The Social Lives of Forests

THE SOCIAL LIVES OF FORESTS

Past, Present, and Future of
Woodland Resurgence

EDITED BY SUSANNA B. HECHT,
KATHLEEN D. MORRISON, AND
CHRISTINE PADOCH

The University of Chicago Press   Chicago and London

The University of Chicago Press, Chicago 60637
The University of Chicago Press, Ltd., London
© 2014 by The University of Chicago
All rights reserved. Published 2014.
Paperback edition 2016
Printed in the United States of America

23 22 21 20 19 18 17 16   2 3 4 5 6

ISBN-13: 978-0-226-32266-7 (cloth)
ISBN-13: 978-0-226-32268-1 (paper)
ISBN-13: 978-0-226-02413-4 (e-book)
DOI: 10.7208/chicago/9780226024134.001.0001

"False Forest History, Complicit Social Analysis: Rethinking Some
West African Environmental Narratives" (Fairhead and Leach) orig-
inally published in *World Development* 23, no. 6 (2005): 1023–36.
"Mutant Ecologies: Radioactive Life in Post–Cold War New Mexico"
(Masco) is an edited version of chapter 7 from his book *The Nuclear
Borderlands: The Manhattan Project in Post–Cold War New Mexico*
(Princeton University Press, 2008). *Handbook of South American
Archaeology*, chapter "Amazonia: The Historical Ecology of a Do-
mesticated Landscape" (Erickson) 2008, with kind permission from
Springer Science+Business Media B.V. This contribution was edited
for length. "Amazonia 1492: Pristine Forest or Cultural Parkland?"
(Heckenberger et al.) originally published in *Science* 301, no. 5640
(September 2003): 1710–14 (DOI: 0.1126/science.1086112).

Library of Congress Cataloging-in-Publication Data

The social lives of forests : past, present, and future of woodland
resurgence / edited by Susanna B. Hecht, Kathleen D. Morrison,
and Christine Padoch.
    pages : illustrations ; cm
Papers of the conference held at the University of Chicago
May 30–31, 2008.
    Includes bibliographical references and index.
    ISBN 978-0-226-32266-7 (cloth : alk. paper) — ISBN
978-0-226-02413-4 (e-book) 1. Forests and forestry—Social aspects—Congresses. I. Hecht, Susanna B., editor of compilation.
II. Morrison, Kathleen D., editor of compilation. III. Padoch,
Christine, editor of compilation.
    SD387.S55S63 2014
    577.3–dc23

                                        2013021533

♾ This paper meets the requirements of ANSI/NISO Z39.48–1992
(Permanence of Paper).

CONTENTS

ILLUSTRATIONS

Tables

Figures

ACKNOWLEDGMENTS

The conference that was the germ of this book initially unfolded when Susanna Hecht and I were fellows at the Center for Advanced Study in the Behavioral Sciences (CASBS) at Stanford, sitting under the oak trees, eating lunch, and bemoaning the lack of effort on the social histories and ecologies of inhabited landscapes. We were doing exactly what we were supposed to do at such a place, which exists to provide adequate time to think, away from the usual scramble from class to committee meeting. Susanna's insistence on the critical importance of and inattention to woodland resurgence made me take a closer look at my own data from South Asia, where the dominant narrative of environmental change is one of progressive, inevitable forest loss through time, a trope that, as it turns out, only partially reflects actual vegetation trajectories. The complexity of the relationships between humans and forests, it was clear, required a range of disciplinary approaches, and so the idea of an international and interdisciplinary conference was born.

The ideal venue for this conference was the newly formed Program on the Global Environment (PGE) at the University of Chicago. The director of PGE, Mark Lycett, invited us to hold the inaugural conference for his program on May 30 and 31, 2008. We thank him for his confidence in the project and for the significant financial and staff support provided by PGE. Planning for the conference was greatly facilitated by adding Christine Padoch to the editorial team. Christine and I joined Susanna one weekend in Princeton, where she was a fellow at the Institute for Advanced Study; there we brainstormed about what might be the best approach and developed a set of names. We would like to thank all of the presenters and participants who made the conference such a success, including those who were unable to contribute papers to the volume. Ecologists, historians, geographers, anthropologists, archaeologists, biologists, and others interacted congenially and shared ideas, approaches, and excitement in a way suggesting the importance of this topic and the value of interdisciplinary approaches. Staff at the Center for International Studies (CIS) and PGE did a great deal to make the conference work, and we would like to thank Dean Clason and Jamie Bender for extraordinary work on hospitality and organization.

After the conference, we solicited several additional papers. This, as well as the many other commitments of the editorial team, slowed down production of the book, and we thank all of our authors as well as our supportive and long-suffering editor at the University of Chicago Press, Christie Henry, for their patience. Christie attended the conference and listened to us promise her a rapid turnaround, something that did not happen. The delay did, however, allow us to think more carefully about the project, and ultimately the book was better for it. Part way through the editorial process, Christine Padoch became the director of the Forests and Livelihoods Programme at the Center for International Forestry Research (CIFOR) in Indonesia, another commitment through which work on the volume was refracted. Susanna Hecht held a Guggenheim, and various funds from the University of California Committee on Research, and I received a Mellon New Directions Fellowship, all of which both took time away from and simultaneously helped facilitate work on this project. Additional funding from the Center for International Studies at the University of Chicago supported publication costs. Last, but certainly not least, we extend profound thanks to our two editorial assistants, Meleiza Figueroa and Christine Malcom, for helping us pull the massive manuscript together from what was otherwise looming chaos. Their painstaking work was essential in transforming a disparate batch of papers into a polished manuscript.

After twenty-eight chapters, it might seem that there would be no more to say, but for me at least, the completion of this volume represents more a beginning than an end. I am indebted to Susanna for her insistence on the importance of woodland resurgence and on broader ways of thinking through forest transitions. She demanded as well that we work harder to get out the word that humans and forests have intertwined histories and that both social and biological approaches are required to understand these socio-natural worlds. Christine's extraordinary grasp of the field—really, many fields—and of contemporary policies and politics significantly enhanced our range of topics and contributors; her wise counsel kept us on track. Finally, I am grateful to both Susanna and Christine for their embrace of the historical dimension so central to my own thinking. They, in turn, helped bring me into the present—and the future—of forests, people, and their interconnected social lives.

Kathleen D. Morrison
Chicago, IL

1 ✳ From Fragmentation to Forest Resurgence: Paradigms, Representations, and Practices

SUSANNA B. HECHT, KATHLEEN D. MORRISON, AND CHRISTINE PADOCH

In the popular consciousness, the dynamics of tropical ecosystems embody a single narrative of catastrophic deforestation and land degradation. Emphasizing clearing, fragmentation, and a unidirectional narrative of forest loss, this apocalyptic vision has held sway over the last several decades. Indeed, not only historical accounts but also most of the scientific literatures on tropical development and the emerging (and enormous) literature on fragment ecologies partake of this view. The approach in development studies and conservation biology, rooted as it is in "islands" of conservation, and inevitable fragmentation of remaining forests outside of parks and reserves, has, however, ignored a major countertrend. There are now widespread and complex processes of forest regeneration and woodland recreation throughout the tropical world (Angelson 2007; Barbier, Burgess, and Grainger 2010; de Jong 2010; Farley 2007; Grainger 1995; Grau et al. 2008; Klooster 2003; Palo 1999; Rudel, Bates, and Machinguiashi 2002), processes that have as yet been underanalyzed. In fact, reports from Malthusian "poster children" like El Salvador and the Sahel show significant, surprising, and very rapid woodland resurgence in areas where, to quote ecologist John Terborgh (1999), "Nature has been extinguished." Similarly, recent historical analyses have called into question dominant narratives of catastrophic forest loss as the universal trajectory of human occupation in the tropics (Denevan 2001; Posey and Balée 1989; Barbier, Burgess, and Grainger 2010; Grainger 2008; Heckenberger and Neves 2009; Heckenberger, Petersen, and Goes Neves 2001).

What this book outlines is the dynamism of the ecological "matrix": the world of working landscapes that connects old growth fragments, rejuvenating them through working landscapes, daily life, and livelihoods in the creation of a society of nature. These landscapes are home to more than two billion people, so this nexus is not trivial. While half the world's population now resides in urban areas, urbanites fundamentally depend on nonurban landscapes, and especially in the tropics, on forested landscapes for water, materials, foods, and, increasingly, carbon mitigation. Forested landscapes are as much a human habitat as are cities and agrarian fields. Like the latter, forests incarnate histories, ways of seeing and feeling, traditions, formal and informal legal regimes, livelihoods, meanings, and now global economics, geopolitics, and, increasingly, energy and climate politics.

Forest recovery, more widely known as the "forest transition," has taken on a life of its own as a cottage industry for explaining woodland resurgence. Woodland resurgence has been documented from the United States and Europe, and has historically been associated with urbanization and industrialization and the shift from a rural to an urban economy supported by intensified agriculture in the late nineteenth and early twentieth centuries. Some authors have sought to universalize this model of the forest transition as a repeating narrative of urbanization that reduced human numbers in the countryside (Mather 1992; Bellemare, Motzkin, and Foster 2002; Foster 2002; Foster, Motzkin et al. 2002). Rural populations fell, trees regrew, farming in the eastern and southern United States mostly relocated to the larger scale agro-industrial zones of the Midwest, with specialty crops moving to California and the Southwest, as well as abroad to Chile and Mexico. Depopulation, relocation, and urbanization and enclosures of the late nineteenth and twentieth centuries seemed more or less to embrace this process. This model cannot, however, be extrapolated to the twenty-first century for the developing world. Indeed, historical evidence reveals some of the complexity and diversity of relationships between urbanization, agricultural strategies, and forest cover.

The processes we see in the developing world now are a good deal more complicated and involve areas that can be densely inhabited yet increasingly forested, a model at odds with the Euro-American case. This transformation involves significant revision and analysis of the history, meaning, natures, and politics of these landscapes as well as of forest transition theory itself. These are the topics of this book, a body of work that, we hope, will help recast understanding of the nature/culture/development nexus. Focusing on the history of forest landscapes, our authors explore these landscapes as places of significant human action, with complex institutions, ecologies, and economies that transformed these areas in the past and shape them today and into the future in unique ways. Currently, the development processes associated with local, global, and regional capitalisms give rise to the socio-ecologies of new woodlands and structure protection of older historical forests. Policy contexts ranging from authoritarian capitalism to neoliberal economic reforms, social movements, globalized commodity and labor markets, as well as the effect of environmental discourses, whether focused on identity, consumption, branding, or planetary change, have dramatically altered the rules of the game in inhabited landscapes. These contexts and discourses should be altering the analytic frameworks for understanding forests more fully than they have to date. This book tries to redress this analytic gap.

At the same time, studies of long-term human and forest histories suggest that past relationships, too, have been far more complex than generally assumed, again calling basic analytical frameworks into question. There is now an emerging literature and an ever larger empirical base and sets of case studies that are capturing the diversity of these processes and their contingency. While some processes, like the Chinese-led expansion of rubber forests described by Menzies and Fox, involve vast agro-industrial simplifications of highly diverse agro-ecosystem and swidden forest systems, other systems, like that of Açaí in eastern Amazonia, are replacing relatively low diversity pasture systems with increasingly diverse forest systems. Forest recovery and management for both local needs and global commodities by women in Africa defy the preconceptions about an inevitable downward ecological spiral. Meanwhile, the legacy of globalized logging operations provided the skilling necessary for managing timber in secondary successions to serve local markets. Longer term studies supported by research in archaeology and paleoecology not only show that forest loss is not a simple, unidirectional process, but also that many forests long thought to be "pristine" actually have long histories of human occupation and use. The real point is that the received ideas about what is going on in forests in the developing world fail to capture this complexity, requiring serious rethinking of our narratives about them.

Woodland resurgences in the developing world are complex and diverse. They have different historical roots, occur in diverse cultural matrices and ecologies, and reflect modern and modernizing processes as well as traditional socio-natural regimes. Complex tenurial regimes, emerging regional and global markets, war, new ideologies and territorial identities, institutional rivalries, ecological paradigms, and competing authorities vie for the political spaces that forests have become. Hallmarks of globalization, such as international migrations and their remittances, energy markets, environmental movements, and global commodity markets now shape non-urban spaces in a new kind of rurality. Woodland recoveries in inhabited landscapes, as discussed in this volume, provide intellectual and practical contexts for the intersection of social theory, historical, matrix, and political ecology, and development transformations. The studies we have selected have broad implications for understanding a crucial yet often "invisible" global process on the forefront of environmental and social analysis in development, one of importance across all tropical regions on earth. What is also clear is how highly interdisciplinary such research has—necessarily—become, as the authors, studies, and bibliography in this volume make abundantly clear.

From Lost Paradise to Social Reconstruction

We will not summarize the chapters of individual sections here, since each section has its own short introductory essay, but rather highlight several essential themes. First, we stress the *complexity* of forestry recovery processes. Forest clearing has been understood through relatively few explanatory models (Malthusian pressures, markets, policy distortions, faulty tenurial regimes, environmental values), and analyses have been largely focused on frontier zones, at the peripheries of modern states, and often in places with competing sovereignties or political economies. The overwhelming tendency of tropical researchers trained in the natural sciences has been to view land-cover change through the Malthusian optic of population increase, and this perspective has a tenacious hold on explanatory paradigms of forest clearing. It is as if the mere presence of humans could account for forest clearance, with population serving as a simple proxy for the intensity of impact. Such a view, of course, ignores the role of land-use practices, settlement forms, institutions, local ecologies, and all the other complex relations that mediate forest cover and human populations. Viewing most of the tropics as functionally empty, ahistoric spaces of wildness with minimal human impacts in the past, researchers have assumed a primordiality in which the shaping forces were planetary processes like climate change and evolutionary dynamics. This construction, which omits human action from "natural" processes, can only admit humans as agents of disturbance, clearing and degrading forest lands from their presumed original conditions. The view of "old growth" as the only type of high-quality forest occluded many of the effects of human action on forest systems, including those that created the kind of ecosystems scholars were, in fact, observing. Such misrecognitions not only produced problematic ecological frameworks, but also worked to erase human histories from forest landscapes, too often leaving contemporary forest residents at the mercy of models and policies ignorant of their long-term roles in shaping these landscapes. At the same time, areas that have had recognizable human impacts were often assumed to be degraded, or not even counted as forests at all. In this Malthusian context where the very existence of human occupation necessarily leads to degradation and loss, few possibilities other than set-asides, like biological reserves or national parks devoid of people, could be imagined as countering the inevitable effect of human destructiveness or contagion.

What this deeply Malthusian position has done is to create categories of the "wild" and the "wrecked"; this has reinforced the invisibility of social processes that affect and are inscribed in the institutions and politics of forests everywhere on the planet. The reality, as many of our authors show, is that highly diverse processes produce woodlands, including—most surprisingly from the Malthusian view—forests in inhabited places. Moreover, what have often seemed to be primordial forests in fact turn out not to be so if we examine the historical and archeological records (Erickson, chapter 15; Heckenberger et al., chapter 24), and especially when combined with biological evidence (Lentz and Lane, chapter 13; Morrison and Lycett, chapter 11). Tragically, restoration efforts that fail to acknowledge the role of humans in co-constructing forests (and other vegetation associations) seem to doomed to fail in the absence of these historical agents, as Neumann notes in chapter 3. Indeed, even the forest ideologies discussed by Adam Smith (Albritton Jonsson, chapter 4) and contemporary dynamics like the Cold War, the environmental implications of which were thought to lie elsewhere (as in immolation of the planet), rather than a process instrumental in creating "wildness" (see Peluso and Vandergeest, chapter 5 and Masco, chapter 6, in this regard), have powerfully shaped understandings of forests and their larger engagements. Ecological matrices and successional structures themselves have been, and are now, deeply influenced by regional histories, politics, and institutions.

What lies under, behind, or in the past of these forests are their long-term legacies of human action: they are artifacts of human ideas about nature, expressions of economies, and places mediated by institutions and their materialized practices as much as they are habitats evolving in response to evolution and planetary processes. In this light, chapters such as those of Lentz and Lane, Erickson, Heckenberger et al., and Morrison and Lycett go far in recasting the categories for understanding what are often taken as "natural" environments. What is especially trenchant from these studies is that our Western categories are very weak at understanding much softer divisions between the wild and nonwild and even the urban and the wild. The historical ecology of domesticated landscapes so typical of the historical occupation of indigenous and traditional groups suggests much vaster scales of extensive changes in landscapes, the durability of these changes, and the reality that many of them create enriching habitats and matrices that support a wide variety of biodiversity. Trajectories of change, furthermore, emerge as complex and multidirectional, with both forest loss and regeneration apparent in paleoenvironmental records. It is imperative, then, to incorporate this revised understanding of human-forest dynamics and history into analyses of more recent cases, which continue to bear the marks of this history in terms of structure and diversity of human effects as well as global processes.

If complexity is one theme of this volume, another is that of *invisibility*, a problem mentioned across the board in the essays collected here. This ecological blindness can be due to: (1) lack of agreement about categories, especially what comprises "forest," and the values and meaning placed on different forest types, such as degraded or secondary; (2) the problems of limited empirical historicisms that only address the immediate landscape structure and do not understand its deeper historical construction; (3) a misreading of forest trends (see especially here studies by Grainger, who suggests that the trend globally has been toward *increasing* woodlands, as well as the classic work of Fairhead and Leach [1996]—both show that most conventionally trained scientists were missing the directionality of the forest trend); or even (4) woodlands "out of place"—in urban areas or supporting urban economies (Crane et al., chapter 28, this volume). The immediate outcome of this blindness (or perhaps amnesia) is that it makes it easy to miss forest cultures, ignore the forest politics and policies that might enhance such woodlands, and fail to valorize the knowledge that inheres in the practices that produced these landscapes. Even the conservation value of forests misapprehended as ruined, secondary, or overly cultural can be seriously underestimated. These kinds of woodlands are often significant sites of biodiversity maintenance, livelihood support, and carbon uptake and storage. This invisibility is an outcome of both systems of thought and inattention to specific forms of empirical data, which our authors explore in detail. The conventional understanding of forests perhaps needs to be revised in light of the emerging rural environmental politics of the twenty-first century, which will be engaging agro-fuels, carbon economies, and food and livelihood dynamics in new ways, and within which forests will be a significant zones of contention.

A third theme is that of *forest history*. In fact, social lives and forests express a broad array of histories of empires, cuisines, commodities, and markets; institutions that produced practices that have unfolded in medium-to-deep time, shaping the landscapes we now confront. The critical element here is that the forests we see are rooted in and contingent on human actions and social configurations in the past. They are divergent in their ecological, institutional, and political content and outcomes, even as they—and their complex histories—constitute today's modern woodlands. Thus conservation areas can range from toxic nuclear dumps to the realms of Indonesia's ethnic cleansing, displaced Africans or to heritage landscapes of the Mediterranean. Pointing to the inconsistent application of the easy equivalence between human presence and landscape degradation, Neumann (chapter 3) highlights the ways in which ideologies that prefer human stewardship for authenticity in European landscapes erase history and displace local populations in their contrary imposition of "wild" nature in African landscapes and in non-European contexts more generally. The range of institutions that have been deployed in woodlands have complex historical roots as well as reflecting modern politics, as many of our authors point out. What forest histories suggest overall is how contingent and divergent forest structures are, what contested terrain woodlands have become, and why these socio-natural places remain among the least legible to the modern state (cf. Scott 2008). These issues are especially important as woodlands become critical sites of environmental and development politics in the present and future.

Forest as Biome: Social Lives and Natural Processes

Both forests and humans have a dual identity, at once biological and artifactual, products of intertwined human and nonhuman histories. As noted, we approach forests from multiple disciplinary angles, with contributors to this volume crossing traditional boundaries that often separate the sciences from humanities and social sciences. This critical intersection highlights another theme of this volume, what might be called the "biotics" of the social lives of forests. Human-modified landscapes can range from systems so complex they are seen as untrammeled, to the vast monocultures now transforming southern China and Southeast Asian forest lands. Matrix, historical, and successional ecologies can be usefully placed within the contexts of forest histories, and more attention to the theoretics, as well as uses, of such landscapes can help enhance their importance as sites supporting conservation, livelihoods, and places mediating some forms of planetary change. These are important arenas for socially and ecologically informed policy. A more generalized landscape approach that integrates different kinds of successional ecologies into both urban and conservation landscapes advances a vision for more resilient landscapes under what will be much harsher planetary dynamics. The habitual dismissal of inhabited forests as "degraded" may well have blinded ecologists to some important applied intellectual and practical possibilities in places they have given up on. These are, after all, socio-*natures*, and much about their dynamics remains outside the purviews of many disciplines, since they neither fit into categories of agriculture, forestry, or wild lands, even though they retain some of them all. If even apparently "untouched" forests in the heart of the Amazon turn out to be sites of past human settlement, then clearly ecological models that ignore that history must be seen as incomplete, at the very least. The tardy recognition, however, that so-called "degraded," secondary, or occupied forests also require biological and social analysis, however, may prove to be a greater failing. It is exactly the poor fit into our categories that makes social lives and the forests produced by them the most challenging arena for integrated analysis from the social and natural sciences, one that will be critical for the twenty-first century. What is especially important is that it seems these types of forests are actually *increasing*. We do not know how long this trend will last, and in general it is hardly on the radar, but given the importance of forests as integrating spaces for many environmental services, livelihoods, household security, economic products, and especially climate mediation, they should be at the top of the agenda.

This book contains many surprises and challenges a great deal of conventional wisdom. The editors hope that our own passions for these surprising ecosystems and their importance for global futures will make the "social lives of forests" an intellectual starting point for the environmental politics of the twenty-first century.

PART I
Conceptual Frameworks

Rethinking Social Lives and Forest Transitions: History, Ideologies, Institutions, and the Matrix

SUSANNA B. HECHT

How do we understand the natural world and what it means? This section brings together a range of disciplines from history to ecology to reflect on "thinking about forests." All authors emphasize the interaction between humans and landscapes in physically constructing biota, and many emphasize the overwhelming importance of ideologies about forests, imagined histories, iconography, institutional arrangements, and competing knowledge systems in structuring how we understand the socio-natures of forests. This first section sets up much of the analytic apparatus for understanding the later pieces in the book. There are several themes that emerge: first, that narratives and ideologies of nature shape our understanding, practices, and politics about nature. The classic paper by Fairhead and Leach on false forest history and the narratives of environmental decline describe how received ideas about the negative environmental impact of Africans on their landscapes are empirically wrong, producing policies and understandings that are enormously at odds with the reality: these farmers actually produced forests around their settlements, and it is their work that encourages the formation of secondary forest thicket in the savanna. In a related analysis, Neumann points out that the European Union explains Mediterranean biodiversity as a function of human management of heritage landscapes, while African populations are seen as antithetical to biodiversity, even though humans evolved with African biodiversity and shaped its landscapes over tens of thousands of years. Going further into the historical ideologies of forests in the temperate zone, Albritton Jonsson revisits the debates over Caledonian forests in the Scottish enlightenment, and examines more specifically how political economist Adam Smith positioned himself vis-à-vis these controversies. Albritton Jonsson highlights several modern echoes: (1) the idea of conjectural histories of primal forests undone either by Roman legions (colonialism) or peasants; (2) the idea that restoration would return Scotland's forests to their ancient and authentic state under the aegis of aristocrats and natural historians; (3) forests as embodiments of a moral order;

and (4) competing narratives of nature—on the one side, iconic landscapes fraught with instability needing to be managed by the crown, aristocrats, or natural historians, on the other the self-regulating action of harmonious and equitable trade, taking timber as a commodity and not a symbol. Although these debates roiled, Adam Smith rejected the symbolism of landscapes for a world of unsentimental exchange and mutual advantage, and lauded the international markets that brought Norwegian woods for the construction of Edinburgh.

Another prominent theme involves placing current forest dynamics in their social histories, a topic also elaborated in many chapters elsewhere in the book. The forest landscapes we think of today as empty, wild, and "natural" usually have a humanized prehistory that is often less far in the past than we imagine. These pasts often include conjectural histories deployed to obscure the institutional, political, and often violent transformations and institutional transitions that underpin them. Further, environmental events that we think of as having one result often turn out to have many others. Take, for example, the Cold War: Masco describes the "rewilding" of Los Alamos's hyper-toxic nuclear waste sites, which have been reinvented as pristine wild landscapes, albeit ones monitored with robotic bees that collect radioactive pollen. The "jungles" of Indonesia underwent an ethnic-enviro cleansing of counterinsurgency to become the empty wildlands mediated by new forestry and conservation institutions, reoccupied by government-supported ethnicities and coteries, and in some areas, definitively switched to agriculture. The Cold War as environmental politics was usually imagined as nuclear immolation, not the creation of an ersatz "Ur" nature and a suite of new institutions.

The next group of analysts raises key questions about the physical nature of forests and their landscapes. Grainger points out the central problem of classification and data sources in discussions of forest trend. Questioning what "forests" are is not a trivial exercise, and this question will become ever more contentious in debates over forest "authenticity" and the value of forests in carbon offsets. The central issue is that the narrative of explosive and endless deforestation that has held sway for the much of the last fifty years—and so dominates thinking about the tropics—may have missed the countertrend. Understanding these processes requires rethinking scholarship and the institutions that support forests, as well as seriously monitoring the direction of forest trend. My piece focuses on how forest resurgence, often described as the forest transition, is explained in several bodies of theory. I end with a discussion of the debates regarding the social natures of forests in the emerging politics of environmental services, emphasizing that twenty-first century rural politics will revolve a good deal less around the "agrarian question" than ever before. Instead, environmental enclosures, as "re-wooded" or inhabited landscapes, now vie with conservation and "re-wilding" set-asides and with efficiency forests (plantations) for the capital flows associated with the emerging carbon economy.

Further shifting the focus away from the imaginary Eden is the path-breaking work by Vandemeer and Perfecto. Recent studies of land change ecology have relentlessly focused on forest fragmentation and extinction dynamics in landscapes, and generally argued for large-scale conservation set-asides. But the emphasis on forest "islands" overlooks the reality that populations in "islands of conservation" are sustained by migration through the ecological matrix of complex, usually inhabited, ecosystems. Thus, they argue, the attention should shift onto shaping the caliber of the matrix. They suggest that a model of rural development that embraces ecological as well as social complexity within a matrix framework will, in the end, serve conservation and development purposes far better than a model of an imag-ined wild on one side and industrial agriculture on the other.

Finally, Chazdon and her coauthors describe the various ways in which successional pathways are modified by human action. Management pro-duces considerable variation in abundance and composition of remnant trees. Hunting or lack of it affects seed predators, often dramatically chang-ing the density of canopy tree and palm seedlings in secondary forests, com-pared with hunting-free zones. Thus, ironically, some major canopy trees do better in hunted areas, because their predators are reduced, and these differ-ences ultimately impact the rate and composition of tree recruitment. The location of forest patches in the surrounding landscape has major implica-tions for colonization and genetic diversity. Human activities shape forest regrowth in a variety of ways, with both positive and negative effects on the diversity and rate of forest regeneration.

2 ＊ False Forest History, Complicit Social Analysis: Rethinking Some West African Environmental Narratives

JAMES FAIRHEAD AND MELISSA LEACH

Introduction

This chapter examines social science analyses that are being used to explain environmental degradation and inform policy responses to it. We focus on two cases pertinent to an exploration of how applied social science knowledge about people–environment relations is produced. They exemplify the type of social analysis often brought to bear to explain environmental degradation in Africa. These assumptions have credibility in large part because they are stabilized within "narratives" (Roe 1991); that is, logical stories that provide scripts and justifications for development action. But once dissected from the reality they seek to construct, these explanations reveal how the applied social sciences can be used to lend weight to popular Western perceptions about African society and environment—a mythical reality. By stripping away the explained from explanations of it, our cases pave the way for rethinking people–environment relationships in this region. We advance alternative assumptions that better fit the facts. These cases involve Guinea's forest margin zone, where the explanations of land-use change suggest that local community institutions were once more able to control environmental resources than they are today, explaining forest loss in terms of "institutional breakdown." An armory of purported factors is marshaled for explaining such social rupture, the results of which seem so evident in a degraded landscape. These include socioeconomic change, increasing mobility, the weakening of traditional authority, more individuated farming, new economic and cultural aspirations of the young, the alienation of local resource control to state structures, and the emergence of "anarchic" charcoal, fuelwood, and timber businesses for urban markets. Migration is added to these arguments: "forest peoples" with supposedly forest-friendly lifestyles were disrupted by the influence of "savanna peoples." Overlaying all is the specter of Malthusian population growth. Foreign observers today date such socio-environmental disruption to the notorious regime of Guinea's first republic (1958 to 1984) under Sékou Touré, imagining the colonial period as a golden age, while nationals look to the precolonial period to find so-called good society and environment. As if to make the point, one scholar forced the social-environmental Eden back to that period documented by the thirteenth-century Arab geographers (Zerouki 1993).

The production of history serves many ends, and social scientists have been complicit in producing a view of current history as one of increasing detachment from a harmonious past. Treating this past as the template for the resolution of today's tensions, they have imagined links between social and environmental conditions in ways that inform policies that now marginalize inhabitants from what little resource control they have.

Case 1: Forest Islands of Kissidougou

THE DEFORESTATION NARRATIVE

Kissidougou looks degraded. The landscape is largely savanna, especially open in the dry season when fires burn off the grasses and defoliate the few savanna trees. Nonetheless, surrounding each of the prefecture's villages are patches of immense semi-deciduous humid forest. Scientists and policymakers consider these "islands" and streamside gallery forests as relics of an original, much more extensive, dense woodland cover. Inhabitants have, they assert, progressively converted forest into "derived" savanna by their shifting cultivation and burning practices, preserving only the forest around their villages to protect their settlements from fire and wind, for shade for tree crops, for fortifications, and to provide seclusion for ritual activities. They argue that today's climate would support forest cover and infer from the presence of "relic" forest islands that it once did:

> At origin, the forest between Kissidougou and Kankan was . . . a dense, humid, semi-deciduous forest. The trigger of degradation is the farming system and the fragility of climate and soils in tropical regions. Some primary formations still exist, however, in the form of peri-village forest islands and gallery forests on the banks of water courses. These forest islands show the existence of a dense forest, which is today replaced in large part by degraded secondary forest. (Programme D'aménagement Des Bassins Versants Haute Guinée 1992, 6–7)[1]

Deforestation provokes problems at several levels. At the local level it leads to soil degradation and renders farming less productive and sustainable. At the regional level deforestation is thought to have caused irregularities in Niger river flow and in rainfall. In addition, it is contributing to global warming.

Social analysis has always been deployed for explaining this problem. In the early part of this century, the celebrated French colonial botanist, Auguste Chevalier, considered migration and trade during the post-occupation period to be responsible for an increase in fire-setting from a previous, less forest-harmful level (Chevalier 1909). He thought inhabitants conserved the forest islands for cultural reasons (Chevalier 1933). In 1948, Adam echoed the idea that the Mandinka were a "savanna" people who had migrated southward into the forest zone and created savanna there (Adam 1948).

More recently, professional social scientists focused on environmental issues in Kissidougou for international or bilaterally funded programs. One team, responsible for the Niger River protection program, illustrate this focus:

Our questions sought to explain the deterioration of the environment, viz: erosion and soil impoverishment, the drying up of water sources, the origin and nature of forest destruction. the origin of perverse use of bush fire. . . . We infer a strong relationship between soil erosion, environmental degradation and the break-up and impoverishment of socio-economic structures and relation. . . . The more a community is in equilibrium . . . the healthier is the nature of its relations with the environment. In these communities, the existence of the living is above all justified by good management of what the ancestors have left to them. This is inscribed in the collection of laws, concrete and abstract, rational and irrational, which, once disturbed from the exterior, can be the cause of a deterioration which manifests itself in social, religious, political and economic institutions, and the environment. (Programme d'Amenagement des Hautes Bassins du Fleuve Niger, n.d., 4–7)

In a second study on local burning, the author inventoried cultural traits around the practice of fire. "We have tried to retrace the transition from traditional to 'modern' practice" (Zerouki 1993, 1). The author argues that "modernity" is responsible for disrupting the once successful integration of fire control within diffuse sets of intra- and intervillage village social, cultural, and political relationships. He finds that "[d]egradation seems to be recent" and that "it accelerates with the development of an urban network . . . and population growth" (Zerouki 1993). A coresearcher on this same study expands on the causes of such "dysfunctioning":

According to inquiries from elders . . . and by IFAN in 1968, the whole region was covered with forest about 99 years ago. In the Samorian period, war chiefs used fire for better visibility and for encampments. Since independence, there has been demystification of sacred forests, installation of wood mills, and brickmaking, nomadic farming and herding. Fire, and runaway population aggravate processes of vegetation degradation. (Fofana et al. 1993, 19)

A study of an area at the eastern border of Kissidougou claims that:

The degradation of forests continues in an accelerating fashion. . . . Peasant exploitation is correctly identified as the principal factor of destruction. The social reasons for fire setting in hunting are . . . closely linked to growing tendencies of commercialisation and monetarization. . . . This underlines the loss in importance of traditional organizations of hunters which, to date, are marked by an anti-commercial character. . . . Traditional structures for natural resources, most often of pre-Islamic origin, incorporate conservation aspects. Some still operate . . . but a change is beginning to show itself . . . which implies a dissolution of traditional regulatory structures. (Stieglitz 1990, 54, 70, 77)

Demographic change is also faulted by most analysts. Ponsart-Dureau, for example, an agronomist advising a nearby project, considers that:

Around 1945. the forest, according to the elders, reached a limit 30 km north of Kissidougou town. Today, its northern limit is found at the level of Gueckedou-Macenta, having retreated about 100 km. . . . Demographic growth forces the villagers to exploit their land completely and to practice deforestation which disequilibriates the natural milieu. (Ponsart-Dureau 1986, 9–10, 60)

Thus in different ways, each of these analyses contributes to a narrative now as prominent in Kissidougou's education and administrative circles as it is in social science analyses: Kissidougou had an extensive forest cover that was maintained under low population densities by a functional social order whose regulations controlled and limited people's inherently degrading land and vegetation use. The breakdown of such organized resource management under internal and external pressures, combined with population growth, led to the deforestation so evident in the landscape today.

Observers invariably consider degradation as a recent, ongoing, and aggravating problem. The social and economic changes are (similar to "runaway demography") always seen to be accelerating out from a "zero point."

Policy has followed from the assumptions of this narrative and has changed little since its first elaboration in the colonial period. The first emphasis is

on the reduction of upland farming—seen as inherently degrading and becoming more so under greater individualization and population growth—
in favor of wetland farming. Upland agriculture must be rationalized and
intensified through "model" agroforestry systems and reorganization of
tenure and fallow systems. Second, policies have focused on bushfire control through externally imposed prohibitions, regulations, and practices
(e.g., early-burning). Third, policies have attempted to control deforestation through prohibitions on the felling of protected tree species and
through the reservation of certain forest patches. Fourth, there are attempts
at forest recovery through tree planting in village territories. Uniting these
policies are technology packages, such as inland valley swamp development,
tree planting from nurseries, and organizations in resource management.
In Guinea's colonial and first republic periods, the degradation narrative
justified removing the villagers' "control" over resources in favor of the
state. In bushfire, upland use, timber-felling, and forest reservation policies, modern government administrations took over resource tenure and
regulated local use through permits, fines, and at times, military repression.
More recently, emphasis has shifted toward reconstituting broken community control over resources: *Gestion de terroir villageois* approaches that
provide for village-level planning of bushfire, upland and forest use, participatory" tree planting, and conservation of forest islands in favor of the
community.

THE COUNTERNARRATIVE

Examining how vegetation has actually changed in Kissidougou is a necessary first step in evaluating these analyses. Fortunately, a number of historical data sources make this possible.[2] Existing aerial photographs for Kissidougou clearly show the state of the vegetation in 1952–1953. These provide
incontrovertible evidence that during this recent, supposedly most degrading period, the vegetation pattern and area of forest and savanna have in fact
remained relatively stable. Changes that have occurred do not involve forest
loss; rather there are large areas where forest cover has increased, and where
savannas have become more, not less, woody. Forest islands have formed
and enlarged, and in many areas, savannas evident in the 1950s have ceded
to secondary forest vegetation.

    To examine vegetation change further back, we reviewed descriptions
and maps of Kissidougou's landscape made during the early French military
occupation (1890s to 1910) and indicators of past vegetation and practices
from oral history. Villagers suggest, quite contrary to "outside" interpretations, that they established forest islands around their settlements and

that it is their work that encourages the formation of secondary forest savannas. In twenty-seven of the thirty-eight villages we investigated, elders recounted how their ancestors had founded settlements in savanna and gradually encouraged the growth of forest around them.

Earlier documentary sources from the 1780s to 1860s do not suggest extensive forest cover; indeed, just the opposite. Harrison, traveling to Kissi (c. 1780, see Hair 1962), and Seymour (1859–1860) in Toma country southeast of Kissi, describe short grass savannas and an absolute scarcity of trees in places that now support extensive humid forest. Sims (1859–1860), speaking of the area just to the southeast (between Beyla and Kerouane) writes: "There are no trees; the whole country is prairie; for firewood the people have to substitute cow dung, and a kind of moss which grows abundantly in that country." This picture of less, not more, forest cover in the nineteenth century is supported by early oral history data. All the above-mentioned villages claiming foundation in savanna were established during or before the nineteenth century. Several village foundation stories in the south refer to conflicts triggered by the scarcity of construction wood, bizarre given the present forest and thicket vegetation.

It appears that social science analyses in Kissidougou have been providing explanations for forest loss that has not actually taken place. As we suggest, there are other ways of conceiving of these relationships that better fit and explain vegetation history.[3]

The first reconception involves recognizing that local land use can be vegetation enriching as well as degrading. It can increase the proportion of useful vegetation forms and species in the landscape according to local values and productivity criteria. Thus, villagers encouraged the formation of forest islands around their villages for protection, shelter for tree crops, sources of gathering products, and concealment of ritual activities. They achieved this through everyday use of village margin land (for thatch, cattle-tethering, which reduces flammable grasses, and household waste deposition, which fertilizes the forest successions beginning to develop), and with deliberate techniques like planting forest-initiating trees and cultivating the margins to create soil conditions suitable for tree establishment. Much farming is concentrated on land that farmers have improved through habitation, gardening, or fallow improvement through intensive cattle-grazing, seed-source protection, the multiplication of savanna trees from suckers, or distributing forest-initiating creepers. These knowledge systems and practices are found in all Kissidougou's ethno-linguistic groups. There seems to be little basis for distinguishing between forest and savanna people.

A related reconception concerns the character of natural resource management "organization." Environmental management always depended less on community authorities and socio-cultural organizations (which might be "threatened" by social change), than on the sum of a diffuse set of relations—a constellation more than a structure. Indeed, the maintenance of long-term productivity is, in many cases, built into short-term production patterns; whether carried out for oneself, one's household, or one's compound, these improvements interact with others—spatially or temporally—so that the combined effect on resource enrichment is greater than the sum of their parts. Thus, the fires set in the early and mid-dry season by hunters to clear small hunting grounds, and by others to protect property and fallows, create barriers to more devastating later fires. The small tree crop plantations that people make and protect behind their kitchen gardens add to the village forest. For much "resource management" there is no need for village or higher-level management structures to "regulate degrading pressures." Village authorities do intervene in certain activities—for example, managing early-burning around the village, protecting palm trees, imposing cattle-tethering dates, and coordinating the fallow rotations of farmers' contiguous plots in some Kissi areas. Village and higher-level organizations also exert control over external factors that influence the agricultural environment, such as in negotiating with Fula pastoralists or representatives of the forest service.

In this context, socioeconomic change has translated into landscape enrichment. Villagers have adapted forest island quality to changing socioeconomic conditions and commercial signals—managing them as fortresses during precolonial warfare, extending them for coffee planting when it became profitable, and abandoning coffee in favor of fruit trees and gathered products as prices fell again. Urban employment opportunities, youth emigration, and more individual economic opportunities have contributed to changes in farming organization, but today's smaller farm households use and improve fallows as large compound ones did earlier, and modern women's individualized, commercial food cropping in upland gardens engages practices that upgrade soils and vegetation (Leach and Fairhead 1995). There have been many social and economic changes, but these changes are expressed in the landscape largely through land use and management priorities, not through organizational "breakdown" and vegetation degradation. Relationships between demographic and environmental change are very different from the rapid population/growth-deforestation relationship upheld by the policy narrative.

Evidence does not support the idea of dramatic population growth. Census data suggest that Kissidougou's rural population has increased by only 70 percent since 1917. Growth concentrates around urban Kissidougou and major road axes, but in many areas population has remained almost stagnant. Precolonial evidence shows that certain areas had nineteenth-century rural populations significantly higher than today and suffered radical depopulation during late-nineteenth century wars. Oral accounts, explorers' reports, early sixteenth through eighteenth century documents that mention the region, and broader regional history and archaeology suggest that Kissidougou had relatively high farming populations from the sixteenth century and long before. There is as little evidence for dramatic population increase in the present century from a low precolonial baseline as there is for dramatic forest loss.

In this context, Kissidougou's forest increase might be supposed to relate to population stagnation or decline, but this argument, however, depends on the assumption that local land use converts forest and forest fallows to savanna, and thus that more people means more forest loss. A counternarrative better fits evidence of local land-use practices and vegetation history: from earlier extensive savanna, there has been a broadly positive relationship between the peopling of this region and its forest cover. First, settlements are associated with the formation of forest islands, so more villages mean more forest islands, with greater multiplication of settlements and forest islands during the nineteenth century, when dispersed settlement was a survival strategy, than in the twentieth, when population growth was accommodated through the expansion of existing settlements. Second, greater population density helps control fire by providing the necessary labor and by filling the landscape with more places (upgraded fallows, plantations, settlement sites) that people need to protect. In certain cases, protected sites easily enable the entire exclusion of fire from the territory. The districts where upland savannas have recently ceded to dense forest fallow vegetation correlate broadly with the areas where population has grown. In contrast, low population densities make fire prevention impossible and are a major factor in the persistence of wild fire in the north.

Viewing socio-environment relations in terms of landscape enrichment-through-use by diverse resource management constellations responding to changing incentives better explains demonstrable vegetation and population history. Policies conceived within the degradation narrative have sometimes undermined these relations. In removing local control over resources, they have often interfered with local management. In the north, for example, external fire control and prohibition prevented villagers from practicing their sequenced management strategies[4] and rendered village and plantation protection more difficult. Removal of local resource tenure has reduced villagers' abilities to profit from past enrichment activities (e.g., in selling their forest island trees for timber) and incentives for further landscape enrichment. The implementation of repressive environmental policies has taxed rural populations for supposedly harmful activities that were, in fact, benign or beneficial. More recent approaches, which focus on decentralizing resource control by establishing village-level organization and management plans, actually risk undermining the existing flexible, diverse resource management relations. When initiated by state agencies with considerable foreign support and predefined ideas about environmental dynamics, real decentralization can be undermined. Finally, the high investment in "redressing" Kissidougou's supposed environmental degradation carries opportunity costs in terms of other, more pressing, rural development problems.

Vegetation history and its counternarrative of landscape enrichment entail different policy implications, emphasizing support to proven local practices. There are clearly many techniques and land uses that increase forest cover, and that could provide an effective basis for external support. In working with the local ecology of fire, soils, vegetation successions, and animal dynamics, these integrated vegetation management practices are more locally appropriate, integrated with the social matrix, and thus more cost-effective in terms of labor than are the forestry "packages" generally proposed by outside agencies. Given that farming in the region is not inevitably degrading, environmental policy may look to support as well as to rationalize and regulate agriculture, specifically to support upland farming practices that improve soils and fallow vegetation. Fundamentally, the more important priority is to create the enabling policy and socioeconomic conditions where local resource management can act effectively.

Case 2: The Ziama Forest Reserve

THE DEFORESTATION NARRATIVE

Traveling south from Kissidougou, one enters the Upper Guinean forest region. This region is populous, with only two significant intact forest blocks.[5] Covering an area of about 120,000 hectares, Ziama was designated a colonial forest reserve in 1932, an international biosphere reserve in 1980, and is now the object of a major World Bank–financed conservation project. Policy narratives concerning Ziama reproduce those of Kissidougou with one major scale exception: Changes in the status of a major forest block are at stake, and the conservation concern is partly global.

The Ziama forest is considered to be under considerable threat as an important relic of once greater forest cover. As table 2.1, drawn from an International Union for Conservation of Nature (IUCN) report on Ziama, indicates, forest cover in this part of Guinea is now only 20 percent of what

Table 2.1. Area of humid forest in
forest Guinea at different times

Period	Area (hectares)
At origin	1,930,000
c. 1958	1,300,000
c. 1980	1,075,000
1986	397,000

Source: République de Guineé (1990).

it was "at origin," and the report emphasizes that forest area is declining rapidly. Apart from the loss of biodiversity, this reduction is said to cause drying of the local and regional climate, increasing forest loss in a vicious cycle that threatens regional agriculture.

Regional studies and administrative perceptions are based on social analysis of deforestation and encroachment in the remaining Ziama reserve. The most detailed and explicit version of the analysis is found in a socioeconomic study commissioned by a conservation project (Baum and Weimer 1992). The assumptions it forwards are not dissimilar to Kissidougou's: growing populations of immigrants and indigenous farmers, who have lost "traditional" values and organizational forms, who are clearing forested land.

As in Kissidougou, a contrast is drawn between a forest people, the indigenous Toma (Loma), and a savanna people, in this case the Konianke (Mandinka), whose immigration and savanna ways threaten the forest. Thus we read of the Toma that they are "largely fixed in their customary conceptions and habitual mode of life" (Baum and Weimer 1992, 12). The authors explain that the Toma "historical and social evolution as a people in a forest environment . . . favors a tendency to contemplation and sobriety" (ibid.). These attitudes supported a traditional society in harmony with the forest, so the argument goes.

But the Toma have lost their forest ways: "The forest has largely lost its customary importance, in favor of an essentially agricultural use of space" (Baum and Weimer 1992, 13). The authors are surprised to find that women manage the principal crop, rice, reinforcing the idea that the Toma have only just learned to farm. This view builds on colonial perceptions that Toma had "a very primitive agriculture, centred on pluvial rice based on forest clearings" (Portères 1965, 688 and 726).

Changes in the Toma agricultural economy are linked to the opening of the area to markets and the need to feed growing populations. Both trends are linked to the immigration of Mandinka people from the north, also central to the explanation of the area's demographic evolution. The authors present a picture of a long-term, very gradual peopling of the Ziama region through the immigration of Toma people and then brusque changes as Mandinka arrive. It is said that there were two villages present in the reserve when it was designated in 1932, Boo and Kpanya, having 542 and 370 persons at that time. Boo, which now has a population of some 1,600, is said to have had a population of 500 when it was founded, giving the impression that the forest might have been lightly inhabited for long periods by forest people. Immigration into the region is reported to have risen by four-to-sevenfold in sixty years. This rapid population growth is seen to have created severe land pressure in the areas neighboring the reserve. Assumptions about carrying capacity under shifting cultivation are used to argue that population: land ratios are now "fully saturated." This largely accounts for farmers encroaching on the reserved land for farming.

This narrative concerning a last block of "pristine" natural forest, threatened by recent socioeconomic change and population pressure, provides a powerful justification for conservation. It also entails guidelines for conservation policy. "Original" forest is easily defined as a global or regional heritage and its conservation by global and regional guardians as a moral imperative. Within earlier, colonially derived approaches, the reservation of such a forest was justified with minimal regard for local interests for reserved land and resources. In Ziama, "policing and patrolling" approaches characterized early forest conservation. More recently, emphasis has been placed on the need for participation of and acceptance by local populations if conservation is to be sustainable. Local resistance to and failure to respect the reserve are seen in terms of land shortage and economic pressure, and the presumed policy needs are for socioeconomic development and agricultural intensification in the marginal area around the forest, accompanied by restricting land tenure to reduce current and future pressures on the reserve.

THE COUNTERNARRATIVE

Once again, historical data show how vegetation, population, and society have changed in this region, and the extent to which the assumptions are ill-founded. In the Ziama case, detailed descriptions come from the writings of several highly educated American-Liberians who visited what is today the forest reserve in the mid-nineteenth century (Anderson 1870; Starr 1912; Seymour 1859–1860). What they saw and described in no way conformed to the image of sparse Toma hunter-gatherer populations living in harmony with an isolated high forest. The two "enclave" villages, now situated within the forests covering the wide Diani river plain then lay in savannas. The Ziama mountain massif, now considered the heart of the primary forest, was either bare rock or covered "with cane grass and scarcely any tree but the palm" (Seymour 1859–1860, in Fairhead et al. 2003, 342). From the top of the massif, Seymour describes the plain as "covered with small bushes and grass, and it gives the country the appearance of an old farm, with palms standing scattered all over it" (Ibid).

The region had large populations, by all accounts significantly larger, not smaller, than today. Anderson (in 1874) considered Kpanya as "very large" (when his account described 2,500 people as small) and Seymour (in 1859–1860) estimated Boo to have 3,600 inhabitants. In addition, as the elders of the villages describe, these large villages had many smaller dependent settlements that no longer exist. Seymour and Anderson describe large savanna farms of rice, maize, and cassava stretching as far as the eye could see. It was also commercially prosperous. Seymour noted fifty looms and five blacksmiths in Boo, and found some women wearing jewelry worth $20 to $30 at that time. A little further north, at Kuankan, people walked several miles from the mountains to the plain to sell firewood. As Seymour noted, both enclave settlements had daily and weekly markets, and these traded in foodstuffs, livestock, cash crops (such as cotton and kola), and artisanal goods of every description. The region was not economically or geographically marginal, but central to busy and long-established forest-savanna trade routes.

In the mid-nineteenth century the Ziama area clearly did not fit the images that today's policy narratives construct for it. Unsurprisingly, then, its subsequent history also overturns the conventional narrative's image of unilineal population increase and forest destruction. The story that explains how this region became a "primary forest" reserve within only 130 years of being heavily populated savanna turns, instead, on the wars from 1870 to 1910 (Fairhead and Leach 1994). Sustained military conflict first with Mandinka groups and then with the colonizing French caused major depopulation and economic devastation. It is this, not low "baseline" precolonial population densities, that explains the region's sparse populations at the turn of the century. On the abandoned settlements, fields, and fallows, forest grew. By 1932 the French colonial administration recognized secondary forest worthy of reservation. The rapid forest growth suggests that earlier intensive farming and savanna maintenance did not cause irreversible damage to forest vegetation and could indicate a positive legacy of earlier management practices. By the early 1980s, conservationists were unable to distinguish Ziama's forest regrowth from primary forest. Populations since 1932 have not grown by the 400 to 700 percent suggested in the socioeconomic study. Using the study's own statistics, in the forty-one villages in the vicinity of the reserve, populations have increased by only 80 percent since 1932, or by 120 percent if recent influxes of Liberian and Sierra Leonean refugees are taken into account.

The assumptions that social science researchers are using to understand the nature and change of people–environment interactions in Ziama are at odds with a demonstrable counternarrative centering on warfare, depopulation, forest regeneration, and land alienation. These better encompass the experience and attitudes of today's Toma inhabitants, whose houses display portraits of ancestors who were killed during the wars. This mismatch of narratives underlies the failure of the reserve administration to build any constructive relationship with local inhabitants. Development activities around the reserve are insufficient to calm this conflict and prevent "encroachment." Achieving sustainable conservation, let alone of a participatory nature, remains a distant goal.

When today's inhabitants "encroach," they are reclaiming ancestral lands, a once peopled and prosperous domain, now politically alienated from them. Recognizing this suggests alternative, more fruitful guidelines for policy. Historically grounded claims to land and political authority need to be recognized and seriously addressed through conservation arrangements that cede tenurial control to local landholders within the context of management agreements.

The Regional Narrative and Its Alternative History

The stories concerning vegetation change and its social causes that inform policy in Kissidougou and Ziama are examples of a broader narrative that contains assumptions also written into national, regional, and international policy documents.

Thus it was lamented in Guinea's agricultural development policy strategy that:

> The north of forest Guinea (Beyla, Kissidougou and Gueckedou) is no longer a pre-forest region, but an 'ex-forest' or 'post forest' region. . . . This degradation of the natural environment . . . is the result of rural societies little adapted to the rapid structural, demographic and economic changes of this century, and above all, these last years. . . . The problem today is the decline of traditional control of the orderly exploitation of space and its resources, which has not managed to follow or adapt to the recent, rapid change in the rural world. This management becomes insufficient given a brutal increase in population [and] a progressive loss in the power of traditional control. (Republique de Guinée and Jean 1989, 8)

Focusing on the population component, a recent World Bank policy review argues that:

> Traditional farming and livestock husbandry practices, traditional dependency on wood for energy and for building material, traditional land tenure arrangements and traditional burdens on rural women worked well when population densities were low and population grew slowly. With the shock of extremely rapid population growth . . . these practices could not evolve fast enough. Thus they became the major source of forest destruction and degradation of the rural environment. (Cleaver 1992, 67)

This, it is argued, leads to vicious spirals of shortening fallows, land depletion, yield declines, and subsequent migration to marginal lands and forests. Environmental crisis results less from the overall effect of population pressure on resource availability than from the multiple effects of population pressure on the institutions seeking to control resource access and use.

Given how inapplicable current narratives are to local situations, the Kissidougou and Ziama cases suggest more fundamental examination of the origins and purposes of the regional narratives and how these depend on—and expose—the field of Western imagination concerning African society.

In particular, they show that stereotypes of the colonial era are alive and well in the applied social sciences. Whether they are used to justify policies of external repression or policies of social reorganization and "participatory" development, the narratives justify and make imperative a role for the outsider in the control of rural resources. The broader assumptions that the regional narratives contain can be summarized as follows:

1. African vegetation was once primal, consisting of a climax vegetation. Against this old growth natural vegetation, one can judge levels of degradation.
2. African society can be seen in terms of a traditional functional order. Such order was once harmoniously integrated with natural vegetation (e.g., as epitomized in the idea of a forest people and a savanna people). African farming, land- and resource-use practices generally degrade the original vegetation and are limited only by functional social organization (regulation and authority). From environmental degradation one can diagnose the social ills of organizational dysfunction.
3. African rural populations only increase and do so fast. Population increase is environmentally and socially damaging.
4. African society is essentially sedentary and subsistence oriented with anticommercial sentiment. Money, mobility, and trade are modern and

lead to socio-environmental dysfunction. African history consisted of the continuous reproduction of tradition until it began to become modern, whether with markets and mobility or via colonial intervention.

The links forged between these assumptions mean that vegetation change carries very profound moral messages. "Original climax vegetation" and "traditional functional society" provide fundamental baselines, so that whether the concern is about society or the environment, it is possible to judge that something is wrong and justify outside intervention.

These assumptions, stabilized within social science analysis, are destructive and ultimately have limited policy relevance. "The hard fact," as Sayer (1992) puts it, "is that most aid projects, and especially those in forestry fail." As the Kissidougou and Ziama cases exemplify, misleading narratives are fundamental to this failure. There are counternarratives that better fit the facts of vegetation history and suggest other assumptions that better reflect realities of African environments. The parameters of this counternarrative also fit better with recent developments in ecological and social theory and do not perpetuate the justifications for outside intervention in local resource control.

The old narratives viewed landscapes from the optic of disappearance of natural climax vegetation, but newer strands in ecological theory reject the idea of a single environmental endpoint. When climate scientists suggest that Africa has experienced both long-period, deep climatic fluctuations and sharp climatic variability, the history of vegetation can be seen as one of continual transition, rather than of divergence from a single, once-extant climax. Recent theory suggests that such repeated transitions are likely to be between particular "stable" vegetation states, each determined by a multifactorial complex, rather than by trends in any particular variable (Sprugel 1991; Scoones 1994). Given the multiplicity of interacting factors, shifts reflect historical conjunctures of ecological factors. From this viewpoint, there is no basis for identifying a region's fundamental, archetypal vegetation. Vegetation is in continual transition, and its trajectory is determined by the legacy of past vegetation paths and present ecological conditions.

Ideas of environmental optima dovetailed neatly with ideas of static social "maxima"—of tradition and structure—so prized in colonial anthropology. But notions of society with fixed social structures maintained by functional adaptation and regulation are challenged by recent social theories emphasizing social action and capacities to shape and determine rules. Such continual restructuring through social change challenges the notion of a baseline traditional societal state. That African social forms have been in constant transition dovetails with the view of dynamic vegetation change.

There is no baseline for how society values vegetation, and therefore no basis for the moral argument that indigenous values once preserved a more natural ideal. Vegetation values shift with social, economic, and political changes of a conjunctural nature. These are also socially differentiated: the high forest and wildife priorities of today's global conservation planners are very different from the bush fallow properties of today's Toma.

In the West African context, these social and economic transitions have taken place within a long historical context of movement and migration, agriculture and commerce, and political and religious turbulence. The relationship between social and environmental change does not turn on the dramatic increase in any of these, but rather on people's responses to changing signals within this broader, dynamic continuity. Thus Kissidougou villagers have adapted forest island form to meet changing needs for fortification and different cash crops. Demographic change, rather than consisting always of unilinear population increase, involves periods of stability and decline, of shocks as well as secular trends. Depending on prevailing ecological and economic conditions, the effects of population growth periods can be positive as well as negative.

In the West African forest margin zones, climatic transition appears to have involved rehumidification since the mid-nineteenth century, following a long, relatively dry phase (Nicholson 1979). Where the combination of ecological factors makes conditions marginal for forests, creating a precarious balance between forest vegetation and fire-maintained savanna, people's activities can make the difference, allowing forest vegetation to develop in grasslands. Where people have socioeconomic or political reasons to create forest, they do so in small patches, triggering transitions in small parts of the landscape, as has happened, for example, in Kissidougou. In open savanna and with low population densities, fire is harder to control, but as populations increase and transitions to forest are provoked in more places, fire is reduced and may eventually be eliminated. Agricultural priorities may mean large areas are maintained as bush fallow rather than allowed to develop into high forest. As populations increase further, fallow periods may need to be shortened and some re-savannization can occur. But if population is removed at that point, and given the legacy of people's previous land-use practices, the area may develop into high forest, as happened early this century in Ziama.

This regional counternarrative provides different, and more appropriate, guidelines for policy. In presenting socio-environmental change in a way that better fits local experience, it provides a more effective basis for dialogue and participatory development work with local populations. In removing the baseline link between social and vegetation form, it removes the justification for external intervention in the organization of resource management to reestablish a lost social order, whether by replacement with external control or by the externally promoted "community reorganization" of recent, more decentralized approaches. It suggests that more important priorities are to create the enabling policy and economic conditions in which local resource management constellations can act effectively to support the diverse existing local institutional forms and to build on the beneficial environmental implications of broader rural development and pricing policies—an approach that now finds support in some regional policy institutions (e.g., Environnement et Développement du Tiers-Monde [ENDA] 1992). Finally, as McNeely argues, "because chance factors, human influence and small climatic variation can cause very substantial changes in vegetation, [the biodiversity for] any given landscape will vary substantially over any significant time period—and no one variant is necessarily more 'natural' than the others" (1994, 3). From this perspective, environmental policy can call on no moral high ground in re-creating the natural (or the social that went with it). It clearly becomes a question of social or political choice about what vegetation forms are desirable at any given time in social history, and about ensuring that conflicting perspectives on this—for example, between local, global, and intergenerational interests—are adequately articulated and addressed.

3 ✳ Stories of Nature's Hybridity in Europe: Implications for Forest Conservation in the Global South

RODERICK P. NEUMANN

Do human activities improve nature or diminish it? Debates on this question in Western science and philosophy are centuries old and unresolved. At least since the publication of George Perkins Marsh's *Man and Nature,* scientific opinion has tilted decidedly toward a view of humankind as generally destructive. Marsh believed that nature produced ecological communities of "almost unchanging permanence of form, outline, and proportion" (Marsh 1965 [1864], 29), which human activities had a tendency to disrupt and destroy. For Marsh—and conservationists who subsequently have drawn inspiration from his great work—human society could only disturb nature's perfect balance. No clearer illustration of this perspective can be found than the original title he had proposed for his masterwork, "Man the Disturber of Nature's Harmonies" (Lowenthal 1965, xxiii).

Combined with the neo-Malthusianism of much environmentalist thought, Marsh's idea of humans as destroyers of nature's balance was transformed into a powerful discourse shaping all manner of interventions into rural land uses in the colonial and postcolonial territories of European empire. In the presence of great uncertainty and absence of verifiable empirical evidence, so-called "environmental orthodoxies" or "degradation narratives" came to dominate scientific understanding of environmental change (Leach and Mearns 1996; Fairhead and Leach 1996; Stott and Sullivan 2000). State-led nature conservation practices that began in the Americas around the time of *Man and Nature*'s 1864 publication, and eventually spread throughout the colonial world, emphasized the separation of humans from nature and the creation of territorially bounded, fortress-style protected forests, wildlife reserves, and national parks.

In recent decades, ideas of wilderness, primeval nature, and stable climax communities have given way to ideas of nature-culture hybrids, socially produced nature, and second nature (Smith 1984; Castree and Braun 1998; Zimmerer 2000). Some biological conservationists have attacked these developments as politically dangerous, arguing that such hybrid and relational perspectives on nature justify ecologically destructive practices (Soulé and Lease 1995). The European Union (EU), however, has embraced hybridity in their biodiversity conservation strategy. According to the European Environmental Agency (EEA), Europe's biodiversity is the product of centuries of human interaction with nature (EEA 2006). Land abandonment resulting from demographic and socioeconomic shifts in rural Europe is thus "considered detrimental to biodiversity" (EEA 2004, 2). The European Commission's (EC) Directorate General, Environment, thus concludes that "traditional management" has produced high levels of biodiversity in Europe's forests "that might be lost if these areas were abandoned" (EC 2003b, 13). Hybridity rather than duality provides the foundation for the EU's environmental narrative and conservation strategy.

In this chapter I analyze the EU's forest biodiversity conservation strategy as a regionally-based environmental narrative, which I refer to as the EU biodiversity narrative. I position the EU biodiversity narrative within a global one and draw comparisons with other regional narratives, especially Africa. My goal for the chapter is threefold. First, I use the cases of Africa and Europe as a means to reveal the deep philosophical and ideological differences in conceptualizing nature-society relations from one world region to the next. Second, I challenge the exceptionalism of the EU's biodiversity narrative, using evidence from Africa and other world regions. Third, I examine conservation and rural development policies of the EU as a first step toward rethinking the causes of and solutions for biodiversity loss in the Global South.

Biodiversity Crisis: Global and Regional Narratives

Along with climate change, biodiversity, a recently invented term, has become one of the most powerful global environmental narratives. It is an all-encompassing concept that refers to the diversity within the entirety of the world's biosphere, as measured at different levels, including between individuals, populations, species, communities, and ecosystems. Biodiversity thus encompasses genetic, species, and habitat diversity, although it is most commonly measured as species diversity in the scientific forestry and conservation literatures. Although many scientists are convinced that we are experiencing a global biodiversity crisis, there is a great deal of uncertainty involved in quantifying specific trends. Problems include a paucity of baseline data (notably the absence of a comprehensive species inventory), multiple methodologies producing incompatible results, a general lack of time series data, and an inability to determine the degree to which our assessment of biodiversity reflects actual trends in the earth's biosphere or is an artifact of monitoring (Secretariat of the Convention on Biological Diversity 2001; Bowker 2000). Consequently, practitioners have labeled conservation biology a "crisis science," meaning that in a context of great uncertainty, as with certain branches of medicine, the risks of inaction are greater than those of action. Critical observers suggest that there is a deep philosophical problem revealed in the conceptualization of biodiversity within conservation biology; the tendency to equate biodiversity with wilderness in a pristine, prehuman state (e.g., Cronon 1995). At least among environmental historians, this sort of "rigid distinction between wild nature and human communities" is rejected as a model for understanding biodiversity threats in Europe (Ford 2007, 114). Whether driven by these critical insights or not, the institutions of the EU have articulated a narrative of biodiversity loss and conservation that has little to do with the wilderness model of nature.

Given that both the levels of biodiversity and numbers of species threatened with extinction are often geographically concentrated, some conservation biologists base protection strategies on biodiversity "hotspots." Norman Myers coined the term hotspot in 1988 to identify areas in the biosphere that have high levels of species diversity, endemism, and habitat loss. Since the term was introduced, conservationists have identified thirty-four hotspots around the world in both terrestrial and marine environments. Biodiversity hotspots have become a major focus for conservation programs and funding, with Conservation International building its program agenda around the concept, and the World Bank and MacArthur Foundation providing hundreds of millions of dollars of financing.

The hotspot initiative provides a window through which to examine regional variations in the global biodiversity narrative. The following passage on a West African forest hotspot (Guinean Forests) is excerpted from Conservation International's website:

> The Guinean Forests of West Africa hotspot is one of the most critically fragmented regions on the planet . . . slash-and-burn agriculture, has exacted the greatest toll on the region's forests. . . . With human population exploding in the region, fallow periods are becoming shorter.[1]

This is a common, neo-Malthusian degradation narrative about tropical forests that echoes Marsh's man-and-nature duality (although pristine is placed in scare quotes). The website's description of the Mediterranean Basin hotspot, however, moves decidedly in the direction of hybridity.

> The Mediterranean Basin has experienced intensive human development and impact on its ecosystems for thousands of years, significantly longer

than any other hotspot. . . . The agricultural lands, evergreen woodlands and maquis habitats that dominate the hotspot today are the result of these anthropogenic disturbances over several millennia. Paradoxically, grazing and fire can maintain species richness, while in their absence, closed forests are often less diverse.

Conservation International treats this region's hotspot as exceptional and recognizes that biodiversity is anthropogenic. The former observation, Europe's exceptionalism, is a theme that recurs often in the EU biodiversity narrative. The latter observation is so incongruent with mainstream conservation biology models that Conservation International can only consider paradoxical the fact that an anthropogenic forest is more diverse than a hypothetically pristine one. The remainder of the chapter is dedicated to the critical examination of these two observations.

Regional Degradation Narratives: Africa

Colonial forestry experts, trained in the temperate zone, arrived in the Global South with limited understanding either of the complexity and dynamism of its tropical forest ecosystems or of local environmental knowledge and practices. In the absence of experience and knowledge, forest scientists relied on an ideological mix of Marshian ideas of nature's delicate balance combined with a racialized, hierarchical concept of humankind and a faith in the superiority of Western science. Ideology thus colored their interpretations of forest conditions and dynamics. The mere presence of native peoples could be read as an indication of environmental degradation, and foresters repeatedly identified the "native agriculturalist," "nomadic herder," or similar essentialized actor as the primary threat to trees and forests (see Neumann 1998; Davis 2005; Doolittle 2007). The crude portrait of natives as nature's destroyers provided European colonial states (and, later, postcolonial states) with the moral and scientific authority to seize control of land and natural resources. Fortress-style parks and reserves—spaces that require the wholesale removal of human settlement and use—logically follow from this colonial perspective on nature-society relations. It is worth digressing to note that European parks and reserves generally are not constructed on the fortress model, but rather allow settlement and use to continue, albeit closely regulated.

I refer to my earlier research in Tanzania to explore how state-sponsored forestry and wildlife science conceptualized nature-society relations in colonial and postcolonial Africa. In my study at Arusha National Park I found that a history of community displacement and loss lay at the heart of an ongoing political conflict over biodiversity conservation (Neumann 1998). The site was declared a forest reserve toward the end of the nineteenth century. Foresters' earliest efforts were aimed at stopping traditional livestock grazing and pasture burning. Later, the Forest Department declared the collection of all nontimber forest products (NTFPs) off-limits to the neighboring communities (Neumann 1997). The newly independent government designated part of the forest reserve as the country's first postcolonial national park. The postcolonial state expanded the park boundaries several times by annexing African ancestral lands, ultimately creating a wilderness park encompassing most of the forest reserve. Another digression is warranted to note that the EU's biodiversity conservation strategy actually *encourages* "traditional" practices of burning, transhumant livestock grazing, NTFP collection, coppicing, and many other practices that Tanzania's colonial and postcolonial governments banned in virtually all forest reserves and national parks.

I documented a similar history of colonial and postcolonial displacement in southeastern Tanzania (Neumann 2001). The Selous Game Reserve is popularly recognized as Africa's largest remaining expanse of primordial wilderness. It is indeed roadless and unpopulated, but it became a wilderness only recently. Over the centuries many African peoples, agriculturalists, hunters and gatherers, and livestock herders, have occupied this territory. In the 1930s, however, the Tanzania Game Department had difficulty managing agriculture-elephant conflicts. They devised a scheme of enormous geographic extent that required all elephants to be herded into the reserve area and "shot out" everywhere else, and all human settlement to be relocated outside of the reserve boundaries. Ultimately the colonial government relocated 40,000 people in the 1940s in the largest forced resettlement in Tanzania's colonial history. The postcolonial government curtailed all remaining local community access to NTFPs within the reserve in 1989.

The histories of Arusha National Park and the Selous Game Reserve represent typical stories of fortress-style conservation in Africa. Such efforts to protect forests and conserve biodiversity through mass evictions and restrictions on traditional land-use practices are widespread in Africa and the Global South. They are based on neo-Malthusian concerns and claims of irrational and unsustainable resource use. Such orthodox, mainstream views of environmental destruction have come under increasing scrutiny in the postcolonial world. Advances in nonequilibrium ecology and ethnographic studies of African agroforestry and agro-silvo-pastoralism show these claims to be based less on science and empirical evidence than on ideology and political interests. In a path-breaking study, Fairhead and Leach (1996) inverted the widely accepted scientific thinking that indigenous land and resource use had led to the historic and continued destruction of West African tropical forests. Rather, they concluded, *"human settlement has generally been responsible for forming forest islands"* (1996, 79, emphasis in the original). Another study from West Africa found that rare species may be dependent on human-derived habitats and that biodiversity levels for some taxa may be enhanced by land-use practices (Kandeh and Richards 1996). Though these studies appear controversial and paradigm shifting in the context of African environmental history, in the EU's biodiversity narrative, similar declarations of anthropogenic biodiversity are accepted as "normal science."

Forests and the EU Biodiversity Narrative

European Union environmental officials observe that there is a "biodiversity crisis in western Europe" (EEA 2006, 20). The EU Strategy for Sustainable Development, adopted by the European Council in Gothenburg in 2001, included a plan to halt the loss of biodiversity by 2010. The cornerstone of the EU's efforts to protect habitats and thereby halt biodiversity loss is the Natura 2000 network of protected areas, the most significant and comprehensive initiative for nature conservation in European history. Natura 2000 is a pan-EU system intended to create the conditions necessary to conserve habitats and the biodiversity that they support (figure 3.1). The EU's 1992 Habitats Directive (92/43/EEC) and 1979 Birds Directive (79/409/EEC)— the primary legislative instruments for Natura 2000—together require all member states to identify sites for inclusion in the network. The selection and designation of Natura 2000 sites is based on a hierarchical classification system of European habitats, called CORINE (Coordination of Information on the Environment) biotopes.

Through analyzing the CORINE biotopes *Interpretation Manual* (EC 2003a) and other EU documents, an EU biodiversity narrative emerges. First, many of the CORINE habitat types, including forest habitats, are defined as being produced by long-term human husbandry. Furthermore, for many of the forest habitat types, the nature-and-society boundary is blurred and indeterminate. "The distinction between spontaneous forests and long-established formations of artificial origin is often difficult" (EC 2003a, 122). Ecologists of European forests agree that "the concept of natural forest is vague" and that "due to long and widespread human influence, finding and defining the natural forests has proven to be extremely difficult" (Rouvinen and Kouki 2008, 135). Because many highly biodiverse habitats were shaped by human use, and because the nature and society boundary is indeterminate, the EU has emphasized the importance of traditional land management practices for biodiversity conservation. According to the EEA, Europe's biodiversity

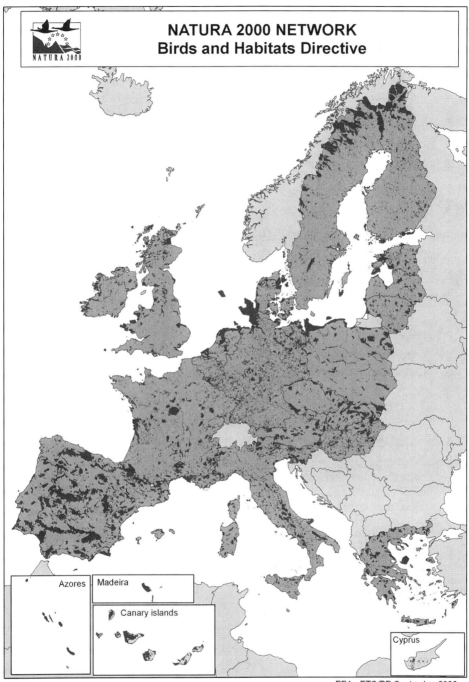

NATURA 2000 NETWORK
Birds and Habitats Directive

Azores Madeira

Canary islands

Cyprus

EEA - ETC/BD September 2006

3.1. Natura 2000 sites in the EU-25

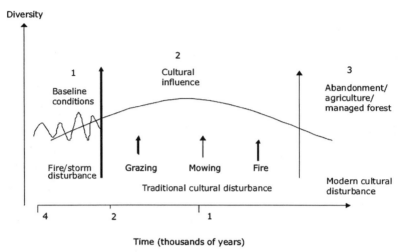

Diversity

1
Baseline
conditions

Fire/storm
disturbance

2
Cultural
influence

Grazing   Mowing   Fire

Traditional cultural disturbance

3
Abandonment/
agriculture/
managed forest

Modern cultural
disturbance

4   2   1

Time (thousands of years)

3.2. Model of changes in biodiversity associated with the development and abandonment of traditional agriculture in Europe. Source: EEA 2006.

peaks under human management of extensive land-use systems (figure 3.2).
As the EU biodiversity narrative has it, Europe's biodiversity, in contrast
with other world regions, is exceptional, perhaps unique, in its hybridity:
"In Europe, more than on any other continent, the influence of human activity has shaped biodiversity over time" (EEA 2006, 19). European forest
ecologists expound exceptionalism consistently and seemingly without the
need for documentation (e.g., Scarascia-Mugnozza et al. 2000; Farina et al.
2005). The problem facing the EU, however, is that the traditional, extensive land-use practices that helped create its biodiversity are disappearing.
The EU has linked biodiversity loss to changes in agriculture and rural occupancy, especially rural abandonment. According to the EEA, "A loss of
biodiversity and heritage landscapes is almost always associated with farm
abandonment" (2006, 36).

Forests are critical to the EU's goal of halting biodiversity loss. Over half
of all Natura 2000 sites include forest habitats, and the EU gives priority to
integrating biodiversity conservation into forest management (EC 2003b).
Coordination of EU forest policy falls to the Ministerial Conference on the
Protection of Forests in Europe (MCPFE). Rescaling forest policy and management from the national to the supranational level for Natura 2000 and
other pan-EU initiatives necessarily rescales forest science. Specifically, the
question "What is a forest?" had been answered previously by twenty-eight
separate national forest agencies. The categorization, measurement, and
bounding of forests has to be standardized across the EU, a challenge made

Table 3.1. Classification of *dehesas* habitat under four EU classification systems

Classification System	CORINE[a] Land Cover	CORINE Biotopes	Current MCPFE[b]	EEA[c] Proposed European Forest Types
Dehesas Classification	Agroforestry area	Grassland	Nonforest (i.e., not included in classification)	Broadleaved evergreen forest

[a]Coordination of Information on the Environment
[b]Ministerial Conference on the Protection of Forests in Europe
[c]European Environmental Agency

more difficult by the EU's use of several vegetation classification systems. For example, the CORINE land cover project uses three categories of forests. The CORINE biotopes project, which is the system used for classifying Natura 2000 sites, uses six. There is a proposed new forest classification system already being used by the EEA that probably will be adopted soon. What are classified as forests under one system are not in the other. As an illustration, the vegetation formation, *dehesas*, a woodland habitat of high biodiversity found on the Iberian Peninsula and managed in an agro-silvo-pastoral system is defined very differently in each classification system (see table 3.1).

Interestingly, the MCPFE also classifies forests by "degree of naturalness," of which there are three grades: forests undisturbed by man, semi-natural forests, and plantation forests. The great majority (88 percent) of forests in Europe, including those with relatively high levels of biodiversity, are classified as semi-natural. It is worth examining the description of the former two grades to highlight how the boundary between nature and society is negotiated within European forestry.

> Forests undisturbed by man—Forest showing natural forest dynamics, such as natural tree composition, occurrence of dead wood, natural age structure and natural regeneration processes, the area of which is large enough to maintain its natural characteristics and where there has been no known significant human intervention or where the last significant human intervention was long enough ago to have allowed the natural species composition and processes to have become re-established. (EEA 2008, 93)

The question "What is a natural forest in Europe?" is answered with a tautology. A natural forest is one that has natural forest dynamics, natural tree composition, natural characteristics, or, if it is left alone long enough, natural species composition and processes. Semi-natural forests are a hybrid of nature and husbandry.

> Semi-natural forest—Forest of native species, established through plant-ing, seeding or assisted natural regeneration. Includes areas under inten-sive management where native species are used. . . . Naturally regener-ated trees from other species than those planted/seeded may be present. May include areas with naturally regenerated trees of introduced species. Includes areas under intensive management where deliberate efforts, such as thinning or fertilising, are made to improve or optimise desirable func-tions of the forest. (EEA 2008, 94)

Forests are considered to be semi-natural if they have native trees that are planted, or if they have nonnative trees that are not planted, but "forest" presumably does not include nonnative trees that are planted.

Natura 2000 and other EU programs have made biodiversity central to forest management and planning in the EU, a priority shift that has caused some skepticism among forestry agencies in member states (Bengtsson et al. 2000). Perhaps in response to this skepticism, the European Commis-sion found it necessary to publish an interpretation guide, *Natura 2000 and Forests*, specifically to address the "widespread misconception" that the EU was creating "a system of strict nature reserves" (EC 2003b, 12). The docu-ment lays out a set of assumptions that will guide forest biodiversity policy.

> The premise of this document is that Europe's natural heritage has been transformed by centuries of human use. (EC 2003b, 7)

> The majority of sites of [Natura 2000] have been influenced by human cul-ture for hundreds of years. In many cases, it is this very human influence that has contributed to development of an ecologically valuable habitat. (EC 2003b, 12)

> In the case of forestry . . . the conservation of biodiversity often depends on the maintenance of human activities. (EC 2003b, 14)

What does the shift to biodiversity conservation mean in terms of existing silvicultural practices? For some biotopes, threatened species require forest habitats that are actively managed through coppicing, pollarding, and other traditional, extensive practices such as transhumant livestock grazing (EEA 2008). According to the EC's interpretation manual, "Certain forms of forest management can have positive effects on biodiversity" (EC 2003b, 21). Forest

3.3. A shepherd and his flock in Spain's Sierra de Cazorla. The EU views the maintenance of such traditional silvo-pastoral systems as critical to biodiversity conservation
in Mediterranean Europe.

"naturalness" and biodiversity are thus often inversely related in the EU: "In
general it can be concluded that those sites with the highest biodiversity . . . are
probably the ones that have undergone the most intense change" (EC 2003b,
95). The overall message of EU documents is clear. Forest biodiversity in Europe is a hybrid of nature-culture interaction, and no forest communities
need worry about being evicted wholesale to enforce Natura 2000.

Indeed, rather than the mass evictions that have marked the history of
biodiversity conservation in Africa, EU biodiversity conservation policy is
aimed at *maintaining* so-called traditional, extensive land-use practices. According to the EAA, "The current concerns about biodiversity loss, including
within forests, are closely connected to the transformation of the traditional
landscape" (EEA 2008, 12). The EU literature often singles out Mediterranean Europe as an area with both high rates of biodiversity and high rates
of rural abandonment and general decline in traditional land management.
Consequently, the EU forestry strategy urges the "maintenance of traditional
management of those silvo-pastoral systems with high levels of biodiversity
which may be lost if these areas are abandoned (for instance, in the Mediterranean regions)" (EC 2003b, 24) (see figure 3.3). Hence, EU biodiversity
conservation strategy includes rural development policies, including direct
payments to land managers, designed to keep people in the countryside.

Chief among revenue transfer instruments is the Common Agricultural Policy (CAP). The EU treated CAP to a series of reforms over the past

twenty-plus years, such as introducing opportunities for farmers to receive financial support for activities other than farming. In 2003 the environment became central to CAP and farm payments are now based on their historic levels, provided farmers "undertake to comply with a suite of EU directives (including the birds and habitats directives) and keep their land in 'good agricultural and environmental condition'" (EEA 2006, 35). These are referred to as "first pillar" payments, which were originally intended as agricultural productivity catalysts, but are now tied to environmental conditions. "Second pillar" payments fall under so-called agri-environment schemes and less favored area schemes, which member states can use for promoting the maintenance of "environmentally friendly farming systems" (EEA 2004, 12). With as much as 50 percent of the Natura 2000 sites in Mediterranean Europe dependent on (often economically marginal) extensive farming practices, such policy measures are critical.

> Where the continuation of positive farming practices is uneconomic to farmers, however, the active management of important habitats needs to be supported by additional measures. Agri-environment schemes and other CAP policy instruments are likely also to play a key role, therefore, in maintaining the conservation status of the future Natura 2000 network. (EEA 2005, 97)

For example, farmers in areas designated as "less favorable" are eligible to receive per hectare payments under the second pillar of CAP while continuing to receive conventional first pillar payments. In sum, as a key part of its biodiversity conservation strategy, the EU is subsidizing economically uncompetitive traditional land-use practices on marginally productive lands.

Discussion and Conclusion

Is Europe's biodiversity exceptional in its hybridity? Using similarly limited evidence—broad extrapolations from archeology and geographically and temporally limited ecological case studies—the answer is no. Indeed, much of this volume can be read as a refutation of European exceptionalism (see especially the chapters by Heckenberger et al. and Erickson). Studies from virtually all regions suggest that humans are key in shaping biodiversity at multiple levels, from habitat diversity to genetic diversity. Recent research in Africa is illustrative (see Beinart 2000 for a review). In addition to the West African studies highlighted previously, studies in East Africa have determined that the region's savanna is not pristine, but rather shaped by millennia of livestock herding and burning (Homewood and Rodgers 1991;

Moe, Wegge, and Kapela 1990; Shetler 2007). Given that all evidence points to the evolutionary origin of *Homo sapiens* in this region 200,000 years ago, conservation biologists should be at a loss to explain how the region's biodiversity managed to remain free of anthropogenic influence. Indeed, the only logical conclusion to be drawn from archeological, anthropological, and ecological findings is that humans have created and shaped the landscapes and biodiversity of East Africa no less than we have in Europe.

The exceptionalism at the core of the EU biodiversity narrative has been constructed without contrasting evidence from other world regions. Also absent is an explicit theoretical argument that would explain why the genesis of Europe's biodiversity should be exceptional. In the absence of inductive theory or careful hypothesis testing of the role of humans in the production and/or destruction of the biodiversity we observe, narrative becomes key in explaining its form, distribution, and quantity. A study of Southeast Asia's forest narratives demonstrates how explanations of the origins of biodiversity hinge on whether events are read as the beginning or ending of the story. "If we see [the swidden cycle] as starting with a forest cut, the forest itself gets naturalized. If we see it as the last in a series of stages starting from planting crops, the forest is a product of the process of fallowing after planting" (Peluso and Vandergeest 2001, 767). From this example we see how temperate-based or temperate-trained experts may be reading the landscapes of Europe from the latter perspective and other landscapes from the former perspective. That is, in Europe biodiversity is humanized and incorporated into narratives of culture and history. In the tropical South, biodiversity is naturalized and incorporated into a narrative of an undiscovered primordial wilderness.

Conservation policy in the Global South is increasingly out of step with the findings of contemporary archeological, geographical, anthropological, and ecological research. The contrast with the EU conservation policy is striking. Across Africa, colonial and postcolonial states have forcefully evicted people from their ancestral lands in the name of biodiversity conservation, while in Europe EU conservation efforts are directed at keeping rural people in place. In Latin America and Southeast Asia, states have implemented penalties and modernization programs to halt traditional swidden agriculture. African pastoralists have faced decades of state pressure to destock, fence, and sedentarize. Meanwhile in Europe, agroforestry, agro-silvo-pastoral systems, and extensive grazing regimes are all considered "high nature value" land uses that are subsidized to remain in place against all social, economic, and demographic odds. Why are traditional extensive farming practices in the tropics inherently threatening to biodiversity, while in Europe these practices are inherently constructive of a biodiverse landscape? If any question remains about the extreme regional contrasts in conservation policy, imagine that Maasai pastoralists, rather than being rounded up by military and police forces and having their communities razed, were subsidized to continue traditional grazing and burning practices in the East African savanna.

In both Africa and Europe, most conservationists and policymakers act according to the respective regional environmental narratives. The EU biodiversity narrative has inverted the neo-Malthusian perspective on the population-environment relationship that has dominated biological conservation theory and practice in other regions. At the very least, such an inversion opens the possibility that population pressure will be questioned as the sole or even principal explanation for biodiversity loss in the Global South. Combined with paradigm-shifting new findings—including many reported in this volume—from archeology, geography, history, and anthropology on the long history of human occupation and manipulation of purportedly "pristine" tropical forests, the EU biodiversity narrative could lay the groundwork for an alternative story of biodiversity and alternative conservation models for Africa and elsewhere. If a key goal of forest management is to slow and halt global biodiversity loss, then the EU biodiversity narrative suggests that the Marshian vision of pristine nature first needs to be reassessed region by region.

4 ✳ *Adam Smith in the Forest*

FREDRIK ALBRITTON JONSSON

In the eighteenth century, the Scottish elite discovered the problem of deforestation. This was something of a belated revelation, given that Scotland had lost most of its woodlands in prehistoric times. But Scottish natural historians and agricultural writers mistook the slow rhythm of climate change and prehistoric clearing for a moral narrative about invasion and neglect. Indeed, they looked at climate change itself as an anthropogenic event. They imagined that the destruction of the great ancient forest of Scotland had made the land cold and wet. Peat moss was a sign of moral failure. Schemes of afforestation were acts of restoration. This creative anachronism proved fertile, sowing the seeds of modern Scottish conservationism (Lindsay 1977; Smout 2000; Smout, Macdonald, and Watson 2005).

The story of eighteenth-century Scottish anxieties about deforestation also sheds new light on the origins of modern economics. The alarm coincided more or less exactly with the rise of classical political economy. Adam Smith wrote and revised *The Wealth of Nations* in the midst of these conservationist campaigns. How did Smith respond to such environmental anxieties? By contrasting *his* forest with that of the Scottish natural historians, it becomes possible to see the Scottish Enlightenment as a contest between different forms of economic expertise vying for primacy. Their disagreement centered on the character of the natural order. Where natural historians stressed the instability of the natural world on the periphery, political economists emphasized its properties of self-regulation and internal harmony. Natural historians used the argument from instability to insist on the importance of long-term management of forest resources by the state and aristocratic landowners. Political economists instead treated timber as a common commodity without political or patriotic significance. The anxiety about forest clearance thus served as a wedge that helped drive apart the enlightened sciences of natural history and classical political economy. The same social and political forces that mobilized Scottish elites around the problem of deforestation rendered it invisible in Smith's *Wealth of Nations*.

A Denuded Land

The field of classical political economy emerged at a moment of increasing environmental pressure. At the end of the eighteenth century, British society was pushing up against the limits of the advanced organic economy. Yet the path forward into a mineral-energy economy was not clearly perceived either by politicians or savants. What may seem like a perverse indifference to the power of steam and mechanized industry becomes more intelligible when we link such attitudes to revised accounts of the pace of industrialization. These show that only in the second quarter of the nineteenth century did industrial production begin to take on a dominant role in the British economy. A genuine optimism about the prospects of the industrial economy may not have surfaced until the middle of the nineteenth century and the popular success of the Crystal Palace exhibition of 1851 (Wrigley 1988; Caton 1985; Richards 2003; Pomeranz 2000; Williams 2003; Matthew 2000). From this perspective, the notion that the early political economists failed to predict the industrial breakthrough turns out to be an anachronistic judgment. For Adam Smith and T. R. Malthus, improvement of agriculture, the division of labor, and the liberalization of trade offered the main possibilities for growth within the ecological limits set by the organic energy economy (Wrigley 1988).

When we do away with the idea of the Scottish Enlightenment as a preliminary step toward an industrial economy, we also become more attuned to how environmental anxieties shaped intellectual concerns in the period. A prime example is the widespread interest in planting and forest history among the Scots. When Samuel Johnson and James Boswell made their famous tour of the Scottish Highlands in 1773, they were traveling through one of the least forested regions in Europe. General Roy's military survey (1747 to 1755) indicated that no more than 4 percent of the total land surface of Scotland was covered by woodland. The Englishman Johnson sneered that "Few regions have been denuded like this, where many centuries must have passed in waste without the least thought of future supply" (Johnson and Boswell 1984, 39; Smout 2000). Yet Johnson's comment was hardly fair. Indeed, General Roy seems to have seriously underestimated the woodland cover. The true number for the period was likely closer to 8 percent. Scottish landowners and sylviculturists had already begun to take an interest in afforestation by the early decades of the eighteenth century. Among the more spectacular initiatives was that of Archibald of Monymusk, whose men planted as many as two million trees before 1754 in northeast Scotland, and the Dukes of Atholl, who between 1730 and 1830 succeeded in planting over twenty-one million trees on their estate in the central Highlands.

Another measure of how seriously Scottish elites embraced the planting agenda was the enthusiastic support offered for William Boutcher's forestry manual *Treatise on Forest Trees*, published in 1775, the year of Johnson's travelogue. No fewer than 441 subscribers pledged to buy one or more copies of his book. The supporters included a very large portion of the Scottish aristocracy and gentry as well as many professionals in Edinburgh (House and Dingwall 2003; Smout, Macdonald, and Watson 2005; Boutcher 1775).

The mission to plant new forests was cast as a peculiarly aristocratic responsibility. Great landowners like the Earl of Haddington regarded tree planting as an act of paternalist foresight that would increase the value of one's patrimony while serving the greater public. In a similar spirit, the economist James Anderson encouraged the gentry to embrace forestry as a slow, but particularly secure, form of investment in his 1771 newspaper campaign. Afforestation here represented a practical version of conservative ideology, which combined an orientation toward landed investment with an ideal of long-term organic growth. Anderson and his compatriots also emphasized the duty of landowners to plant woods on marginal soils and at high altitudes. Plantations were a means of redeeming wastelands and fulfilling a providential plan of optimal land use within the nation. Trees were civilizing agents that transformed the wilderness into a valuable investment. Something of this rhetoric can be traced back to the naturalist surveys commissioned by the British government on the Annexed Estates confiscated from rebellious Highland landowners after the 1745 rebellion. The natural historian John Williams proposed a scheme of extensive oak plantations on the Annexed Estates. This plan was in turn linked to increasing worries about the timber supply of the nation. During the Napoleonic Wars, the fourth Duke of Atholl converted his mountainous estate into a massive plantation of larch trees, intended as a substitute for the dwindling oak stands of England. Planting thus became connected with the desire for strategic autarky in wartime (Anderson and Taylor 1967; Smout 2003; Hamilton 1761; Anderson 1777; Albion 2000; Albritton Jonsson 2013; National Archives of Scotland 1770–1771. File E 727/46/22, 12).

Such patrician patriotism did little to satisfy the local needs of the poor. In the Gaelic culture of Highland cottagers and subtenants, the wood was a deeply ambiguous place. It was the gateway to the world of the spirits and the refuge of *ceatharnaich-choille*—"forest warriors" or brigands. To be an outlaw was to be *fo choill*—"under the forest." The forest also served as a poetic symbol of kinship, particularly in the sense of clan solidarity. The fatherly chieftain was compared to a tall tree towering over his people. However, in economic terms, it is not clear that plebeians particularly welcomed the wave of planting across Scotland. Indeed, there are signs that planters and plebeian tenants were locked in a quiet struggle over land use towards the end of the eighteenth century. John Williams disparaged traditional forms of land use by alleging that usufruct led to a tragedy of the commons. He recommended that all oak plantations in the Highlands be enclosed. Private property was the best means to conservation:

> For while they get their wood for nothing, every one will destroy all he can, that he may not have a worse share than his neighbour; and it is well seen what horrid havock they make, now they have the cutting of it themselves. (National Archives of Scotland, John Williams Memorial of 1770–71. File E 727/46/7/6, 11)

Young plantations were often damaged by grazing cattle and goats. Since the middle of the seventeenth century, cattle droving had been a central feature of the Highland economy. Even the poor would often keep a cow or goat and feed it on the commons. Black cattle grazed the lower slopes while sheep roamed in higher elevations. During the long wars with France, the price of cattle and the value of timber and coppice bark increased simultaneously and therefore exacerbated the conflict over land use. The most recent calculations hint at a modest expansion of forest cover in the period to about 9 percent of the land surface. However, contemporary perceptions were more pessimistic. John Sinclair's survey from 1814 gave a much lower estimate, around 5 percent (Newton 2000; Lindsay 1977; Smout 2003; Graham 1812; Smout 2000; MacInnes 1996; Smith 1982; Smout, Macdonald, and Watson 2005).

Resurrecting the Forest

The proponents of afforestation regarded their work as a patriotic form of restoration. They were merely returning Scotland to its ancient and authentic state. Over the period 1760 to 1815, a host of Scottish naturalists and agricultural improvers amassed a speculative history of the Scottish woodlands to prove this point. They drew on a wide range of sources, including ancient history, oral tradition, and botanical surveys. A recurring claim in these reports was that buried roots, stumps, and sometimes entire trees submerged in the peat bogs of the Highlands and Hebrides constituted remnants of a vast pine forest. In reality, these stumps were often four thousand years old, but the eighteenth-century naturalists tended to date them as Roman or more recent in origin.

Reverend John Walker led the way with his tour of the Hebrides in 1764. An early advocate of Linnaean method in Britain, Walker's fieldwork proceeded on the premise that providence had made nothing in vain, and that every local ecosystem, no matter how seemingly barren and impoverished, held hidden riches visible to the naturalist's gaze. On his tour of the Hebrides, Walker approached the islands with the attitude of a theological detective, looking for the loving hand of God in little known native plants and other local resources. While visiting Colonsay in the south of the archipelago, Walker observed a plantation of coppice woods near the harbor. These ash trees had grown "above 40 feet high" (Walker and Mckay 1980, 117) in less than thirty years. Evidently, when planted close together in sheltered conditions, trees could resist the damaging winds and salty foam from the Atlantic. After a long journey through treeless regions, the little copse seemed a revelation. Popular belief had it that no trees could grow in the Hebrides. In contrast, Walker saw the Colonsay ash trees as part of a conjectural history of Scottish woodlands. He noted that peat bogs across the region contained numerous remnants of trees. On this evidence, the timber famine of the islands was man-made: "The Woods are long ago extirpated by time, by the Cattle, and the Consumption of the Inhabitants" (Walker and Mckay 1980, 124). Walker's new history of nature insisted that the decay of woodlands could be reversed through ecological exchange. He called for a public nursery in every larger district of the Highlands, listing pitch fir, larch, silver fir, walnut, and sugar maple among others as desirable additions (Walker and Mckay 1980; Walker 1808).[1]

One frequent hypothesis of the Scottish writers was that the woodlands had been destroyed by foreign invaders in the form of Roman legions, Viking marauders, or even British government troops after the last Highland rebellion. Quite typical in this respect was Alexander Blackadder's report to John Walker's Agricultural Society of Edinburgh in 1790. Blackadder laid the blame on the Roman emperor Severus. After waging a futile campaign in Scotland, the emperor had come to the realization that the Caledonians could not be conquered "So long as they had so many castles—by which was meant their woods and morasses." Consequently, Severus had ordered his legions to destroy the Caledonian people by an act of systematic biological warfare: "Cutting the woods and draining the morasses." In Blackadder's mind, the Caledonian forest signified patriotic resistance and heroic national identity (Smout 2000; Smout, Macdonald, and Watson 2005).[2]

Perhaps the most significant and spurious evidence offered in favor of a primeval Scottish forest came in the form of the epic poetry of James Macpherson. His so-called translations from obscure manuscripts of doubtful authenticity depicted a world of Gaelic warriors of the third century AD. Macpherson set much of the action of *Fingal* in a dense and gloomy woodland that seemed to cover much of his imaginary Caledonia. Thanks to the immense popularity of his poetry, this mythical image may well have taken root among a wide circle of readers (Macpherson 1763).

The conjectural history of forest clearance contained a theory of climate change. Boutcher's treatise on forestry from 1775 presupposed the power of trees to affect ground temperatures. In the preface to the work, Boutcher (1775, viii–ix) proposed that "Making plantations of Forest trees . . . [offered] . . . the most solid foundation for promoting all the different branches of husbandry and gardening." Woodlands provided shelter from winds and cold air, making the land more hospitable to both native and foreign crops. Logically then, a good improver of the land should always begin by planting trees to prepare the way for other schemes of agriculture and horticulture. Through large-scale schemes of planting, patriotic landowners could ameliorate the climate of Scotland as a whole. Boutcher was drawing here on an ancient tradition of medical thought, which linked forests to climate change. But he was also reversing the received notion that forest clearance could make a cold climate more temperate. Whereas natural historians of colonial peripheries tended to see deforestation as a necessary step toward establishing European agriculture and a healthy, temperate atmosphere conducive to settlement, Boutcher's book pointed in the other direction and opened the door to a critique of anthropogenic cooling caused by deforestation (Boutcher 1775; Kupperman 1982; Glacken 1967; Golinski 2007).

In order to demonstrate the variability of climate, the Scottish naturalists turned to oral testimony and folk tradition. Reverend Dr. James Robertson of Callander observed a cooling trend in the uplands of Perthshire. He stressed that deforestation had recently made agriculture more difficult in the hilly districts of the country. John Smith reported a similar trend in Argyll. Very likely, the local informants who supplied this information were simply recalling the harsh winters of the 1690s or later periods of dearth. Yet naturalists and planters grafted such oral traditions to their own discourse about deforestation, thus occluding (*sensu* Leach and Fairhead 2000) the folk memory of the Little Ice Age with enlightened anxieties (Sinclair and Board of Agriculture 1795; Robertson 1799; Smith 1798). This line of research reached a climax in William Aiton's (1805, 1811) rambling *Treatise on the Origins, Qualities and Cultivation of Moss Earth*. Like Boutcher, Aiton wrote with the sanction of the Scottish establishment. His book was published under the patronage of the Highland Society of Scotland, the most powerful agrarian lobby group in northern Britain. Aiton's main thesis was that the great peat mosses of Scotland had been produced by the destruction of the native woodland and that peat itself was a product of decaying wood left to rot in the ground. Through experiments measuring the amount of water absorbed by peat moss, Aiton insisted that the mosses acted to retain moisture and frost, reducing surface temperatures. While most of the peat mosses were located across the Highland periphery, the accumulated effect was to dampen and cool the climate across the Scottish nation (Aiton 1811).

The forest histories from Walker to Aiton present a northern counterpoint to the science-based conservationism of the Indian Ocean in the Enlightenment. Thanks to Richard Grove's (1995) work, we know that an effort to elucidate the ecological effects of forest clearance gathered pace after midcentury on the French island of Mauritius. According to Grove, it was the French official and naturalist Pierre Poivre who first connected deforestation to problems of desiccation and erosion on Mauritius in the late 1760s. Drawing on a metropolitan science of plant respiration, Poivre persuaded the colonial government to implement a policy of conservation, setting aside forest preserves in the mountains in order to halt adverse climate change. For Grove, Poivre's science was produced by a convergence of multiple trends: the ecological pressures of European colonialism were particularly visible within a vulnerable island environment; the power of the French colonial state allowed it to overrule private property claims by planters; and an evocative European literary discourse elevated tropical islands into mythical emblems of Edenic purity that must be protected (Grove 1995).

Nothing but a mirage separated the Scottish history of the forest from that of French Mauritius. By understanding the chronology and causes of deforestation in terms of anthropogenic cooling, rather than Ice Age climate change, Scottish savants and improvers confused history with ecology. Yet their discourse followed tracks roughly parallel to that of Mauritius: climate was history; forests were idylls of the nation; environmental degradation required expert governance; improvement on the periphery should be guided by natural history. The main divergence was political: the state mobilized on a lesser scale in the Scottish case, leaving much of the work of afforestation to the great landowners. Moreover, the authority of Scottish natural history was contested by the rival expertise of classical political economy. Hence, the Scottish case uncovers a set of questions that elude Grove's research. How did Adam Smith respond to the environmental anxieties of his day? What was the place of the forest in the naturalist discourse of Smith?

A Token of Exchange

In the midst of the national outcry about the plight of Scottish woodlands, Adam Smith composed and published *The Wealth of Nations* (first edition 1776). Writing on the heels of Samuel Johnson's scathing remarks about the denuded state of Scotland, Smith certainly was familiar with the debate on deforestation. Yet, the political economist looked at this phenomenon as a simple indicator of progress:

> In its rude beginnings, the greater part of every country is covered with wood, which is then a mere incumbrance [sic] of no value to the landlord, who would gladly give it to any body for the cutting. As agriculture advances, the woods are partly cleared by the progress of tillage, and partly go to decay in consequence of the increased number of cattle. (Smith 1976, 183)

The reduction of woodland was, to Smith, an entirely desirable affair, analogous to the removal of the surplus population of small tenants from the countryside. Timber was a commodity with no sentimental value. Forest clearance constituted an important material dimension of Smith's stadial history. As humanity moved from the stages of foraging through pastoralism and settled agriculture, woodlands naturally contracted. When wood became too scarce to provide a source of heating fuel, the nation turned to alternative energy sources, including coal or peat. Only once in *The Wealth of Nations* did Smith attribute a strong ideological meaning to timber. While discussing the construction of the New Town of Edinburgh, he observed that there was scarcely "a single stick of Scotch timber" (Smith 1976, 183) in that entire section of the city. Instead, the wood had come across the sea from Norway and the Baltic region. The Georgian splendor of the New Town, the supreme material expression of the Enlightenment in Scotland, was thus a token of exchange rather than national autarky. Smith and his successors in the tradition of classical political economy had no patience for arguments about the strategic necessity of conservation. Timber served the same function as grain in *The Wealth of Nations*. Its economic value was shorn of any sentimental and patriotic association with native soil and self-sufficiency in favor of the symbolism of international exchange: peaceful prosperity and mutual advantage (Smith 1976; cf. Malthus 1826; Mill 1848).

But the environment was not absent from Smith's political economy. There was a hidden subtext to his argument, a political ecology underpinning his ideal of commerce. The stability of the natural world served as a model and a warrant for the operations of exchange.

In a Newtonian vein, Smith saw the "natural price" of commodities as a force analogous to gravitational attraction. He also looked to Linnaeus's notion of the self-regulating economy of nature to imagine the equilibrating processes in the exchange economy. In his discussion of the rates of profit on different kinds of capital, Smith spoke of soil fertility as a natural kind of labor, superior in productivity to that of manufacturing. He noted: "The land constitutes by far the greatest, the most important, and the most durable part of the wealth of every extensive country" (1976, 258). Agriculture set the limit to all other forms of growth in *The Wealth of Nations*. The efficiency of mixed husbandry determined the level of the food supply, population growth, urbanization, and the ultimate extent of the division of labor. On this reading, Smith recognized that advanced economies were subject to diminishing returns and would eventually settle into a mature form close to the ecological limits of production. He seems to have interpreted the stagnant performance of the Dutch economy in the eighteenth century as a version of such a steady state. From this perspective, Smith's political economy looks surprisingly like a forerunner of ecological economics, premised on the recognition that the economy formed a subset of the environment (Smith 1976; Schabas 2005; Wrigley 1988; Daly 1996).

Yet Smith employed natural knowledge selectively in his economic analysis. He was all too happy to embrace evidence of benign natural cycles while systematically rejecting evidence of ecological degradation and irreversible decline. *The Wealth of Nations* made little mention of the ecological disturbances and geographic peculiarities that preoccupied so many natural historians of the periphery. When Smith did discuss instabilities, he tended to follow two strategies. Either he assumed that nature was capable of correcting its own errors, as in his analysis of the ungulate irruption of cattle in colonial North America. Or in the case of famine, he employed a political means of rationalizing disaster. Here, he rejected an ecological explanation based on drought cycles and interpreted famine instead as the effect of monopoly corruption and the inhibition of free exchange. This seems to be the pattern that Smith followed in his discussion of the Bengal famine of 1770 (Johns 1999; Mulcahy 2006; Smith 1976).

The same selective view of the natural order offers a final clue about Smith's cool appraisal of deforestation. Simply put, climate had no history in Smith's economic theory. He was familiar with the eighteenth century commonplace of anthropogenic climate change from half a dozen learned works in his own library, including the natural historian Pehr Kalm's (1772) *Voyage to North America*, which he discussed explicitly in *The Wealth of Nations*. Yet Smith never made explicit use of the idea, not even in his account of deforestation as a civilizing movement. Instead, his text presupposed a climate uniform over time and distributed regularly in space. This geometric and uniformitarian view of the natural world resembled greatly the theory of nature embraced by Smith's close friend, the geologist James Hutton. The function of climate was to produce a natural basis for exchange. Different climate zones generated varying natural advantages of agriculture across the world, which in turn stimulated international trade. It was a simple, geometric notion fit for an enlightened deity (Smith 1976; Kalm 1772; Mizuta 2000; Hutton 1797, 1788).[3]

From this brief discussion, the full extent of Smith's indifference to forest clearance should be clear. The myth of the primeval Scottish forest assumed a world susceptible to violent ecological swings, natural disasters, and fundamental alterations. It also dictated that state and civil society must be mobilized to counteract these movements when necessary. Last but not least, the myth gave natural historians a central role to play in guiding the work of planning and restoration. Such a chain of reasoning was anathema to Smith for all the reasons mentioned. Yet his silence on a subject of such widespread controversy may have had the ironic effect of sanctioning the division of labor between natural history and political economy. As Smith's fame grew and his mode of economic analysis began to shape government policy and intellectual habits across civil society, the issue of ecological degradation on the periphery was left unattended for others to exploit. By rendering the problem of the Caledonian forest invisible in *The Wealth of Nations*, Smith may have unwittingly helped to consolidate the economic authority of natural historians on the margins of the nation and empire.[4]

5 * Jungles, Forests, and the Theatre of Wars: Insurgency, Counterinsurgency, and the Political Forest in Southeast Asia

NANCY LEE PELUSO AND PETER VANDERGEEST

Introduction

War affects forests and forestry. Forests have been the sites of many forms of political contest. While they have been the mantle of the poor, they have also been shelters of resistance and theatres of war. Forest-related institutions and the ideologies that govern them, all hallmarks of modern states, were also produced in Southeast Asia in the crucible of wars. Emergencies and insurgencies of the twentieth century, ethnic wars, and cold wars in the tropics shaped institutions and practices because jungles were the front lines. The direct impacts of modern war—napalming trees, blasting at wildlife, ubiquitous land mines, carpet bombing, and massacres—are well known. Yet less well known and barely documented (Kuletz 1998; Peluso and Harwell 2001; Peluso 2003b; Barber and Talbot 2003; Kaimowitz 2005) are the indirect, often invisible, and especially institutional effects of war, and an array of "medium hard" technologies of power: forced relocation (strategic hamlets); state-sponsored and spontaneous colonization of forest and indigenous territories (transmigration); criminalization of traditional practices (laws and decrees against the use of fire, shifting cultivation, etc.); exclusions of local populations from their traditional resource terrains with new legislation (national parks and forests); and the transformations of bureaucratic/ military structures and surveillance for the management of forests. These all have taken place within contexts of ethnic and racial conflicts and politics in and over forests.

In this chapter, we focus on the institutional dimensions of war in the creation of forest politics, policies, and representations. How did conflicts, especially insurgency and counterinsurgency in the Cold War, produce policies of exclusion of traditional, often politically suspect, ethnic or racialized populations; programs of colonization, such as resettlement into contested zones; and the transformation of peopled jungles into management regimes of militarized forests, social control, economic production, scientific forestry, conservation, and surveillance? What had been zones of jungle warfare were "tamed" through violence, lawfare, and bureaucracy in the context of the political contests—Communist, capitalist, or Islamic—over Southeast Asian nation-states after World War II.

We argue that war, insurgency, and counterinsurgency contributed to resurgence of forests in many areas as local agrarian-forest populations were moved out in order to undermine rebel supply and support. This included resettlement of traditional forest populations away from the woods into far more controllable spaces, such as strategic hamlets, or through directed colonization. It also included the imposition of forest regimes on spaces still occupied by people, whose continued residency turned them into criminals. New institutional contexts—modern forest ministries—emerged at this time for controlling access and rights to resources, and defined the kind of forests and activities that could be carried out in them (national forests, national parks, buffer zones, etc.). At the same time, these ministries consolidated spatial control of insurgent areas through violence and the instruments of a militarized forest management. Finally, the new forestry involved a sharp redefinition of the nature of agriculture and forests in policy, practice, and satellite imagery that was enormously at odds with the complex and interactive mosaic of forests and agriculture that had been the historical norm in forest systems in much of Southeast Asia. The practices and logics underpinning forestry, and the ensemble of laws, local institutions, and actions enacted in this period, produced in new ways what we have called elsewhere "political" forests (Peluso and Vandergeest 2001). The wartime interventions of national states in these regions were explicitly designed to serve emergent national political ends.

Because war makes states and states make war, it is important to understand that modern state forestry and its territories were an outcome of the violent politics through which new national states were made after World War II in Indonesia, Malaysia, and Thailand through their political associations with various emergencies and insurgencies, staged in forest areas then called jungles. Here, wild lands and insurgent peoples had to be transformed from the inchoate to the legible, and from exotic minorities into docile citizens, and here sovereignty over rebel territories had to be definitively claimed. The making of state forests has always involved the political act of allocating jurisdiction over territories to scientific forestry agencies, regardless of local histories or circumstances. Creating or maintaining state and state-authorized claims often involves violence, such as the forced removal of resident populations with competing claims or the armed prevention of perceived encroachments on state-claimed forest territories by local livelihood uses (Hecht and Cockburn 1989; Guha 1989; Poffenberger 1990; Peluso 1992; Vandergeest and Peluso 1995; Sivaramakrishnan 1997; Laungaramsri 2001; Bryant 1996; Rangan 2000; Scott 1998).

The period of political violence from the 1950s through the 1970s in Southeast Asia and in other tropical forest regions represented a transformation for forests in three distinctive ways. First, the jungles of wartime were shifted ideologically, discursively, practically, and institutionally to forests (see, e.g., Slater 1996; Peluso 2003b; Sioh 2004). This process involved casting the term "jungle" as a space of relatively unassimilated, barbaric, essentially tribal, people living in untamed nameless spaces of nature, or as threatening, colonially defined "aliens," such as Chinese, and contrasting this with both the calming, controlled aspects of settled agriculture and the idea of scientific (forest) management of the "intractable wild," whether woodlands or people. Shifting from inchoate jungles to scientifically manageable and legible forests further distinguished *forests* from *agriculture* politically, spatially, and in terms of agency jurisdiction (see also Dove 1992; Sivaramakrishnan 1999; Sivaramakrishnan and Agrawal 2000).

Second, state responses to jungle-based populations and threats to national sovereignty involved stabilizing and racializing categories of minority and majority ethnic groups in Southeast Asia in terms of their perceived allegiance (or lack of it) to new national states, in the fragile and fragmenting moments of early post–World War II state formation. Further, to the extent that religion is closely associated in Southeast Asia with the making of race and ethnicity—particularly the elision of "Muslim" and "Malay"[1] in Malaysia and Indonesia—forest-based radical Islamist conflicts were less demonized than Communist ones, because they were seen as less of a territorial and cultural threat at that time.

Third, technologically sophisticated surveillance grew in remote areas in the context and course of war. Insurgency stimulated the military technologies for surveilling jungles; these technologies later articulated with the intentions and needs of forest managers. The expense of such technologies precluded their early development for forest monitoring until the timber industry became an important part of national economies (Leigh 1998). After insurgencies were quelled, forest conservation and management benefited from the technologies developed for jungle warfare and counterinsurgency. These three processes created new ideas about forests and new practices that reshaped ecologies through both the practice of forestry and its institutions. We explore these ideas through forest politics in Indonesia, Malaysia, and Thailand.

Background: Forest-Based Emergencies and Insurgencies in Southeast Asia from World War II to the Cold War

The Japanese occupation of much of Southeast Asia (1942 to 1945) generated forest-based political resistance, first to the Japanese military and occupying forces, and subsequently to colonial powers returning to the region after World War II. Both periods of political violence involved occupation, war, and revolution, resulting in forest destruction and major population movements (Godwin 1940; Soepardi Poerwokoesoemo 1974; Kathirithamby-Wells 2005). During these wars, colonial state forestry continued in many parts of Southeast Asia, but the mission of state agencies was to contribute to the war effort. In Indonesia, the wartime Japanese government put the Forestry Department under the Department of War. The legacies of war included extensive timber cutting and the production of other crops, such as castor plants (*Jatropha*) for motor oil, all directed to strategic purposes. People were forced to move into some forests to cut them down, or moved spontaneously into forests to hide from the Japanese. The destruction of forestry infrastructure was also a part of the allies' scorched earth policy during the occupation (Soepardi Poerwokoesoemo 1974).

After World War II, the British returned to Malaya and the Dutch to Indonesia. Thailand was never formally colonized, and as a Japanese ally, it was not formally occupied during the war. In Indonesia, the Dutch faced immediate resistance. The Indonesian Revolution (1945 to 1949) affected forestry mostly in Java, especially the teak forests, which were harvested for fuelwood to power trains; in addition, teak timbers helped build railroads and roads. In colonial Malaya, the British declared an emergency in 1948, as the Malayan Communist Party (MCP), which had led an armed, jungle-based resistance to the Japanese during the occupation, became increasingly militant. Decolonization was delayed as the British responded to the MCP with counterinsurgency tactics that became a model for subsequent counterinsurgencies around the world. By the mid-1950s the MCP was all but defeated, and Malaya became independent in 1957. In 1963, Singapore and the Borneo states of Sarawak and Sabah joined Malaya to form the Federation of Malaysia. In Sarawak, armed resistance to the Federation of Malaya by Communist forces was supported internationally by Indonesia's President Sukarno, producing another jungle-based border conflict, the so-called *Konfrontasi*.

During the Malayan and Sarawak emergencies, half a million Chinese forest squatters were moved into camps called New Villages (Stubbs 1989, 286; Sioh 2004). In Malaysia, the forestry department developed silviculture models specifically for land where forest had been cleared for farms. The Malayan Uniform System had been initially developed as a technique for regenerating deforested areas cut by the Japanese and peasants resettled by Japanese during the occupation. The Malayan Emergency (1948 to 1957) also allowed the forest department to promote this technique as a scientific rationale for the expulsion of Chinese and other cultivators who had "occupied" forest villages and were suspected of supporting MCP insurgents (Wyatt-Smith 1947, 1949; Ali 1966). These scientific practices were both silvicultural management techniques and forest department actions that later facilitated their claiming the land as political forests.

Cold War Confrontations

The anticolonial insurgencies were important and set the stage for the most widespread type of political violence affecting scientific forestry and forests in Cold War era Southeast Asia: jungle-based or mountain insurgencies. Most of these were inspired by Maoist revolutionary ideas. The infamous slogan, "Let the countryside surround the cities," was meant to mobilize peasants and other rural subjects (jungle-based or not) to rise up and take over the cities where the governments of the newly emergent nation-states were based.

Not all forest-based insurgencies were Maoist. Regional Islamic rebellions in Indonesia (1950 to 1957), particularly in western Java, and parts of Sumatra and Sulawesi, were launched by Islamist militants desiring an Islamic state in newly independent Indonesia. Forest-based rebellions were largely associated with the *Darul Islam* movement and *Tentara Islam Indonesia* (the Islamic Army of Indonesia), referred to in shorthand as DI/TII. These rebellions were supported internationally by the United States, Britain, and other Western powers for their anti-Communist stance, even though they were challenging the Indonesian state and its leader, Sukarno.

Many of the jungle wars took place in what were, at the time, heavily forested border or difficult-access mountainous areas or swampy lowland forest. These forests had been long occupied and farmed by swidden cultivators, settled agriculturalists, and hunter-gatherers; many households combined all three types of activities. Central state control had not yet been effectively established, either on paper or in practice on the ground. The political threats posed by the insurgencies to colonial (in Malaya and Sarawak) and post-colonial states motivated these states to devote tremendous resources to strengthening their controls on these areas, both by "winning hearts and minds" (creating national subjects through engaging them in state programs of agriculture and schooling), and by turning "wild jungles" and "remote mountains" into zoned and controlled state territories.[2] In each region, the ways that governments asserted control over these peopled, uncontrolled, violent spaces they called jungles shaped the future practice of state forestry and the forests themselves.

A Clash of Civilizations: The Spaces of Insurgency and Ideologies of Jungles

The major episodes of political violence on which we base our arguments include: The 1948 to 1957 Malayan Emergency in Peninsular Malaya; the Communist Party of Thailand's insurgencies from the mid-1960s to the early 1980s; the violence between Indonesia and Malaysia (Sarawak) in Borneo as part of Confrontation (1963 to 1966); and the complex Communist and Islamist insurgencies in Indonesia and Sarawak in the 1960s and early 1970s. Besides those directed toward the Islamist forest-based rebellions (Nasution 1965), some eight years (1966 to 1974) of counterinsurgency operations were carried out in Sarawak and West Kalimantan.[3]

The political violence that engaged these insurgencies was not simple resistance to the imposition of colonial or post-colonial forestry, but was committed to what we call "alternative civilizational projects"—the potential to create polities such as Communist or Islamic states that differed from the actual political forms and orientations of the new nations that emerged in the wake of World War II and the Japanese occupation.

Many of the groups involved in forest-based political violence discussed here intended to build different forms of states based on Marxist, Maoist, or Islamist ideas. However, while some insurgents enlisted, attracted, or forced people living in forests to engage in violence or provide physical support for their forest-based fighters, these incidents of political violence and opposition did not emerge primarily from the desires of forest villagers themselves. Rather, students, organizers, party members, combatants, and others from urban and more integrated rural areas went "down" to the countryside or into the jungles and mountains, to stage the insurgencies.

Many insurgents saw themselves as alternative nationalists and freedom fighters. This contrasts with depictions in the press and the political science literature of forest-based violence and rebellion as the actions of recalcitrant tribal peoples in jungles, or remote mountains, in what Scott (1998) calls "non-state," inchoate spaces. Many of the topographic characteristics that Scott argued made these spaces ideal as non-state spaces in fact made them ideal for organizing insurgent states—for example, having difficult terrain to access, tree cover protecting them from overhead surveillance, and crops such as manioc that could not easily be confiscated. A central feature of the Maoist model was the "base area" where the party could gain territorial control, and from which they could expand to eventually surround and capture the cities (McColl 1967). Sites with the following characteristics were sought for base areas: some previous political or revolutionary activities (implying support from local populations); relatively easy access to major political targets; weak or ambiguous political control by states (thus they were often located at provincial or national boundaries); a terrain with cover (forested mountains were a common type); and the potential for economic self-sufficiency (meaning support from local populations) (McColl 1967). These desired characteristics fit the circumstances of insurgents in Indonesian Borneo, Java, and Sarawak who operated from border areas, mountains, and jungles.

As in China, the Communist parties in Southeast Asia were largely urban based, initially, and decided to move to the countryside and jungles after they were made illegal or violently attacked. The association of Communism with urban labor and intellectuals, as well as the adoption of Maoist revolutionary strategies, all contributed to the idea in Malaya and Thailand that Communism was not inherently Thai or Malay, but Chinese. These associations were a first step in the racialization of counterinsurgency.

The situation was somewhat different in Indonesia, however, because the Communist Party (PKI) and affiliate organizations were legal, not to mention key components of Indonesian politics; that is until March 1966 when Indonesia's second president, Suharto, criminalized Communist, Socialist, and other left parties and organizations. Although Communism was best organized in urban areas of Indonesia, they moved into the countryside through various recruitment pushes and actions (*aksi*) to engage workers in the forestry and plantation industries in Java and Sumatra, as well as peasants all over the country.

These insurgent, alternative civilizing projects had the potential to succeed as their successes in Indochina (e.g., Vietnam) and China clearly show. In our cases, further violence and land-use policies divided jungles into permanent political forests devoid of *legal* occupants/residents, territories under the aegis of forestry agencies, and settled, controllable agricultural zones whose people were under the authority of civil administrations and territories largely managed by agricultural departments. Forests construed without authorized residents were the opposite of highly peopled, and thereby suspicious, dangerous jungles in many ways.

Counterinsurgency, Evictions, Colonization, and Political Forestry

People were forced or encouraged to move out of forests because settled agricultural villages were crucial to counterinsurgency operations. Ultimately, the goals of states engaged in counterinsurgency at this time coincided with those of political forestry: they hoped to transform "jungles" from populated untamed, dangerous entities, into more orderly, managed forests with agricultural areas in villages next to them, whose people were located and locatable neatly and securely (Slater 1996; Peluso 2003b). One key aspect of counterinsurgency was thwarting insurgents' attempts to incorporate rural people into their movements by cutting off food and supply line links between insurgents and forest residents. This meant destroying physical access to prevent political education, empathetic support, and recruitment.

State actors reclaimed or expropriated peoples' land in the name of national security, then reallocated it in new forms of property. Local people in insurgent areas were made more dependent on central states through various forms of violent displacement and through incorporation into agricultural development schemes or programs. The landscape effects, property rights, and management goals of counterinsurgency varied. Sometimes counterinsurgency strategy intended to produce empty forests without settled people, privileging forest resurgence, protection, or management of woodlands. At other times forests were sites of directed colonization to be replaced with permanent agriculture.

In these ways, the transformation from "jungles" to forest and forest agriculture to forest resurgence was violently articulated with nationalist statecraft in terms of both forest territories—political forests—and the making of politicized, minority, national subjects.

Reorganizing Space and Reconstituting the Nation: Taking the Forest out of the Jungle

Cultivating Southeast Asian populations of forests and rice fields and mountain empires and lowland polities historically had multifaceted trade, social, and cultural interactions. Although Scott (2009) has most famously argued that forested uplands of Southeast Asia were constituted predominantly with those who resisted lowland governance, we are looking at a much more recent, and long formed landscape, where social and territorial interactions were complex and often ethnicized (see, e.g., Vandergeest and Peluso 2006a, 2006b). In the Cold War period, taking the forest—a management space— out of the jungle involved the rhetorical categorization and deployment of the idea of jungles as wild places occupied by wild people. This idea was important to both insurgency and counterinsurgency; jungles, like mountains, resided at the very edges of state power, which was why insurgents chose them for their base areas. If insurgents saw themselves as agents of an alternative nationalism, states characterized them as either primitive wild people or political renegades. These representations were meant to lend legitimacy to state projects aimed at controlling these regions.

The desires of the authoritarian governments to cleanse jungles of their undesirable primitives or political oppositions articulated conveniently with early environmental ideologies that tended to frame forests as uninhabited spaces of nature that needed to be set aside and away from human activities for conservation or biodiversity reasons. These political ideologies, like the political goals of counterinsurgency, dovetailed with goals of political forestry and scientific forestry, meant to transform jungles into orderly, spatially distinct, managed forests and controlled agricultural areas. Local people in areas occupied by insurgents were demonized and made dependent on central states through various forms of violent displacement and/or incorporation into agricultural development schemes or programs. Forest property rights, national jurisdiction, and various minority-majority questions were being addressed on national, rather than provincial, scales as earlier polities were transformed into the terrains of the nation-state, and as national forests and national parks came into being. Forests have a long and important role in the making of modern imagery of nation-states, including the United States, Brazil, and Germany, but this imagery involved idealized nature, not places fraught with warfare. In Southeast Asia, insurgents, "natives," and other empires had to be excised from forest histories.[4]

Racialization and the Political Forest

Ethnic and racial politics involved essentializing minority populations along the lines of biocultural (including genetic) as well as political difference. Three basic categories were deployed but differently inflected. "Tribal peoples" and "Chinese" were constructed as national minorities, while "lowland groups" were depicted as national majorities, whether or not the demographics supported this view. In some places, both tribal peoples and Chinese were designated additionally as alien races, based on ideas about the naturalness of national boundaries and migration patterns imagined historically back to times long before the emergence of the nation-state (Anderson 1984). Majority and minority status were connected to alleged political loyalties. In areas where majority peoples lived in both upland and lowland environments, uplanders, being more remote from state power, were more suspect than lowlanders. Association with Communist parties in Malaysia, Indonesia, Thailand, or with Islamists in Indonesia also rendered individuals living in any of these environments suspect.

Tribal peoples, such as Hmong (in Thailand), Karen, various groups of Dayaks, and Orang Asli were refashioned as "backward," but paradoxically also as warlike, fierce, and violent. Notably, they were seen to have potential to use their tribal warfare skills to support the insurgents' political violence against the mainstream states. The dark side of what Tsing (1999) called the "green development fantasy," in which forest-based, tribal peoples are represented as noble savages with ecological sensibilities, is the shift in perception during times of insurgency, such that the noble savage becomes the savage headhunter needing to be civilized.

Complicating this issue of national loyalties from the inside of "weak state" territories was the problem that tribal peoples' purported jungle knowledge was important to both the conduct of counterinsurgency operations as well as to insurgents. Indeed, many insurgents took similar negative views of jungle-based people, looking down on them as did national state leaders. Leaders and theoreticians in the insurgent parties and groups were neither tribal minorities nor forest-based. Their treatment of jungle-dwelling people under Communism, had it succeeded in Thailand, Indonesia, or Malaysia, may have been equally coercive in removing them from forests as was that of the national states that succeeded in defeating these insurgencies.[5]

Even more suspect to the states were rural people of Chinese background who lived their lives with legal categorizations such as Overseas Chinese, Indonesian Chinese, Sino-Thai, or after the Chinese dialect they spoke most frequently, Khek, Foochow, or Teochieu.[6] In Peninsular Malaya, the Communist Party (CPM) consisted primarily of urban and Chinese intellectuals, a feature that made Chinese in rural areas also suspect to state authorities.

The third group of racialized/ethnicized subjects, "national ethnic majorities," presented a more complex picture, especially in Indonesia. The Javanese constituted a national majority in Indonesia, but other people of Sumatran, Sulawesi, and Balinese heritage were considered ethnically different from Javanese but not *tribal*. Racialization of these politics was complicated because Communism was legal until early 1966 and a powerful discourse in Indonesia until 1965. Membership cut a broad swath across minority and majority populations. The ethnicization of Communism as Chinese was an explicit and important political strategy in West Kalimantan. Another confusing factor in thinking about the Indonesian case was the Islamism. This minority, radical, Islamist position was held mainly by people associated with ethnic majorities in Java, Sumatra, and Sulawesi.

Counterinsurgency operations resettled people in at least three ways that articulated with the goals of political forestry. First, rural Chinese in Malaya, Sarawak, and Kalimantan were forced out of forests or jungles when suspected of supporting insurgents. They were removed into controlled settlements, known later as strategic hamlets. Next, tribal peoples, such as the various hill tribes in Thailand and Dayak settlements in West Kalimantan, were also moved into consolidated villages to facilitate their "protection" and surveillance.

The third strategy involved government-planned and spontaneous colonization by ethnic majorities. Colonists moved into conflicted or post-conflict zones in large-scale resettlement schemes (also called colonization or transmigration) because governments assumed the colonists, being ethnic majorities, would not support insurgents (Uhlig 1984; Soemadi 1974). Javanese and Sundanese, among others, were moved to West Kalimantan. They were (unbeknownst to them) given land in areas that had been emptied by the expulsion of Chinese out of those so-called jungles, which had been inhabited rural areas previously. Some 100,000 rural Chinese were evicted, leaving hundreds of thousands of hectares—more than could possibly be claimed by or allocated to locals for immediate and continuous use. Transmigration converted massive amounts of jungle to rubber and oil palm production, transferred property rights from customary to private, and in many cases changed the definition of local. This was because provincial authority was transformed by national resource politics.

In 1967, the first national Forest Law was established. This law created a *national* forest for the first time; previously, forest areas had been under provincial governors' authority. National minorities now had different relations to these national forests. Lines on maps were drawn to allocate national forest concessions in West Kalimantan and elsewhere. No recognition was made on these forest maps of the local villages enclosed within these forests—nor of their prior land and resource claims. Thus the construction of Indonesia's political forests was concurrent with counterinsurgency against local Chinese and Communists, after the 1965 massacres in Bali and Java.

In Thailand, very large areas of land already demarcated and gazetted as reserve forests were occupied by millions of spontaneous migrants, who were not part of official resettlement programs (Uhlig 1984; Hirsch 1990; Vandergeest 1996). Although this was clearly in violation of forestry law, these settlements were condoned and even encouraged by local authorities, who saw the movement of people into these areas as a way of decreasing forest cover for insurgents, as well as a counterinsurgency strategy aimed at land-poor farmers susceptible to insurgent propaganda. The government also set up planned settlements in forest areas and sponsored the movement of lowland Thai farmers into these areas. Agricultural concessions were thus promoted as much as or more than forest restoration.

In forest areas where majority ethnic people were resettled, or near international border areas, governments described majority populations as needing military protection. This, of course, consolidated the regional expansion of national states (de Koninck and Dery 1997; Dove 1985). Former soldiers and police were resettled into areas considered dangerous or ongoing security threats (*"rawan"*) in Indonesia, notably in West Kalimantan. This practice has been used by victorious sides after many wars (e.g., in the United States after the Revolutionary War, in various Central American countries after the 1980s and 1990s wars; and in the Soviet Union [Brown 2003], China [Menzies 1994], and Vietnam [McElwee 2004]). These ex-combatants, and the new army bases built to accommodate new forces stationed there, provided a powerful visible symbol of the *national* Indonesian state in the former jungles of West Kalimantan (Peluso 2009).

Colonists were expected to clear forests for permanent agriculture, changing the insurgent environment and the region's ecological makeup and providing a state-dependent national and nationalist bulwark against antinationalist rebellion. Regardless of which strategy was adopted—and in some places, such as West Kalimantan, all three were deployed—the intention was to divide forests and agriculture into separate territorial domains in order to isolate insurgents from the cover provided by the jungle and to nationalize forests as colonist zones, conservation areas, and forest concessions.

Deploying Military Resources in Jungle Emergencies

A key way that militaries supported forestry was through the intensified surveillance and mapping of forest areas. In Thailand, the Royal Survey Department became in many ways an arm of the US military, which used aerial photos to produce a series of 1:50,000 maps of all forested areas starting in the 1950s, periodically updated based on new aerial photographs. The maps were shared with other government departments, in particular the forestry department, where they became the base maps for forestry work and to demarcate reserve forests. By the early 1970s, more than 40 percent of Thailand was demarcated on such maps as reserve forest, with virtually no ground checking into local forest use. Similar stories about mapping can be told about Malaysia, Sarawak, and Kalimantan (see, e.g., Harper 1997, 21; Barr, Brown, and Casson 1999).

A second significant technology of counterinsurgency access was road building. Roads had two purposes: to provide easier military access for military operations to facilitate conversions and capitalist projects, and to draw existing populations more into the sphere of central state rule through their greater involvement in production of commercial crops. Between 1960 and 1980, total road length in Thailand tripled (Hirsch 1990, 50). Road building facilitated "spontaneous" migration, as land-poor farmers flooded to the forests to grow economic cash crops (maize, cassava, sugarcane), also promoted by the government (Uhlig 1984; Hirsch 1990). Roads were often built through reserve forests (Uhlig 1984).

Rural development in these areas was an explicit form of counterinsurgency. In Thailand, for example, a key program was the Accelerated Rural Development (ARD) scheme, supported by USAID where insurgency was most active. This insurgency period was also the period that the United Nations Food and Agriculture Organization (FAO) was promoting its Forest-for-Development model (Westoby 1987). Forestry for development was generally preceded by enclosure/reservation of forests and the coerced exclusion of rural people from forests. Militaries also transferred other technologies and organizational cultures to forestry departments. Helicopters, for example, started out as a technology of war, but soon became a technology that assisted foresters monitoring forest cover and rural settlement. The organizational structures and institutional patterns of forestry had long imitated the military, as reflected in uniforms, the territorial structure of forest management, the rotation of foresters to avoid their becoming too attached to the people in their districts, and the arming of forester enforcement units. Finally, in some areas, forestry department and militaries worked together to both control and profit from forest exploitation, as in Indonesia, where timber concessions were allocated to PT Yamaker—a military-held timber concession—along the Borneo international border between Indonesia and Malaysia.

We finish this section with a few comments on how forest-based violence contributed to the separation of jungle into forest and agriculture. One way was simply that forests became dangerous places for farmers and forest product (previously called "jungle produce") collectors. For example, on the Indonesian side of Borneo, people were afraid to make individual *ladangs* (swidden fields) in the forest for fear of being mistaken for rebels. Many people were afraid to run into combatants (government or oppositional) in the forested areas and stayed out. Insurgents also suspected villagers. Forest villagers were also expected to engage in unusual activities in the forest (becoming "expert" guides for Indonesian troops, for example). And after timber concessions were allocated to the military, they were afraid to complain or act if they lost access to land, trees, or other forest products.

Even after the insurgents no longer posed a serious challenge to urban-based states in the region, security concerns continued to shape the practice of professional forestry. The fear of further insurgencies helped motivate the reshaping of property rights to land and forest products, the practices around forestry, and the use of military personnel as private guards for forest enterprises. These continued to put foresters, militaries, and big extractive businesses into close connection, and to shape what happened to the forests.

Discussion

The role of political violence and the kinds of forestry practices it generated differed substantially from the legal and institutional processes we have described in previous papers on the making of professional or scientific forestry and forests in Southeast Asia (Peluso and Vandergeest 2001; Vandergeest and Peluso 2006a, 2006b, 1995). Political violence in the form of insurgency and counterinsurgency is a specific kind of forest-based violence. Applied in the jungles of Southeast Asia that did not adopt Communist rule (i.e., Malaysia, Thailand, and Indonesia after 1965), forestry for development was also a strategy for counterinsurgency, nation-state building, and the production of forests and forestry. The ideologies and institutional practices associated with the 1970s conservation era's romantic notions of preserving rainforests, primary forests, or pristine forests were both preceded by and enabled by this earlier, critical period, largely through the consolidation of the principle that cultivated jungles should be transformed into non-overlapping zones of uninhabited forests on the one hand and agriculture on the other. Forgotten or ignored in most contemporary conservation discourse, political violence was depicted as having taken place in previously pristine forests, not peopled jungles. Where nation-state control was solidified, jungle discourses have largely disappeared. But use of the term jungles to refer to tropical forests recurs when anti-state political violence occurs in border or other marginal and contested forests, such as in recent and ongoing conflicts in the Philippines (Mindanao), Burma/ Myanmar, and West Papua.

During Cold War insurgencies and wars, the jungles of Southeast Asia represented a variety of frontiers: national boundaries; the edges of civilization and national state hegemony; as well as frontiers of deforestation, often through bureaucratically organized means of colonization, forest concessions, and conservation. Contemporary state forestry and the institutional structures and ecologies of the political forests were produced through the claiming of actually inhabited forests and the separation of agricultural activities from forest farming or agroforestry. Even community forestry efforts were based on assumptions that agriculture could be separate from forests (Walker 2004). We have also shown that counterinsurgency through forestry was not only concerned with territorial control, but also produced racialized state subjects connected with forests and forest territories in political and politicized ways. Finally, forests were far more militarized than is normally appreciated.

In sum, our major argument has been that it is difficult to understand the shape, resurgence, or social lives of forests today without understanding their connections to political violence—and that violence itself must be understood in more nuanced ways. The emergencies and insurgencies of the Cold War era were qualitatively different than the violent colonial conquests and the structural violence both moments generated. The outcome is that, except in conditions of war, forests are no longer jungles.

6 ✳ *Mutant Ecologies: Radioactive Life in Post–Cold War New Mexico*

JOSEPH MASCO

Given the multigenerational technoscientific focus on global nuclear war in Los Alamos, New Mexico, the firestorms of spring 2000 could not have come from a more unexpected source. On May 4, 2000, the US Forest Service lost control of an effort to reduce the regional risk of forest fire by burning underbrush in Bandelier National Monument. Over the next week, winds gusting to over 60 miles per hour turned the "controlled burn" into a raging firestorm, forcing the evacuation of Los Alamos County. In addition to displacing over 25,000 people, the fire forced Los Alamos National Laboratory to close for the first time in its history, leaving scientists to worry about ongoing research projects, the security of their experimental data, and the safety of nuclear materials. Meanwhile, thousands of residents of northern New Mexico fled the region, fearing the health effects of the smoke and ash. By the time the fire was contained, some 48,000 acres had burned, making the "Cerro Grande" blaze the most damaging forest fire in New Mexican history. The town of Los Alamos was the hardest hit, with hundreds of homes reduced to smoldering ruins. Almost one-third of Los Alamos National Laboratory's 27,000 acres burned, causing more than $300 million in damage. Significant areas of Santa Clara Pueblo and San Ildefonso Pueblo were also scorched, damaging water supplies as well as sacred sites. Concerns about how the charred mountain landscape would handle the torrential summer rains kept residents in fear of flooding for months afterward, hampering rebuilding efforts and extending the scope of the tragedy.

A forest fire of this magnitude is terrifying, and undeniably traumatic for all involved. One of the most striking cultural aspects of the Cerro Grande fire, however, was how many New Mexicans experienced the blaze as nothing less than a nuclear apocalypse. During the midst of the crisis, conversation among many New Mexicans focused on how the melted ash of burned Los Alamos homes and the neighboring forest was like the melted sand created by the first atomic bomb, which was detonated by Los Alamos scientists in the deserts of central New Mexico in July of 1945. Similarly, some laboratory scientists took pains to calculate the heat of the Los Alamos firestorm so as to compare it to the firestorms created in Hiroshima and Nagasaki in August of 1945. Hundreds of Santa Fe residents, fearful of "radioactive" smoke and ash from the fire, fled the area. When asked how people were coping with the disaster, one laboratory employee and Los Alamos resident put it succinctly: "All of us are thinking of Hiroshima. Now we know what that was like." But while four hundred Los Alamos families tragically lost their homes, and residents of northern New Mexico breathed smoke-filled air for weeks, and the regional environment was heavily damaged, miraculously, no one was killed or suffered a serious physical injury during the fire. So what does it mean that residents immediately negotiated their charred mountain landscape by referring to themselves as "survivors" and evoking the atomic bombings of Hiroshima and Nagasaki (events in which some 200,000 people lost their lives) as a point of reference? What does it mean that, ten years after the Cold War, so many New Mexicans came to inhabit a post-nuclear landscape in May of 2000 and to understand a most terrible forest fire in decidedly nuclear terms?

Local reactions to the Cerro Grande fire gave expression to the traumatic residues of the Cold War nuclear project, while articulating broad-based ecological fears about the long-term nuclear effects of Los Alamos National Laboratory. The smoke from the Cerro Grande Fire, for example, was immediately mapped as dangerous in New Mexico not because of the usual health concerns about smoke inhalation, but because it was taken by many to be radioactive. The plumes of smoke created the effect, for many, of atmospheric fallout from a nuclear explosion, as the particulate matter within the clouds—from destroyed buildings and from radioactive ecologies within the laboratory—generated regional concerns about radiation effects. The billowing smoke, thus, not only drew visible attention to Los Alamos National Laboratory as a threatened site, but also revealed a widespread regional concern about the ecological effects of a half-century of nuclear research (Fresquez, Velasquez, and Naranjo 2000). The fire demonstrated the willingness of many New Mexicans to experience any ecological event involving Los Alamos National Laboratory through the lens of nuclear trauma. To understand the nuclear fears evoked by the Cerro Grande Fire, I suggest we must first understand the historical production of nuclear natures and recognize the cultural processes that have evolved since 1945 to manage life within a radioactive biosphere. For reactions to the forest fire reveal not only a specific apocalyptic sensibility in northern New Mexico, but also a structural change in how nature is now constituted and experienced. This effort to normalize nuclear nature has been a multigenerational process in New Mexico, one effecting concepts of social order, ecology, health, and security, producing multiple experiences of the nuclear uncanny in the aftermath of the fire.

In this chapter, I interrogate the production of nuclear natures in the New Mexico context, arguing that one of the most profound aspects of the Manhattan Project has been to put in motion changes in specific social and biological ecologies that are highly mutable. To understand these new formations, however, we need a political ecology that can forward multigenerational reproduction. Assessing the scale of this biosocial process requires us to unpack contemporary assumptions about the bomb as a technology while interrogating the historical terms and parameters of the Cold War nuclear program as a global ecological experiment. For reactions to the Cerro Grande fire revealed not only widespread nuclear fear, but also repression about the historical production of nuclear nature. Thus, it is important now to ask: How have Americans come to understand their place within a radioactive environment since 1945? What does a shift in knowledge about nuclear nature tell us about the long-term transformations of both the nation-state and the state of nature in the nuclear age?

Radioactive Natures: Life in the Wildlife/Sacrifice Zone

At the end of the Cold War, the US nuclear complex formally occupied a total continental landmass of over 3,300 square miles, involving thirteen major institutions and dozens of smaller production facilities and laboratories (O'Neill 1998, 35). These production sites were predominantly located in isolated, rural areas as a complex form of domestic development. In 1943, huge new industrial economies were created in Oak Ridge, Tennessee, Hanford, Washington, and Los Alamos, New Mexico, and later in Aiken, South Carolina, Amarillo, Texas, Idaho Falls, Idaho, Rocky Flats, Colorado, and at what became the Nevada Test Site (see Hales 1997; O'Neill 1998). It was in these mostly rural, nonindustrial locations that nuclear materials were mass produced, nuclear weapons were built and tested, and nuclear waste was stored, fusing local ecologies and local communities with the American nuclear project. The internal logics of nuclear development required deliberate acts of territorial devastation, producing an archipelago of contaminated sites stretching across the continental United States from South Carolina to Nevada, from Kentucky to Washington, and from Alaska to the Marshall Islands. This "geography of sacrifice," as Valerie Kuletz (1998) has called it, is currently estimated to entail an environmental restoration project costing between $216 to $400 billion for those sites that can, in fact, be remediated, and it is likely to cost more than the Cold War nuclear arsenal itself (see Schwartz 1998; US Department of Energy [DOE] 1995a, 1995b). Nuclear security has required complex new forms of internal cannibalism, as both the biology of citizens and the territories of the state encounter an array of new nuclear signatures after 1943.

In the post–Cold War period, the US nuclear complex has implicitly recognized these transformations through a new type of territorial reinscription. On October 30, 1999, for example, Secretary of Energy Bill Richardson announced the formation of a 1000-acre wildlife preserve within a 43-square-mile territory of Los Alamos National Laboratory (LANL). The new White Rock Canyon Preserve was singled out by the DOE as a "unique ecosystem," one that is "home to bald eagles, peregrine falcons, southwestern flycatchers, 300 other species of mammals, birds, reptiles, and amphibians, as well as 900 species of plants" (US DOE 1999, 1). As Secretary Richardson explained:

> How fitting that we are here today at Los Alamos, the place that witnessed the dawn of the atomic age. . . . In places of rare environmental resources, we have a special responsibility to the states and communities that have supported and hosted America's long effort to win the Cold War—and we owe it to future generations to protect these precious places so that they can enjoy nature's plenty just as we do. Los Alamos's White Rock Canyon is such a place, an able bearer of New Mexico's legacy of enchantment. After today, it will be more so as we celebrate the reunification of land and community.

We celebrate the reunification of land and community. The wildlife preserve as a concept forwards a claim on purity, marking specific ecologies worth preserving as precious resources in a state of nature. What can such a claim mean, however, in the context of a US nuclear site? Richardson's appeal to a "legacy of enchantment" as well as to the reunification of land and community in New Mexico comes after a decade of intense environmental politics concerning the Cold War legacies of nuclear weapons work at Los Alamos. The post–Cold War period began in New Mexico with the near simultaneous announcements of a moratorium on nuclear weapons tests and the designation of 2,200 contaminated sites within Los Alamos National Laboratory, requiring an estimated cleanup of over $3.3 billion (US DOE 1995b, xiv). While many New Mexicans discovered the scale of Cold War nuclear research at Los Alamos through its environmental costs, community groups throughout northern New Mexico began mobilizing for health studies as well increased surveillance of water, soil, and air quality (see Athas 1996; Makhijani and Schwartz 1998). The reunification of land and people proposed by the "wildlife preserve" recognizes the unique cultural investments of Pueblo and Nuevomexicano communities in the area now occupied by Los Alamos. However, the discourse of "preservation" enabling such recognition can only do so by ignoring the long-standing practices of environmental ruin, informing past and present research at the laboratory.

This ideological project to link the national security offered by the atomic bomb during the Cold War to sustaining the biodiversity of US territories, however, forwards a deep structural contradiction. The effects of nuclear production have transformed the global environment, making the biosphere itself a postnuclear formation. Since the trace elements of atmospheric fallout are now ubiquitous in soils and waterways, flora and fauna, the "nature" of wildlife as a concept has changed in the nuclear age. If exposure is now a general condition—a question of degree rather than kind—then what does it mean to promote such images of survival in the midst of contamination?

This recuperation of "nature" within post–Cold War debates about the environmental and health dangers of nuclear production articulates a new form of state territoriality. In the continental United States alone, the DOE has recently transformed over 175,800 acres of land by legislative fiat from industrial nuclear sites to wildlife preserves. Carved out of the vast security buffer zones established around nuclear sites, most of these areas were fenced off in the middle of the twentieth century and isolated from human contact during the Cold War. Consequently, these sites were among the most heavily fortified wilderness areas in the world. By presenting these sites as untouched in over fifty years, the DOE seeks to redefine the value and object of that military fortification, replacing nuclear weapons systems with biodiversity as the security object of the nuclear state. This suturing together of wildlife preserve and national sacrifice zone has become an expansive post–Cold War project.

At the Savannah River Site, which produced plutonium and tritium for the US nuclear arsenal, 10,000 acres (of the 200,000-acre nuclear facility) became the Crackerneck Wildlife Management Area and Ecological Reserve in 1999 (US DOE, Savannah River Operation Office 1999). Celebrating some 650 species of aquatic life found on the site, the DOE presented a remarkable image of biodiversity to the public. However, DOE representatives failed to mention that the unusually healthy alligators and rather large bass fish found at the Savannah River Site are also unusually radioactive (Associated Press 1999). Their bodies contain Cesium 137, a byproduct of nuclear material production on the site, which is home to five nuclear reactors. The Savannah River Site now presents a uniquely modern contradiction: the site maintains a massive environmental problem in the form of 34 million gallons of high-level radioactive waste, a multi-millennial challenge to the future, but it has been rescripted by the nuclear state as an ecological reserve preserved, as the DOE notes, for "future generations."

At the Idaho National Engineering and Environmental Laboratory (INEEL), 74,000 acres are now included in the Sagebrush Steppe Ecosystem Reserve. The DOE has devoted this preserve to the protection of some 4,000 species of plants and 270 species of animals—including the ferruginous hawk, the pygmy rabbit, and Townsend's big-eared bat (US DOE, INEEL 1999). Inaugurating the reserve, Secretary Richardson remarked:

> The Department of Interior estimates that 98 percent of intact sagebrush steppe ecosystems have been destroyed or significantly altered since European settlement of this country. Because the INEEL has been a largely protected and secure facility for 50 years, it is still home to a large section of unimpacted sagebrush habitat. Our action today will help preserve for future generations one of the last vestiges of this important ecosystem.

INEEL–a largely protected and secure facility. With fifty-two nuclear reactors and eleven gigantic tanks filled with 580,000 gallons of high-level nuclear waste, INEEL is redefining the definition of "protected" and "secure"—as well as "impact" and "risk"—for distant future generations. Townsend's big-eared bat and the pygmy rabbit may have gained new state recognition via the reserve, but their new status is primarily a bureaucratic one that does not address the mobility of animals, ecosystems, and radionuclides between territories identified as wildlife reserves and nuclear production sites.

The hard insight informing these new wildlife preserves is that isolation from human traffic provides an enormous ecological benefit: Human contact is more immediately toxic for many ecosystems than are radioactive materials. The dual structure of the DOE wildlife reserve/sacrifice zone seems to argue, however, that nuclear materials can be kept in place and that the border between preserve and wasteland can be effectively patrolled over millennia. This logic is trumped most convincingly at the Hanford Reservation in Washington State, which produced plutonium for the US arsenal from 1945 to 1992 and is now recognized as the most seriously polluted site in the United States. The DOE has recently devoted 89,000 acres of Hanford's 540 square miles to preserving the long-billed curlew, Hoover's desert parsley, and Columbia yellow cress (US DOE, Pacific Northwest National Laboratory 1999). However, mulberry trees on the Hanford Reservation have been showing increasing amounts of strontium-90 over the last decade (Lavelle 2000); and the Russian thistle plant has recently created a new kind of environmental hazard, the radioactive tumbleweed (Associated Press 2001). The Russian thistle shoots its roots down 20 feet into the earth, sucking strontium-90 and cesium into its system from contaminated areas. The head of the plant eventually breaks off to become a windblown radiation source. Hanford now spends millions of dollars each year managing this form of contamination and has crews armed with pitchforks patrolling the reservation in trucks to wrangle the radioactive weeds. This inability to enforce the distinction between wilderness and wasteland was further dramatized at Hanford in 1998, when fruit flies landed in liquid radioactive material and carried contamination far and wide over the next weeks, requiring nothing less than a $2.5 million dollar DOE cleanup operation (Stang 1998).

Radioactive tumbleweeds, contaminated fruit flies, and toxic alligators— these are all survivals of the Cold War nuclear project, as well as new forms of nuclear nature. Adjacent to each of the DOE wildlife preserves, however, are sites that are not just minimally radioactive according to federal standards, but rather present such profound environmental hazards that they will need to be fenced off and monitored for, in some cases, literally tens of thousands of years. These sites represent Cold War survivals of another kind. Despite the rhetorical and institutional effort to find areas of "purity" within the ecology of the nuclear complex, the broader context involves a massive, state-sponsored territorial sacrifice during the Cold War that has been wildly productive in specific areas. The US nuclear complex could not have produced 70,000 nuclear weapons from 1943 to 1992 without favoring industrial production over environmental concerns. Just as the current background radiation rate normalizes the atmospheric effects of aboveground nuclear testing as an aspect of nature, the new wildlife zones offer an image of nature created through nuclear politics and radioactive practices. The wildlife preserve is thus an exception that proves the rule within the US nuclear complex. Despite the new bureaucratic recognition of the ferruginous hawk, the pygmy rabbit, and the larkspur, the divisions between normal, abnormal, and pathological are being redefined in these nuclear sites, as contaminated nature is recognized to be not only valuable and robust, but to greater or lesser degrees, ever present. In other words, the experimental projects that produced and now maintain the bomb have collectively turned the entire biosphere into an experimental zone—one in which we all live—producing new mutations, as we shall now see, in both natural and social orders.

Environmental Sentinels, or the Militarization of the Honeybee

The radioactive future of the Cold War nuclear complex is already mutating in the post–Cold War period, producing a complex mobilization of future generations, technoscience, and state institutions. The DOE has not only offered up zones of conservation to future generations but also acknowledged that as many as 109 sites within the nuclear complex are too contaminated to remediate. The challenge of what do to with these radioactive sites over decades, centuries, and (in some cases) millennia, is now articulated through a new discourse of environmental surveillance and control known as "long-term stewardship." The US DOE (2001) defines the project in this way:

The Long-Term Stewardship Program will maintain and continuously improve protection of public health, safety, and the environment at a site or portion of a site assigned to DOE for such purposes. This mission includes providing sustained human and environmental well-being through the mitigation of residual risks and the conservation of the site's natural, ecological, and cultural resources. Mission activities will include vigilantly maintaining "post-cleanup" controls on residual hazards; sustaining and maintaining engineered controls, infrastructure, and institutional controls; seeking to avoid or minimize the creation of additional "post-cleanup" long-term stewardship liabilities during current and future site operations; enabling the best land use and resource conservation within the constraints of current and future contamination; and periodic re-evaluation of priorities and strategies in response to changes in knowledge, science, technology, site conditions, or regional setting. The Long-Term Stewardship Program will coordinate activities to identify and promote additional research and development efforts needed to ensure this protection and to incorporate new science and technology developments that result in increased protection of human health and the environment and lower costs.[1]

Sustained human and environmental well-being through the mitigation of risk. The Long-Term Stewardship Program approaches the radioactive and chemical legacies of Cold War nuclear production as a bureaucratic, as well as technoscientific, problem. Promising an increasingly intimate interaction with contaminated sites, the Long-Term Stewardship Program hopes to minimize future environmental effects by systematically deploying as yet undeveloped technologies. This is a utopian program that imagines perfect management of Cold War nuclear waste and contaminated sites for millennia—despite the prior fifty years of environmental neglect.

Creating "sustained human and environmental well-being" in a post-nuclear environment, however, requires a complex new form of governmentality. For Foucault (1991), governmentality is the focus of the state on policing its population to improve the health and well-being of its citizens.

In long-term stewardship, the logic of national security is inverted: the threat of foreign arsenals and armies are replaced by an internal discourse of contamination and territorial colonization. In this context, governance means protecting citizens from the industrial effects of the nuclear security state, thus redrawing the lines between policing and welfare. However, it is not clear how environmental well-being can or will be defined. The DOE cannot return ecosystems to a preindustrial, prenuclear state. Rather, "clean-up" here means meeting US regulatory standards, which are dependent on expected land use. The hope of the Long-Term Stewardship Program is that through surveillance and applying cutting-edge science to the environmental legacy of the Cold War, a kind of ecological stasis can be achieved in the near term, as science improves over time to solve the problems posed by radioactive contamination and waste. However, in recognizing that some sites are too damaged to treat effectively, the program also reveals that the Cold War maintains a powerful claim on a deep future. With budget projections currently made out only to the year 2070, the DOE estimates that the program will require $100 million *per year* simply to maintain the 109 long-term stewardship sites for an indefinite future (US DOE 2001, 108).

If the wildlife zone is one new form of nuclear nature, the long-term stewardship site is another, with an equally deep claim on future generations. Indeed, in orienting scientists, technologies, and communities around long-term stewardship sites, the DOE is also creating long-term stewardship communities, producing entirely new ecosocial orders. To make this point, we do not have to look thousands or even hundreds of years into the future. One long-term stewardship site in Los Alamos is known as Area G, which has been the laboratory's primary nuclear waste site since 1957. Area G is a 100-acre facility located on one of the finger-like mesas that make up the Pajarito Plateau. Low-level radioactive waste (consisting mostly of objects contaminated during laboratory operations), as well as significant quantities of plutonium 239 and uranium 238 from nuclear weapons research, is stored in 500-foot long pits and in deep shafts. While inventories have been carefully documented since 1988, few records were kept for the period from 1957 to 1971, and the records for the period from 1971 to 1988 are poor. The incomplete knowledge of what is in Area G is important because just to the east of the site is the town of White Rock (population 6,800), while immediately north is San Ildefonso Pueblo territory. Pueblo members collect plants and hunt game in the shadow of Area G, as well as maintain shrines and sacred sites in the area. A recent laboratory performance assessment concludes that Area G will be completely full by 2044, initiating a new kind of territorial project:

Active institutional control will continue for a period of 100 years (between 2047 and 2146). During institutional control, site access will be controlled, environmental monitoring will be performed, and closure cap integrity will be maintained. After the institutional-control period, it is assumed the site will be maintained by the DOE or its equivalent for as-yet undefined industrial uses. This industrial-use period is assumed to prevail for the 900 years remaining in the compliance period (between 2147 and 3046). (Hollis 1997, 10)

The 900 years remaining in the compliance period (between 2147 and 3046). Evaluating the exposure risks to future populations along a variety of intrusion scenarios, the report confirms that the Manhattan Project inaugurated a new ecological regime on the Pajarito Plateau, one that is now intimately involved with negotiating the 24,000-year half-life of plutonium and other nuclear materials (see Rothman 1992; Graf 1994). Currently evaluating risk only on a 1,000-year time frame, Area G is nonetheless one instantiation of a larger Cold War nuclear legacy that the discourse of long-term stewardship rhetorically seeks to contain using rational technoscientific measures.

The Area G Performance Assessment concludes, "The ability to contain radioactivity locally depends largely on nature, while the ability to prevent intrusion depends solely on man" (Hollis, 1997, 16). It therefore assumes from the start that "current natural conditions will prevail" and "a government entity will maintain the site and control access to it" for the next 1,000 years (Hollis 1997, 16). Both nature and the state are, for the sake of the study, assumed to be stable entities across the next millennium, even as the evidence of the last fifty years shows a dramatic change in both. Indeed, more subtle changes are already shaping the nuclear future of the Pajarito Plateau, offering a new state of nature, more mutant than stable. Plumes of tritium contamination as well as chemical residues from high explosives are already leaking from Area G, demonstrating that the geology of the Pajarito Plateau is more permeable than previously assumed. Thus, even as the performance assessment assumes a forever-vigilant state agency to watch over a stable ecosystem at Area G, environmental surveillance is revealing a more mobile ecological formation. Indeed, surveillance itself has become the basis for new kinds of nature.

Consider the role now played by the Italian honeybee (*Apis mellifera*) at Area G. As a creature that flies over a wide area foraging for pollen and nectar in flowers and then returns to a fixed location (the hive) to produce honey, the honeybee is a natural environmental surveyor. Los Alamos scientists have demonstrated that the honeybee is particularly sensitive to tritium, a radioactive substance used in nuclear weapons to enhance the size of the explosion that is and notoriously difficult to contain. Deploying the honeybee as an environmental tool since the late 1970s, scientists have documented increasing tritium contamination rates at Area G through the 1990s (Fresquez, Armstrong, and Pratt 1997). This instrumentalization of the honeybee takes more than one form at Los Alamos, but in the context of Area G, it reveals a profound transformation in ecological regimes. Neighboring Pueblo communities identify mesa tops as areas of particular cultural importance, containing shrines and sacred sites that participate in a different conception of nature. Pueblo cosmology has traditionally worked, not to deploy nature as a technoscientific object, but to integrate Pueblo members into the local ecology (see Ortiz 1969). Within Eastern Pueblo cosmologies, the bee plays a crucial role in pollinating plants and is both a symbol and an agent for life itself; consequently, pollen figures prominently in ceremonies of purification and seasonal renewal. The Manhattan Project colonized this ecological regime with one that focuses on the technoscientific deployment of nature. The value of the bee in this new context is no longer as a life-giving entity but as a toxic being, marking the transformation of the plateau from a wild space of nature to a new kind of mutant ecology.

While specific animal forms are being deployed—and reinvented—to shape environmental politics in post–Cold War Los Alamos, a more subtle aspect of the Manhattan Project has been to transform regional human populations into radiation monitors. Activist groups spent much of the 1990s pushing for environmental impact studies and increased regulation of the laboratory, helping to produce a cross-cultural regional dialog about the environmental consequences of nuclear weapons research at Los Alamos. Concurrently, LANL scientists, Pueblo representatives, as well as officials from the Bureau of Indian Affairs, each began conducting independent tests of air, water, soil, plants, and animals in the region, not only to define the level of risk to Pueblo citizens living adjacent to the laboratory but also to confirm the accuracy of LANL science. The Pueblos of Jemez, Cochiti, Santa Clara, and San Ildefonso have begun training new generations of youth as environmental scientists to prepare them to take over responsibility for monitoring the environmental effects of the laboratory. Thus, communities throughout the region—LANL scientists, Los Alamos community members, Native Americans, Nuevomexicanos, and antinuclear activists—all now claim the title of "environmentalist," maintaining deeply felt, if asymmetrical, investments in the Pajarito Plateau. However, while each of these populations is committed to preserving the regional ecology, their cultural understandings of that ecology are construed on radically different terms.

As New Mexicans took an increasingly public interest in LANL's environmental standing after the Cold War, many also played the unwitting role of environmental test subjects. New Mexicans did so at two levels: first, as workers at the laboratory who were monitored for radiation exposures on the job, and second as regional populations who (often unwittingly) participated in the Los Alamos Tissue Analysis Program, an effort started in the 1950s to track radiation exposures via tissue sampling. In the late 1990s, relatives of 407 individuals who had tissue samples taken during autopsies in Los Alamos and regional hospitals brought a class-action lawsuit against the laboratory. The multimillion-dollar settlement acknowledged that informed consent was not received from family members during these autopsies. Workers in the laboratory as well as residents of Northern New Mexico have thus been part of a larger environmental monitoring project for decades—similar to the bees—but, in this case, their own bodies have been placed in the role of "environmental sentinel." In this sense, tracking radionuclides through the biosphere and specific bodies in Northern New Mexico has become an expanding project for all concerned (Petryna 2002). The medical knowledge produced by these efforts, however, remains partial and controversial. The fourfold elevated presence of thyroid cancer in Los Alamos discovered in the 1990s might simply be an effect, for example, of the intensity of the screening regime in Los Alamos hospitals (Athas 1996). Nevertheless, while the long-term health effects of nuclear production at Los Alamos remain controversial at the level of technoscience, there is no doubt of the effect they have had on social imaginations in northern New Mexico. Illnesses throughout the region are attributed to the laboratory, revealing another aspect of the nuclear reinvention of nature.

The Social Logics of Mutation

While interviewing Los Alamos employees who believed their health had been damaged on the job, I was told repeatedly about a videotape that purportedly documented hazardous work conditions at Area G. For these workers, the tape held the promise of standing as evidence in future legal proceedings, a means of making visible to the outside world the everyday practices that were usually shielded by gates, security, and the power of the nation-state. A former Area G worker, who was concerned about his health and did not believe in the veracity of the cumulative radiation badge measurements recorded in his Los Alamos medical file, invited me to view the videotape in his home. As I watched, I was confronted with a complex textual record of mutation. The tape was originally made by Los Alamos personnel to document efforts to consolidate space at Area G for the accruing nuclear waste from laboratory operations. The banality of worker job descriptions is soon ruptured, however, when a tractor accidentally punc-

tures a partially buried barrel of nuclear waste. The narrative then shifts
from recording the formal statements of workers during the handling of
the ruptured barrel to informal moments with the work crew playing to the
camera. Eventually, the multiracial workforce splits along racial lines, as
the white program managers don anticontamination gear to test the drum
for radionuclides, while the Nuevomexicano and Pueblo workers remain in
normal work clothes. The manual labor of digging up and moving barrels
of radioactive waste takes place underneath the deep blue New Mexican sky
with a ferocious wind that completely covers workers in dust from the site.
My host claimed that the dust from the waste site might well have contaminated workers, and then explained to me how easily the radiation monitors
could be turned off at Area G to allow such exposures to go unrecorded.

The videotape reveals the difficult work conditions and physical labor
needed to move drums of nuclear waste, but the novel presence of the camera also becomes central to the recording: The workers not only practice
describing their jobs prior to formal taping and then deal with the accident
while being taped, they also mug for the camera. Midway through the video,
my host interrupts to tell me that he knows what happened to Karen Silk-
wood, the Kerr-McGee whistleblower who died mysteriously in a car crash
in 1974. Her organs were sent to Los Alamos for analysis as part of the tissue registry program but were then mysteriously lost. He tells me that her
organs were placed in a laboratory refrigerator, which subsequently failed,
and was then dumped at Area G, packed full of the damaged organs of US
nuclear workers. Area G becomes, in his presentation, not merely an ongoing health threat to current workers but also literally a grave, a site where
the human evidence of radiation exposures is buried as industrial waste.
He hopes that the videotape can help reveal this fact, documenting for an
outside world the ongoing biological sacrifice of nuclear workers. Twenty
minutes into the videotape, the scene shifts to the office spaces at Area G,
where the camera operator discovers and then plays with the mirror function on the video camera to produce a series of special effects. For the next
20 minutes of tape, he entertains his fellow workers—by giving them a
third eye, or merging their foreheads into giant mutant forms, or giving
them tails, while laughing hysterically at the visual results. The videotape
that begins with the serious work of nuclear waste disposal, in other words,
shifts to a literal discourse of mutation, one that visually transforms each
Area G worker into a monstrous being. The Area G workers I spoke with
focused more on the official acts documented in the first half of the videotape than on the cultural logics and fears revealed in the second half. But
the videotape records not only the everyday practices at Area G, the brute
work of moving nuclear waste around and the precariousness of contain-

ment, but also a surreal form of nuclear play that displays workers not as potential mutants but as present ones—linked by tails, misshaped heads, and multiple eyes.

The Area G videotape ends on an equally jarring note, as it cuts from the play of mutation at the nuclear waste site to a garage somewhere in the northern Rio Grande Valley, where a Nuevomexicano relative of the camera operator (who has taken the camera home) stands stiffly and without emotion in the center of the screen, playing ranchero music on an accordion. This eruption of the nonnuclear everyday into the narrative of Area G is a reminder of the multiple cultural worlds informing life in northern New Mexico that are linked both formally and informally to the nuclear project at Los Alamos. The Area G videotape reveals the radical transformation of the region into a nuclear economy: it documents the burying of nuclear waste on the plateau, permanently transforming the ecology of that space. It also documents the mobilization of whole communities that are now devoted simply to monitoring and working with the nuclear waste produced by America's national security regime, and ultimately, it demonstrates the fears of mutation that permeate workers' psyches, underscoring the psychosocial effects of living within a nuclear ecology. These forces are not static, but rather highly mobile, making it impossible to discuss the regional effects of the Manhattan Project without taking into account how material realities fuse with sociocultural logics and nuclear fear. A political ecology of the bomb that investigates the interaction between regimes of nature reveals the American nuclear project to have been ecologically transformative and multigenerationally productive: it has reinvented the biosphere as a nuclear space, transformed entire populations of plants, animals, insects, and people into "environmental sentinels," and embedded the logics of mutation within both ecologies and cosmologies. Cold War fears of nuclear war have, in other words, been replaced by more subtle and multiple forms of life defined by the ambiguities and dangers of inhabiting specific radioactive spaces, mutant ecologies that now present an ever-evolving biosocial, political, and environmental terrain.

7 ✳ Pan-Tropical Perspectives on Forest Resurgence

ALAN GRAINGER

Introduction

The need for empirical study of forest resurgence in the tropics was only recognized in the first decade of the twenty-first century (Kauppi et al. 2006). Previously, clearance of tropical forests was reported so frequently in the scientific literature and mass media that continuing deforestation was taken for granted. Evidence for natural forest regeneration was limited because no one thought to collect it, and international forest statistics were compiled on the assumption of continuing deforestation. This chapter informs our understanding of tropical forest resurgence by placing it within seven wider perspectives. It looks in turn at forest classification, scholarship, theory, institutions, measurement, evidence, and policy, and concludes that natural reforestation is probably happening on a considerable scale and deserves proper measurement.

The Classification Perspective

CLASSIFYING TROPICAL FOREST TYPES

The tropics comprise a humid zone near the equator and a dry zone close to the Tropics of Cancer and Capricorn. It contains a wide variety of forest types that are commonly divided, for convenience, into forests with closed and open canopies. Forest with a closed canopy, termed *closed forest*, is predominant in the humid tropics, where all types of forest are known collectively as tropical moist forest. Some closed forest is found in the dry tropics, but *open forest* is predominant. It consists of varying mixtures of trees, shrubs, and grasses, often known as *savanna woodlands*. Such *natural forests* are distinguished from *forest plantations*.

TYPES OF FOREST CHANGE

Two terms are used to describe negative forest change. *Deforestation* involves "the temporary or permanent clearance of forest for agriculture or other purposes." Forest *degradation*, in contrast, is the "temporary or permanent reduction in the density, structure, species composition or productivity of forest cover" (Grainger 1996a, 62).

There are three types of positive forest change, or *forestation* (Wiersum 1984). *Afforestation* involves establishing forest plantations on previously cleared land that has not been forested within at least fifty years (Mather 1993). *Reforestation* increases tree cover in existing forested areas. It may occur naturally or involve human intervention. Trees are also planted outside forest areas, for example, on farmlands, mixed with crop cultivation and livestock raising in *agroforestry systems* (Nair 1989).

A hundred years of experiments in tropical forests by colonial foresters found that various silvicultural systems, in which forestation in natural forests was intentionally manipulated, were technically effective (Whitmore and Burnham 1975). But since they were too labor intensive to be economic, after 1960 the focus switched to establishing forest plantations (Evans 1982).

CLASSIFICATION AND MEASUREMENT

Formal classification obscures difficulties in monitoring forest resurgence. For example, *open forest* encompasses a wide range of tree density, which varies from place to place, according to climate and soil. This and the inability to establish a definitive prehuman norm (Eyre 1968) exacerbate difficulties in measuring change in tree density accurately by remote sensing (Lambin 1999). In the humid tropics a lot of land is under shifting cultivation, but only part is cleared at any one time. The rest is in varying stages of natural reforestation, pending subsequent clearance. With rotations of the order of six years this *forest fallow* never reaches the same height as undisturbed forest, but after a few years it is difficult to distinguish the two types using satellite imagery.

The Scholarship Perspective

Current recognition of the importance of forestation in the tropics is just the latest stage in the evolution of narratives of global forest trends. The predominant narrative in the 1980s was that tropical deforestation became significant in the nineteenth century and continued unabated (Tucker and Richards 1983; Richards and Tucker 1988). Forest cover in the temperate and boreal zones was stable after earlier forestation had countered the long history of forest clearance in these zones (Sedjo and Clawson 2000). The outlook then grew more optimistic, with claims about forestation outside the tropics (Sedjo 1992). Early in the new millennium, uncertainty about the global forest area trend became more apparent (Mather 2005).

The Theoretical Perspective

CONTEMPORANEOUS NATIONAL MODELS

Of the two main types of theoretical models of forest change that have been developed in the past thirty years, the first model estimated contemporaneous relationships over short periods of five to ten years between deforestation and underlying socioeconomic causes (or *drivers*). In the absence of time-series data for national forest areas in the tropics, mean annual deforestation rates in a large number of countries, taken from the United Nations (UN) Food and Agriculture Organization (FAO) statistics, were regressed against socioeconomic statistics, such as the national population growth rate (for reviews see Lambin 1997). These reached superficially attractive, but simplistic, conclusions (Lambin et al. 2001), often biased by statistical limitations of which the authors were unaware (Rudel and Roper 1997). Only recently have drivers of reforestation received their due attention (Nagendra 2007).

LONG-TERM NATIONAL MODELS

The second type of model, the *forest transition* model, focused on long-term change and envisaged the possibility of forestation. Mather (1992) proposed that national forest cover follows a U-shaped curve over time, switching from deforestation to reforestation at a turning point, which he called the *forest transition*. Initial studies focused on temperate countries, but tropical transitions are now being reported (Rudel et al. 2005; Mather 2007).

An alternative model treats the U-shaped curve as a special case. In the general case the U-shaped curve has two parts: the decline in forest area, termed the *national land use transition*, and the rise after the forest transition, termed the *forest replenishment period* (fig. 7.1). The two are separated by an interregnum of variable extent (Grainger 1995). In this model, forest replenishment does not automatically follow the end of the national land use transition.

One reason for separating the two parts of the curve is that they respond to different market demand curves. The national land use transition curve is merely the inverse of a curve showing how agricultural area expands in response to rising demand for food as a country develops socially and

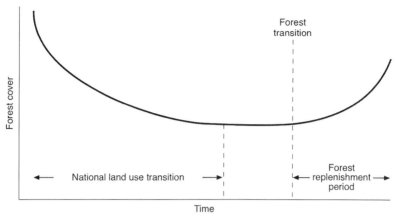

7.1. The national land use transition and forest transition models. Source: Grainger (1995).

economically. It tapers off as the limits of land suitability are reached and farming intensifies following adoption of improved technologies (Drake 1993). In the upward part of the curve, artificial forestation is a response to changing demand for wood and nonmarket environmental services supplied by forests. Natural reforestation may occur if farmland is abandoned as it becomes unproductive or uneconomic.

A U-shaped curve is likely in a *normative scenario* in which deforestation ends once market mechanisms have allocated all land in a country to its optimum use. Much forest would remain. But since land allocation is invariably piecemeal, with marginal farmland cleared before the most productive land is identified, the end of deforestation could be followed by natural reforestation on abandoned marginal land if its soil has not been degraded.

In many countries, however, deforestation does not end until either (a) environmental services collapse and floods and other hazards lead to political pressures on governments to intervene, or (b) wood supplies are restricted and price rises make forest protection and forestation more profitable. A possible lower limit of 0.1 ha per capita was suggested for this *critical scenario* (Grainger 1993), based on the forest area needed to supply mean internal domestic wood demand.

The potential for different forest trajectories was recognized by Mather (1992). The normative and critical scenarios correspond to the *economic development* and *forest scarcity* pathways proposed for forest transitions by Rudel et al. (2005). Their second pathway used the 0.1 ha per capita limit as a benchmark, though it could be exceeded if markets and states break down through poverty or civil war.

The Institutional Perspective

Persistent forest trends or states reflect repetitive human actions or *institutions*. These are "enduring regularities of human action in situation structured by rules, norms, and shared strategies, as well as by the physical world" (Crawford and Ostrom 1995, 582).

Continuing deforestation implies the dominance of agricultural institutions that privilege repeated clearance. Stable forest cover implies the predominance of institutions that privilege long-term forest management for producing timber or preserving biodiversity. These include state institutions informed by those of forestry science or conservation biology, or civil society institutions responding to local needs (Ostrom and Nagendra 2006).

A forest transition therefore requires changes not only in driving and controlling variables, but in how these interact with a shift in the relative dominance of forestry and agricultural institutions. As forestry institutions strengthen, new forest is planted or regrows and remaining forest is better protected. This requires embedding forestry institutions within higher-order democratic institutions (Mather and Needle 1999), so conservationist groups have the same access to policymakers as exploitative groups (Grainger and Malayang 2006).

The Measurement Perspective

Until recently, measurements of tropical forest areas relied on periodic national forest surveys by government forestry departments. Yet only half of tropical countries have had two national forest surveys in the last forty years (table 7.1).

Data on earth resources have been collected by satellites, such as Landsat, since 1972. Scientists have occasionally analyzed satellite images to survey forest in particular countries, but only in the last fifteen years have they attempted to map forest distribution throughout the tropics. Unfortunately, these measurements have limitations, because the resolution of satellite sensors commonly used for broad scale mapping is too low to measure all clearances properly, so sampling with higher resolution images is required. Some groups also sample for reforestation (e.g., Achard et al. 2002) but others do not (e.g., Hansen et al. 2008).

A diversity of *sampling* methods have been used, but Tucker and Townshend (2001) argue that any sampling method could underestimate deforestation, and that only a *wall-to-wall* approach, which compares two entire maps, can give satisfactory results. When Hecht and Saatchi (2007)

Table 7.1. Number of national forest surveys for ninety tropical countries by latest survey date

Latest estimate Recorded estimates	< 1980	1980–1989	1990–1999	2000–	Total
2	1	3	17	22	43
1	3	8	23	3	37
0	–	–	–	–	10

Source: Grainger (2009).

used a wall-to-wall approach by comparing satellite images of the whole of El Salvador in 1989 and 2001 they found that national forest area had expanded by 140 ha.10^3.

The Evidence Perspective

INTERNATIONAL FOREST STATISTICS

The limited amount of actual measurement raises questions about the accuracy of available evidence for tropical forests change and how it has been assembled. The UN's FAO has published compilations of national statistics on tropical forests since 1981 in its Forest Resources Assessment (FRA) series.

FAO depends on its member countries to supply estimates of their forest areas, usually linked to national forest surveys. To turn these into global compilations of national statistics with a uniform reference year (e.g., 2000) FAO generally:

1. Projects forward estimates from the latest survey year to the reference year. The longer the projection period, the greater the likely error involved (fig. 7.2).
2. Assumes that national forest area has declined since the last survey.
3. Assumes that the latest survey is more accurate than previous ones. Back projection from the last survey is used to produce an estimate for the former reference year that is consistent with the latest estimate (fig. 7.2). This often raises earlier figures substantially.

These adjustments can lead to misleading national estimates. For El Salvador, FRA 2000 projected forward a 1978 survey finding of 193 ha.10^3 to give 187 ha.10^3 in 1990 and 107 ha.10^3 in 2000. In FRA 2005, a 2003 survey measurement of 298 ha.10^3 was projected forward to give 292 ha.10^3 for

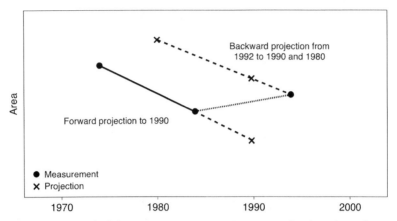

7.2. Projection methods in FAO Forest Resources Assessments, showing a forward projection to 1990 from two earlier surveys (lower line); a backward projection to 1980 and
1990 from a later survey (upper line); and how assumptions of increasing survey accuracy can obscure a reversal in trajectory (dotted line). Source: Grainger (2010).

2005, and backward to give 369 ha.10^3 in 1990. This ignored the possibility
that forest area stopped declining between 1978 and 1989 and started to rise,
as shown by measurements by Hecht and Saatchi (2007).

At pan-tropical scale, estimates of deforestation rates are generally consistent between FRAs, at 15.4, 12.7, and 12.1 ha.10^6.a^{-1} in FRAs 1990, 2000,
and 2005, respectively. However, combining potentially large area adjustments for many countries can lead to inconsistencies between successive
forest area estimates (fig. 7.3).

Thus, according to FRA 1980, Natural Forest covered 1,970 ha.10^6 in
seventy-six tropical countries in 1980 (Lanly 1981), later corrected to 1,935
ha.10^6 (FAO 1982). In FRA 1990, forest area was said to decline from 1,910
to 1,756 ha.10^6 between 1980 and 1990 for ninety countries (FAO 1993). According to FRA 2000, for these ninety countries forest area fell from 1,926
to 1,799 ha.10^6 between 1990 and 2000 (FAO 2001). In FRA 2005, the 1990
estimate rose again, and forest area shrank from 1,949 to 1,768 ha.10^6 from
1990 to 2005 (FAO 2006a).

SCIENTIFIC MEASUREMENTS

Pan-tropical scientific surveys have focused on tropical moist forest. Recent
remote sensing surveys have estimated its mean deforestation rate as 5.8
ha.10^6.a^{-1} for 1990 to 1997 (Achard et al. 2002), and 5.4 ha.10^6.a^{-1} for 2000
through 2005 (Hansen et al. 2008). However, these estimates were based on
sampling, not on comparing forest maps at the start and end of each period.

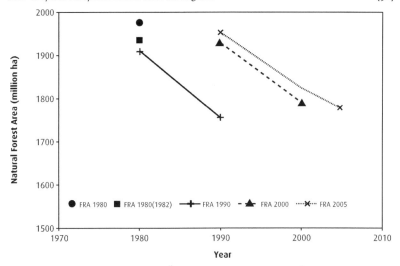

7.3. Trends in natural forest area in ninety tropical countries, 1980–2005, from data in FAO Forest Resources Assessments (FRAs) 1980, 1990, 2000, and 2005. Source: Grainger (2008).

Thus, their small size in comparison with FAO estimates for the whole tropics was understandable.

To see if estimates of deforestation are reflected in the long-term trend in total area, independent estimates of the extent of tropical moist forest were collected and adjusted to cover sixty-three countries containing 95 percent of all tropical moist forest. These included:

1. Five expert assessments from before 1990, including one by this author based on FRA 1980 data (Persson 1974; Sommer 1976; Myers 1980, 1989; Grainger 1984).
2. One using Closed Broadleaved Forest data (FAOCB) from FRA 1990.
3. Two derived from areas of ecosystem types in FRAs 1990 and 2000 (FAOE). These divided FRA estimates of Natural Forest area between ecosystems using proportions in separate surveys.
4. Three based on remote sensing surveys by the TREES Program (Mayaux, Achard, and Malingreau 1998; Achard et al. 2002) and its successor, the Global Land Cover (GLC) Program (Eva et al. 2002; Mayaux et al. 2003; Stibig, Achard, and Fritz 2004; Stibig and Malingreau 2003).

The resulting time series (fig. 7.4) does not conform to expectations (Grainger 2008). A conservative interpretation is that tropical moist forest area has changed little since 1980, within the limits of error. This does

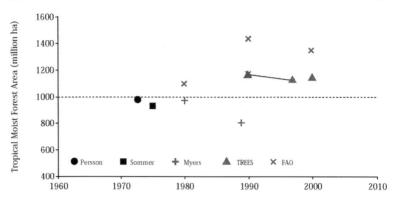

7.4. Estimates of tropical moist forest area for sixty-three countries, 1973–2000. For clarity, the Grainger (1980) estimate derived from FRA 1980 and the FAOCB 1990 estimate are both shown as FAO estimates, and the GLC estimate as a TREES estimate. Source: Grainger (2008).

not mean there has been no deforestation, just that global monitoring is too imprecise to detect it. The GLC estimate of 1,181 ha.10^6 for 2000 is indistinguishable from the FAOCB 1990 figure. It exceeds the TREES estimate of 1,152 ha.10^6 for 1990, but this could be influenced by extrapolation to sixty-three countries. The TREES 1997 estimate of 1,118 ha.10^6 is below that for 1990 but was based on sampling.

Areas for the same year vary, but there is consistency between the remote sensing estimates and between the FAOE figures. Differences between the FAOE and remote sensing figures for the same year (24 percent in 1990 and 14 percent in 2000) are consistent with the 18 percent difference that FAO (2001) found between its estimate of forest area in 2000 and one based on a remote sensing survey. The apparent decline between the two FAOE estimates for 1990 and 2000 reflects the use of different ecosystem classification systems. The Köppen-Trewartha system in FRA 2000 probably misclassified less forest than the Yangambi-UNESCO system in FRA 1990, so the lower figure for 2000 should be more accurate (Grainger 2008).

EVIDENCE FOR NATURAL FOREST RESURGENCE

Another interpretation of figure 7.4 is that decline is obscured because clearance of primary forest is offset by natural forest resurgence elsewhere.

Local evidence for natural reforestation has been found by various remote sensing studies (e.g. Perz and Skole 2003; Naughton-Treves 2004; Hecht and Saatchi 2007). Sampling measurements in pan-tropical remote sensing surveys by FAO (2001) and Achard et al. (2002) estimated the re-

Table 7.2. Trend in the area of forest plantations in ninety tropical countries in 1980–2005 (ha.10³) in FAO Forest Resources Assessments (FRAs) using different definitions of *forest plantation*

	FRA 1980 1980	FRA 1990 1990	FRA 2000 2000	FRA 2005 2005
	A. Using latest definition			
Africa	1,780	3,000	4,573	9,499
Asia-Pacific	5,111	32,153	54,716	16,726
Latin America	4,620	8,636	8,188	8,482
TOTAL	11,511	43,789	67,478	34,707
	B. Using FRA 1980/1990 definition			
Africa	1,780	3,000	4,000	9,159
Asia-Pacific	5,111	32,153	47,869	44,884
Latin America	4,620	8,636	7,953	8,482
TOTAL	11,511	43,789	59,823	62,525

Source: Grainger (2010).

forestation rate in the 1990s as about 1 ha.10⁶.a⁻¹. This was 10 and 17 percent of the deforestation rate, respectively, because FAO's deforestation rate was higher. Figure 7.4 suggests that the ratio could be much higher.

According to FRA 2005, six countries—Bhutan, Cuba, Gambia, Puerto Rico, St. Vincent, and Vietnam—sustained net natural forest expansion between 1990 and 2005 (FAO 2006a), and so seem to have passed through their forest transitions. Four more—Cape Verde, India, Ivory Coast, and Rwanda—did so based on the trend in the combined area of Natural Forest and Forest Plantations. But net reforestation in all ten countries was only 0.46 ha.10⁶.a⁻¹, or 3.6 percent of Natural Forest loss reported in FRA 2000 for that decade. So remaining resurgence is likely to be widely distributed throughout the tropics.

EVIDENCE FOR FOREST PLANTATION EXPANSION

Forest plantations are also expanding but their total area is only equivalent to 2 percent of natural tropical forest area. According to FRA 2005, the area in our set of ninety countries rose from 29.0 ha.10⁶ in 1990 to 32.7 ha.10⁶ in 2000 and 34.7 ha.10⁶ in 2005 (table 7.2). This was equivalent to a rise of 0.6 ha.10⁶.a⁻¹, just over half of the natural reforestation rate reported in scientific studies.

However, there is inconsistency between estimates in successive FRAs. For ninety tropical countries the total Forest Plantation area apparently increased sixfold from 11.5 ha.10⁶ in 1990 in FRA 1990 to 67.5 ha.10⁶ in 2000 in FRA 2000, but then halved in extent to 34.7 ha.10⁶ in 2005 in FRA 2005 (table 7.2).

One reason for this is uncertainty about plantation areas. Government reports usually refer to the area planted with trees, not to that which survives. So FAO routinely deducts 30 percent from reported areas to correct for high mortality (FAO 2001).

The definition of the Forest Plantations statistic also changes over time. The sharp rise from 1990 to 2000 was driven by Indian plantation area almost doubling from 18.9 to 32.6 ha.10⁶. The decline between 2000 and 2005 is explained by the Indian government reallocating most plantations to a new FRA category of *Semi-Natural Forest* (FAO 2006b), and reporting only 3.2 ha.10⁶ of stands with nonindigenous species as Forest Plantations in FRA 2005.

Since FRA 2000, FAO has included rubber plantations as Forest Plantations, since wood from rubber trees is exploited commercially. FRA 2005 registered a sharp fall of 383 ha.10³ in Forest Plantations area in Malaysia because large areas of rubber plantations were replaced by oil palm plantations, which are not treated as Forest Plantations (FAO 2006a).

If the estimate for 2005 is corrected to be consistent with those in FRAs 1980 and 1990, by removing rubber plantations and the reclassification of Indian plantations, Forest Plantation area in ninety countries rose from 11.5 ha.10⁶ in 1980 to 62.5 ha.10⁶ in 2005 (table 7.2).

The Policy Perspective

Forestation has been promoted in various ways over the last century. In dry areas of Africa, in particular, forestation was promoted by colonial foresters, who saw the unregulated spread of farming leading to formation of desert (Fairhead and Leach 1996). After trying and failing for decades to exclude people from tropical forests designated for timber production, in the 1970s tropical foresters reversed this strategy by involving nonforesters in forest protection and restoration. New community forestry (or social forestry) programs extended the role of forest plantations to supplying the needs of rural people for fuelwood, food, fodder, and other products (FAO 1978).

In the 1990s, tropical forestation was encouraged to create new terrestrial carbon stocks and sinks to offset carbon dioxide emissions and curb global climate change. The Noordwijk Conference (Declaration of the Ministerial Conference on Atmospheric Pollution and Climatic Change 1989) proposed that 12 ha.10⁶ of forest plantations be established every year for this purpose. Degraded tropical lands were thought to be ideal sites for these. According to Sedjo and Solomon (1989), 465 ha.10[6] of new tropical forest plantations would absorb the entire net increase in atmospheric carbon dioxide content from all sources. If variation in plantation growth rates in different environmental contexts were taken into account, this could rise to almost 600 ha.10[6] (Grainger 1996b).

Under the Clean Development Mechanism (CDM) of the Kyoto Protocol of the UN Framework Convention on Climate Change, tree-planting and other projects in developing countries can be funded by developed countries to offset their own carbon emissions. The full potential of the CDM has not yet been realized (Olsen 2007). Natural reforestation for carbon sequestration could be more economically competitive with alternative agricultural land uses than forest plantations (Olschewski and Benitez 2005), but is excluded from the CDM.

A Reduced Emissions from Deforestation and Degradation (REDD+) scheme features highly in negotiations to allow developing countries to take action themselves once the Kyoto Protocol ends in 2012. Increasing forest protection could indirectly facilitate natural reforestation, but there are many practical difficulties, for example, in defining historical baselines and preventing the *leakage* of deforestation from one place to another (Miles and Kapos 2008). This could put at risk forests that are low in carbon but high in biodiversity (Grainger et al. 2009).

The REDD+ scheme will also only be feasible if it is compatible with national development aspirations. So when projecting future normative trends in forest area to use as a basis for estimating reductions in deforestation rates, it would be wise to use trends based on forest transition curves that are development compatible and not rely on abstract mathematical exercises.

Conclusions

Growing interest in tropical deforestation since 1980 in scholarship, theoretical research, statistical collection, scientific measurement, and policy has been welcome. But it has led to the neglect of natural reforestation, which compensates in some measure for it.

Natural reforestation is anticipated by existing theoretical models, but if they and our understanding of tropical forest change are to become more sophisticated, it is vital to expand scientific measurements. A World Forest Observatory (WFO), now in the planning stage, will help enormously by mapping the world's forests every year for the first time (Grainger 2009). To achieve this massive advance in terrestrial observation will demand far more sophisticated international scientific institutions, but the maturing policy framework, in the form of the REDD+ scheme, will encourage this, as a WFO could provide independent confirmation of national reports of reduced deforestation rates.

The inability, even of scientists, to collect data on natural reforestation in the tropics is a reminder of how little we know about changes in terrestrial land cover and land use at pan-tropical and global scales. It is time to start paying more attention to this, especially now that budgeting for every tonne of carbon is assuming global political importance.

8 ✳ The Social Lives of Forest Transitions and Successions: Theories of Forest Resurgence

SUSANNA B. HECHT

Ecological Amnesia: Successions, Transitions, Re-wilding, Re-wooding

INTRODUCTION

In the popular consciousness, tropical ecosystems embody a catastrophic narrative of deforestation and land degradation. Emphasizing clearing and a unidirectional narrative of forest loss, this apocalyptic vision has dominated science and popular perceptions of forest trends over the last half century as forest landscapes became objects of intense environmental concern integrated into modern commodity circuits, national political strategies, and theaters of war. This anxiety was coupled with an explosion of research on the extinction effects of forest fragmentation, research that made anything but relatively large forest old growth areas seem like sites of extreme ecological instability (Arima et al. 2005; Barlow et al. 2006; Broadbent et al. 2008; Cochrane and Laurance 2008; Laurance 2008). In this context, forest resurgence and the re-wooding of landscapes seemed unlikely; when observed, successional landscapes were largely dismissed as degraded and "denatured." Although deforestation dynamics remain strong, there is, however, also a dramatic process of woodland recovery in these contexts that requires analysis.

The tropics have been home to great civilizations that have come and gone. Tropical forest regions have been the targets of the modern "development enterprise" (especially since the mid-nineteenth century[1]) and have experienced both intense exploitation and subsequent declines. Colonial forays and contests driven by industrialization and a rise in mass consumption that included latexes, oils, stimulants, timber, and a range of foods, contributed to the history of the tropics. This history shaped the tropics ecologically, demographically, institutionally, and politically. However, this history is also often reworked through a kind of socio-ecological amnesia in which areas that had once been tightly integrated into global political economies, but then slipped into obscurity, were later recast as remote and untrammeled nature. This is the socio-environmental counterpart to the ecological idea of "shifting baselines," which suggests that there is generational loss in knowledge about abundance and diversities in ecosystems

(Papworth et al. 2009). This dehumanizing of socio-natures is important because it places landscapes into an apparently natural, evolutionary history of wilderness: highly diverse systems lose their human imprimatur as they are ideologically and biotically "re-wilded," even if they bear significant social histories.

Three modern cases illustrate this point. The western Amazon (especially the Purús River) was the site of five wars and the source of half the world's rubber at the turn of the twentieth century. The decline of rubber prices just prior to World War I produced out-migration and a forgetting of the region's history. The Purús was reimagined as pristine at the turn of the twenty-first century, a place at the "ends of the earth and the beginnings of time." The largest National Park in the New World, the bi-national upper Purús park, was recently gazetted there (Hecht 2011).

Other global events, such as the Cold War of the mid-twentieth century, were also instrumental in shaping forests. Proxy wars in this conflict took place in the tropical forests of Central America, Amazonia, tropical Africa, and of course, Southeast Asia. Such conflicts had a significant impact in structuring forest "natures" (Hecht et al. 2006; Peluso 1992; Peluso and Watts 2001). Peluso describes how formerly inhabited forests in Indonesia, for example, experienced "environmental" as well as ethnic cleansing as longtime Chinese residents and indigenous populations were forcibly removed from their forest-farming homes. These places were then recast as wild environmental spaces, "empty land" to be controlled by new forest ministries, conservation set-asides, or transformed into permanent agricultural landscapes (e.g., Java's transmigration program for nationalist colonist projects), or industrial tree crops (see Peluso and Vandergeest, chapter 5 in this volume). These forests had an extensive history of settlement—a history inscribed on the forest landscapes and ecologies before they were transformed for political purposes into what are often today considered "wild" forests.

The Cold War helped reshape the "wild" in the United States as well, through the nuclear theater of the contest. Masco (2007) describes how places so contaminated with nuclear waste that people cannot occupy them have been reimagined as exquisite wildness. The toxicity of the landscapes did not change, but the new portrayal of nuclear dumps obliterated their histories and changed their rules of access, transforming them from among the most dangerous "socio-natures" into ahistorical, biotic landscapes, beautifully re-wilded. These twentieth-century episodes do not require much archeology: these "successions" have histories written on paper as well as landscapes.

Forgetting the human agency in wild landscapes, recasting them as lands without history has its counterpart in the invisibility of today's inhabited woodlands as forests. These linked processes of dehumanizing (or re-wilding) some places, while denaturing others, limits our capacity to understand the dynamics of landscapes—their successions and transitions—as socio-biotic processes. Ideas about forests and forest trends, unmoored from their social and ecological matrices, obscure the "social lives" that underpin them. Simple empirical descriptions of forests and modern classifications of what these places "are" ignore many forces (histories, tenurial regimes, knowledge systems, political economies, institutions, discourses, and other cultural arrangements) that impinge on them today or constituted them in the past, shaping the successional substance of their transitions.

FROM TREES TO FORESTS TO SOCIO-NATURES

Like the word nature, forests are what semioticians call a floating signifier. The term means different things to different people. It is vague, highly variable, and it has multiple referents with symbolic power. Thus, although landscapes and forests obviously exist biotically, what exactly constitutes a "true" or "authentic" forest or what even is counted as forest is controversial. Increasingly this is a topic taken up in the debates over forest resurgence, as well as by those engaged in remote sensing work and global modeling exercises (Colson et al. 2009; Comber, Fisher, and Wadsworth 2005a, 2005b; Grainger 2008; Hecht and Saatchi 2007; Putz and Redford 2010). These questions of classification are not just semantic, especially today, when climate and environmental capital flows are mediated through the value—human ascription to landscape—attributed to different kinds of woodlands. It is for this reason also that social scientists refer to the idea of socio-natures, the construction of forests as both cultural and biotic successional entities, and as landscapes understood through representations. These socio-natures include what we say about forests (e.g., the untrammeled wild, timber yield per hectare, the mantle of the poor, percentage woody cover per pixel), as well as political economies, social rules, rights, processes, and institutions that pertain to them. This forest fabric also has power relations that inhere in it: who rules, who can rule, who makes the rules, and who is overruled in access, control, and meaning of such place— what Foucault has called "governmentality" (Inda 2005).

Thinking about forest transitions is enormously helpful for activating a more complex understanding of landscapes because it helps overcome socio-ecological amnesia by placing successions and transitions in palpable and recent histories. It also liberates us both to imagine forest pasts through an anthropogenic lens to apply an ecological optic to today's humanized landscapes and matrices rather than simply dismissing them as degraded, despoiled, and depauperate because they are fragmented (Levins 1979; Perfecto and Vandermeer 2010; Vandermeer et al. 2004; Vandermeer and Perfecto 2007). The processes of successions and transitions taken together—biotic processes and social change—suggests cyclicality rather than simple linearity, a substantive epistemic shift.

Framing forests and forest resurgence in terms of socio-natures is important for seven main reasons, some pertaining to the theoretics of ecology and other more applied dimensions of landscape review. First, the historical shaping of forests through human management and interaction with ecological processes raises questions of resilience that do not come up if places are seen as more or less primordial. The dynamics, capacities, and limitations for landscape recuperation and reconstruction, including human agency, are especially important under the onslaughts of climate change. Second, this approach permits understanding of species distributions and assemblages as anthropogenically and biotically constructed cohorts (Balée 1998; Lentz 2000; McKey et al. 2001; McKey et al. 2010; Rival 2002). The widespread occurrence of oligarchic forests or pure stand forests in the tropics often involve "lost" anthropogenic histories (Clement 1999; Padoch and de Jong 1990; Peters et al. 1989; Rival 2002; Vormisto et al. 2004). Third, historical ecologies can radically reshape our understanding of human interventions in the past, provide insight into current landscape management, and help construct future practices. Fire ecologies in the construction of tropical landscapes and anthropogenic soils are examples of this (Glaser 2007; Glaser and Woods 2004; Junqueira, Shepard, and Clement 2010; Kim et al. 2007; Paz-Rivera and Putz 2009; Peterson, Neves, and Heckenberger 2001). Fourth, when examining recent historical successions, forests differ because societies and their coteries, ethnic and class fractions, men and women differ in terms of what they want (Adeney, Christensen, and Pimm 2009; Brosius 1997; Buckingham and Kulcur 2009; Deere and Leon 2001; Mathews 2008; Padoch 1999; Rocheleau, Thomas-Slayter, and Wangari 1996; Toledo and Salick 2006). These differences provide ways to assess the sociocultural drivers that shape the ecological and social parameters of such systems. This has been a long-standing realm of cultural ecology and political ecology, and if landscapes are to serve a range of livelihood, climate, and biodiversity purposes, such socio-ecological backstories and their institutionalities are worth knowing about. Fifth, forests are crucial to both urban and rural peoples, directly for livelihood supplements and production inputs, and indirectly for environmental services, such as water and greenhouse emissions absorption, local microclimate modifications, and so on. Urban areas are often increasingly forested, serving as biodiversity refugia (Goddard, Dougill, and Benton 2010). How landscapes work in these hybrid contexts represents a major research frontier. Sixth, environmental politics and markets for environmental services, especially carbon, will increasingly structure the institutions, distributional dynamics meanings, and markets for forests and their environmental services in novel ways. With carbon offsets and Reduced Emissions from Deforestation and Degradation (REDD), what is permitted to transpire in forests, who ultimately defines what are "worthy" forests, and who benefits from the commoditization of the atmosphere and the role of forests in atmospheric politics remains an open question (Chhatre and Agrawal 2009; Neeff and Ascui 2009; Okereke and Dooley 2009; Stickler et al. 2009). Finally, history, identity, and long-term landscape interventions now infuse claims for forests and land resources, competing with international environmental or development enclosures for local sovereignties. Indigenous and traditional peoples' territories—extractive reserves, former slave refuges, as well as native lands—are clear examples of such claims, which may engage oral histories, historical and indigenous cartography, historical ecologies, indigenous classifications and knowledge systems about landscapes, and animate a range of institutional configurations that have shaped them (Comaroff and Comaroff 2008; Escobar 2008; Hecht 2009; Hecht and Cockburn 1989; Linhares 2004; Neumann 1998; Posey and Balick 2006). The politics of such sites are likely to become more contested with time, especially vis-à-vis environmental capital flows into forest conservation and offset projects and an emergent set of re-wilding projects that also seem to have produced conservation refugees and controversial resettlement programs (Brockington, Duffy, and Igoe 2008; Dowie 2009; Neumann 1998). A likely issue will be whether the neoliberal model of extensive privately owned lands managed by ecological proconsuls will prevail over re-wooded populated agroecological matrices and collective landscapes.

Forest Recovery: "The Forest Transition" in the North and South

The re-wooding of the United States and Europe was widely associated with urbanization and industrialization and the structural shift from rural to urban economies (Mather 1992; Mather and Needle 1998). There were regional dislocations of agriculture from one region to another through production efficiencies and technical change within various agricultural sectors (Ramankutty, Heller, and Rhemtulla 2010). Trees regrew, or at least they have for a while (Drummond and Loveland 2010). Populations declined in the countryside. Some authors have sought to universalize this

Euro-American model of the forest transition with the modernization narrative that developing countries will repeat the patterns of the first world, a
generally questionable proposition (Perz 2007).[2] Current transitions in the
developing world have several features that differ from the Euro-American
pattern in general: (1) they have occurred extremely quickly; (2) they reflect
exogenous pressures more than endogenous dynamics; (3) rural areas, even
though forested, have high population densities; and (4) they are, as a general rule much more globalized in terms of commodities, financial flows,
and often, labor (Asner et al. 2002; Garcia-Barrios et al. 2009; Hecht and
Saatchi 2007; Rudel et al. 2005). These transformations have involved significant recasting of the livelihood possibilities, meaning, nature, and politics of these landscapes.

These general features are made more complex by varying historical
roots, diverse cultural-ecological matrices, an array of tenurial systems,
emerging regional and global markets, policy interventions, war, institutional rivalries, new ideologies (including environmental ideologies, neoliberal policies), territorial identities, and competing authorities that vie for
the political spaces forests have become. The resurgence of woodlands also
suggests that the largely agrarian annual cropping models, much more typical of the temperate zone than the tropics, used to understand (and model)
this transition, may have overlooked a reality of annual cropping with successional mosaics involving much more heterogeneity in terms of ecological structures, functions, livelihoods, and mediating formal and informal
institutions in peasant landscapes. In many tropical cases, the agroecological matrix may be a more useful analytic category for understanding transition landscapes (Hecht 2009; Perfecto and Vandermeer 2010).

THEORIZING THE FOREST TRANSITION AND
ITS DISCONTENTS

The forest transition has become an active arena for theorizing drivers of
land-use change, as recent special issues of *Ecology and Society* (edited by
ecologists Mitch Aide and Ricardo Grau) and *Land Use Policy* (edited by geographer Tom Rudel) attest. The next section outlines several of the processes both observed and theorized about the dynamics of forest expansion.
Several schema have been elaborated over the last few years (Barbier, Burgess, and Grainger 2010; Perz 2007; Rudel 2002; Rudel et al. 2005). The dynamics of modern deforestation have been fairly completely analyzed at this
point, but forest recovery remains a much more complex process. Loosely
conceived as a U-shaped curve, re-wooding reflects an array of environ-

mental, institutional, economic, and sociocultural arrangements. These are generally divided into state-led responses to scarcity, market responses, globalization dynamics, and other processes described as socio-ecological feedbacks. Finally, there are the smallholder intensification pathways (Lambin and Meyfroidt 2010). These are useful frameworks for understanding these emerging socio-natures, although they neglect some patterns of change that modify our understanding of what are often perceived as incremental processes. One that is largely absent in the models, but widely present in the world, is catastrophe.

Catastrophic Events and Forest Resurgence

Although not particularly noted in the forest transition literature, events that produce massive demographic changes or severe economic disruptions have historically been associated with forest resurgence. Emblematic in this regard is the extensive forest recovery after the population collapse in the New World associated with the arrival of Europeans and their diseases (Mann 2005, 2008; Nevle and Bird 2008). Other unforeseen events like Chernobyl have produced local ecosystem recovery, and problems like nuclear waste dumps and testing sites in the United States (Masco, chapter 6 in this volume) induced a re-wilding. Increasing severity of weather events or other geophysical catastrophes, like hurricanes, tsunamis, earthquakes, and so on, may well contribute to land-use change, either by making areas uninhabitable or through efforts of biotic stabilization. Barbier (2008) has described how Asian tsunami victims began to plant mangroves as barriers from future assaults. The economic breakdown experienced in much of the rural Midwest in the United States and the collapse of Detroit are producing forest transitions.

War is another widespread impact that often produces forest resurgence when agricultural frontiers are inhibited, local populations are driven away, and landscapes become militarized, as occurred in the buffer zones between the Koreas, Central America in the 1980s, Colombia for much of the last thirty years, and eastern Congo and parts of Southeast Asia. While such events may not be particularly good for the mega-fauna, forests have been refuges for insurgencies. Many such areas have been seeded with land mines, as in Angola where the mine situation only allows about 3 percent of the area to be cultivated (Integrated Regional Information Networks 2002), far less than was possible when it was a jewel in the crown of the Portuguese empire. Catastrophic events are part of planetary and human history, and should be incorporated into an understanding of socio-natures.

Socio-Ecological Abandonment and Forest Resurgence

Land degradation associated with soil nutrient declines or erosion can produce forest recovery as land is taken out of agricultural or other uses. Lambin and Meyfroidt (2010) refer to this as a "socio-ecological feedback," where ecosystem services and utility are degraded by past land uses; human use is withdrawn and endogenous processes of ecosystem recovery kick in. Such derelict landscapes are often seen in Amazonia, where about 20 percent of formerly cleared pastures are now in some form of regrowth (INPE 2008). The empty pasture is an ecological story of soil nutrient decline (Fearnside 2008; Hecht 1985), but it is also an institutional one: land claiming, money laundering, institutional rents, or for many small-holdings, a strategy of economic diversification (Bommel et al. 2010; Hecht 1993; Pacheco 2009; Walker et al. 2009). Severely eroded or desertified landscapes are also often targets of state or nongovernmental organization (NGO)–led reforestation exercises that often include the aims of livelihood additions as well as ecosystem service recuperation (Bryan 1994).

State Policy and the Forest Scarcity/State-Mediated Path

Building on Rudel et al. (2005), Lambin and Meyfroidt (2010), as well as Barbier, Burgess, and Grainger (2010) point to "forest loss" as a pressure that stimulates state policies for reforestation, and subsidize forests as woodland resources and services become scarcer. These models focus on how decline in forests enhances demand for marketable forest goods (and change land values of woodlands), while loss of forest environmental services, especially erosion control, water resources, and conservation, become alternative stimuli for state intervention (Barbier, Burgess, and Grainger 2009; Meyfroidt and Lambin 2008). While these pressures are often seen as market driven, forest programs are typically highly subsidized. The expansion of state-promoted plantation forests in China and its regionally controlled landscapes in Asia exemplify this model (Menzies 2007; Xu et al. 1999). Chinese interventions and policies to relieve scarcity also support environmental services, but largely through the establishment of monocultural plantations.

States act through the politics of national sovereignty, national institutions, and territorial control, and they mediate a range of conservation interventions since states are the key arbiter of property rights, provide counterparts to international funds, shape national and local resource management institutions, respond to varying political pressures, and are themselves actors in determining landscape outcomes through other investments like infrastructure and the provision of social services (Hecht 2009; McAfee and Shapiro 2010). States often use forest policy to change tenure regimes, forest access, and regional economies (Peluso and Vandergeest, this volume; Brown 2003; Kant 2009; Ribot 2001). In China, extensive monocrops of rubber and other industrialized tree crops (oil palm, eucalpytus) associated with both national industries and international demand have been key in shifting land from collective ownership and village control into privatized holdings. In contrast to Latin America where claiming land by clearing has been the historical norm, tree planting in Southeast Asia became a means of encroaching on common lands and privatizing them, increasing rural income inequality and small farmer marginalization (Fox and Vogler 2005). These monospecific plantations raise issues about the larger ecological diversity and vulnerabilities of such systems (Aziz, Laurance, and Clements 2010; Trac et al. 2007). Social analysts question the tenurial outcomes and economic exposure that are often unbuffered, as well as the displacements of complex landscapes and the social systems that underpinned them (Greenough and Tsing 2003; Nevins and Peluso 2008; Primack and Lovejoy 1995).

In Latin America, state programs that support agro-forests and native species reforestation have had a long history as part of rural development programs, but increasingly these funds are being linked to state-led payment for environmental services programs. These programs are taking place in very heterogeneous tenurial regimes ranging from private holdings, collective agrarian reform holdings, ejidos, to a range of nonprivate access regimes (Farley 2007; Grieg-Gran, Porras, and Wunder 2005; McAfee and Shapiro 2010; Rosa, Kandel, and Dimas 2005). The extension of global carbon economies and REDD as state-led programs is likely to expand this state intervention further into many forested regions. Equity dimensions to such capital flows are already controversial (Boerner, Mendoza, and Vosti 2007; Chhatre and Agrawal 2009; Irawan and Tacconi 2009; Soto-Pinto et al. 2010).

The Environmental Kuznets Curve as Forest Transition or the "Modernization-Development Pathway" Structural Change

The Environmental Kuznets Curves (EKC)—generalized U-curves—are based on the idea that as countries develop and increase their national GDPs, their environmental indicators improve because natural resource dependency declines, production techniques and the composition of the economy shift, and the components of goods change. This is "ecological modernization," and in principle, people are more willing to pay for environmental goods and enforce regulation. Initially focused on industrial externalities (pollution and potable water), the term has largely been discredited because of the much-enhanced ecological footprint associated with increasing wealth. Still, the EKC has been applied to deforestation and forest recovery. Deforestation in these models is largely explained by endogenous features, including population, population density, GDP, debt, institutional configurations (e.g., democracy, corruption), and policy factors to test whether the curves exist and at what per capita income level the inflection point—the place where environmental trends shift in a positive direction—occurs. Forest EKC modeling efforts generally rely on FAO panel data that do not easily assess successional/anthropogenic forests (Baker 2003; Bhattarai and Hammig 2004; Caviglia-Harris, Chambers, and Kahn 2009; Cole 2003).

The results of the EKC deforestation models are contradictory and short on empirical data. Bhattarai and Hammig (2004), for example, focusing on the institutional dimensions of clearing and EKC, find that a deforestation EKC exists for Latin America, but the inflection point developed in their model is not supported by field data, since Brazil, at the time of their writing, had the highest per capita income and deforestation rate, while in national geographies where populations were far poorer, regional forest resurgence occurred (see Hecht et al. 2006; Klooster 2003). Koop and Tole (2001) find no EKC for deforestation in Latin America, arguing that policy arrangements and institutions, as well as distortions affecting land use, differ so much as to defy useful comparisons. Mills and Waite (2009), running data on thirty-five countries, show no relation between country wealth and forest conservation. Pfaff and Walker (2010) focused on trade and the transposition of deforestation from some regions to others in new patterns of spatial/economic expressions of regional and international markets. As they put it, one person's transition may rely on another's deforestation. This pattern of deforestation displacement is also found in the forest transition in Vietnam (Meyfroidt and Lambin 2008).

Critics of the EKC point to problems with the data on which assertions are based, policy environments, institutions, forms of governance, problems in comparability between countries and between types of ecosystems, and the statistical methods used to address the questions. Another problem with the deforestation/EKC is that structural change is implied but not specified (Barbier, Burgess, and Grainger 2010; Dietz and Adger 2003; Perz 2007) and the forms that this takes make a difference. It is here, especially, where the social lives of forests unfold.

The economic modernization pathway—or development pathway—implicit in the EKC has been a good deal more complicated in the developing world due to the comparatively weak dynamics of industrialization, the low wage structure, rural semiproletarianization and informality of many urban jobs, the multisitedness of households, circular migration, international migration, a broad range of property regimes, the politics of agrarian reform, labor scarcity, and the expansion of local and regional markets for tree products. All of these factors, among others, potentially affect the upward deflection curve. The current dynamics of a forest transition model must be more carefully specified. Dissatisfaction with the endogenous explanations of the EKC version of the forest transition also caused researchers to explore questions of globalization.

Globalization and the Forest Transition: Products, Labor, and Finance

Globalization is one of the defining features of today's macroeconomies. Globalization has substantial impact on the forest transition via (1) global commodity markets for wage foods and niche goods; (2) international fiscal transfers; (3) global information flow; and (although this is not usually included in definitions of the global economy, it is a ubiquitous feature of global policy and discourse) (4) the rise of the environmental economy. This last has had an impact during the past thirty years, and will intensify with the politics of planetary change.

The global system for wage foods (i.e., basic annual food crops such as corn, soy, wheat, and rice) increasingly involves highly productive, internationally uniform intensive agro-industrial systems integrated into global commodity chains though trading monopsonies that deliver basic foods at low cost, although these systems are ecologically controversial and frequently subsidized (Bush 2010; Patel 2010). Most developing countries embrace cheap food policies for their burgeoning urban populations, since high food prices are a source of political unrest. The often lower price of imported food products from a few mega-producers can undermine the returns to local producers, making local agriculture less competitive for these wage goods. This has generally eliminated small-scale farmers as producers of most global commodities except in small regional markets, clandestine markets, and niche markets (Barkin 2002; Bebbington 1999; McMichael 2006).

NICHE MARKETS

Throughout most of the twentieth century, development policy viewed small farmers as key food producers for urban enclaves and national markets. As generic annual crops became increasingly globalized in their production and distribution, the central strategy for claiming value added for

small farms involved moving toward higher value products for elite markets, and enviro-social branding, where prices are stronger for things like
fair trade, organic, and specialty commodity markets for cacao, coffee, teas,
and foods like açai, edible and cosmetic oils like shea butter, and babassu.
Local markets with specific tastes can also be thought of as "niche"—Oaxa-
can corn and chile markets, Bragança manioc in Belém, sticky rices in the
Philippines. Farmers are not growing generics but specialized, named products. Clandestine drug production is a separate version of a niche market.

Credits for tree products are among the few credits available for nonindustrial farmers, or specialized nontraditional exports. The existence of
such niche markets for forest products is a significant element of woodland
support, and mixed agroforestry systems have had an impact on maintaining and transforming local ecologies (Brondizio 2008; Nevins et al. 2008;
Peeters et al. 2003; Perfecto et al. 2007). In a world where social identity is
increasingly signified by consumption patterns, the simultaneous extension of "distinction" with ubiquity lies at the heart of business models and at
much of the externally perceived cultural content of specialized woodland
products (Bourdieu 1994; Comaroff and Comaroff 2008). The practical result is that these systems often help generate what Meyfroidt and Lambin
(2008) call the "small holder tree-based intensification" pathway, producing global commodities as well as supplying household and local demands
for foods and building materials and other livelihood supplements, and,
frequently, habitat for nonhumans.

FISCAL TRANSFERS: GLOBAL AND LOCAL SUBSIDIES AND
HOUSEHOLD LIVELIHOODS

Remittances

Cash payments into rural households from family members working in
cities are a major financial flow into rural areas. About 10 percent of the
world's population now depends on remittances (Inter-American Development Bank [IADB] 2007), and remittance patterns show a strong correla-
tion with forest resurgence (Garcia-Barrios et al. 2009; Hecht and Saatchi
2007). The magnitude of these transfers is globally significant. International remittances alone are over 317 billion, even though these may be
vulnerable to periodic global fiscal problems (Acosta et al. 2008; Ocampo
2009). The remittance economy is about triple the total development assistance worldwide. It also can exceed direct foreign investment in many
small economies. This is the true development fund "from below" as people,
often working at the low-wage jobs in the industrial or oil-flush economies,

or as seasonal agricultural labor, send portions of their wages home. Latin America receives about 63 billion per year, and the rate of growth of this sector of the economy is dynamic, with about 10 percent per year (Orozco 2002). Although clearly it has been affected by the global downturn (the rate of growth is now about 6 percent), this decline is viewed as having bottomed out (Ocampo 2009).

For many countries in Central America, remittances are now a major portion of GDP. For example, remittances count for 20 percent of the GDP of Honduras, 17 percent of El Salvador and Nicaragua's economies, and 12 percent of Guatemala's GDP. Remittances as a percentage of the value of exports are now worth about half for Nicaragua, 60 percent of Guatemala, and 75 percent for Honduras (IADB 2007), eclipsing any national agricultural or industrial export product. The environmental implications of these capital flows remain largely outside of the purview of migration studies. One dimension, however, is that cash supplements permit families to buy basic foods instead of growing them. While this has implications for food sovereignty, the reality is that there has been a retraction in peasant agricultural systems as a consequence of remittances. Given the global magnitude of migration and remittances, the lack of attention to the environmental issues is surprising, reflecting perhaps a similar indifference to the anthropogenic forests that are in part a result of them (Garcia-Barrios et al. 2009; Grau et al. 2003; Hecht and Saatchi 2007; Klooster 2003).

Pensions and State Payments

The effect of international state transfers on forest resurgence was initially analyzed for Puerto Rico, where ecologists and geographers noted significant forest recovery (Aide et al. 1996; Grau et al. 2004; Rudel, Perez-Lugo, and Zichal 2000). Focusing on US social security and other transfer payments, Puerto Rico shifted from a zone of agricultural production into the New Rurality or, as Grau has called it, post- agricultural—a place of retirement, household safety nets, state entitlements, and forest cover returned. Retirees or others with pensions are also feature of environmental enclaves like Costa Rica, privileged parts of Mexico, and parts of Guatemala, but their environmental impact still remains largely unresearched (Kull, Ibrahim, and Meredith 2007). These are likely to follow the suburbanization models recorded by Baptista (2008) in Florianopolis and Foster on Martha's Vineyard (Foster et al. 2002).

In Eastern Amazonia, Padoch and her colleagues noted that multisited households increasingly depended on the impact of pensions and social security payments, and they reduced patterns of deforestation as basic food items could be purchased, even as they invested more time in producing more valuable tree products, such as açaí and shrimp, that could be sold in local urban markets (Padoch et al. 2008). The impact of these transfers from global and national economies provide an entrée into the complexity of the ethnographies and dynamics of forest recovery, and indicate the surprising side effects of policies and practices that have very little to do with forest management per se. The impact of large-scale government or private transfers, however, may be especially important as rural areas become more central to the environmental economy for climate mediation, such as REDD.

Globalized Environmental Economies

Muirist Conservation in the Neoliberal Age. The environmental economy is among the most powerful in shaping modern rural land use, and is likely to increase in importance with carbon-offset projects. While all cultures in the world and throughout history have generated complex epistemes and systems of knowledge about nature, the globalized discourse about "environment" is largely constructed on American Muirist approaches that emphasize wild nature—set areas that largely exclude human use except for tourism and science. Initially promoted for "beauty, health, and permanence," they were part of the Progressive Era's legacy of public goods. In the post–World War II period, the dramatic vistas and accessibility that had been at the heart of early parks in the developing world gave way to biodiversity conservation (especially for charismatic beasts) and, in the 1990s, increasingly environmental services. As part of international lending conditionalities and the rise of Big International NGOs (BINGOS), park creation and management has become a central element in international environmental philanthropy as well as rural policy. From their inception, set-asides everywhere have involved exclusions of traditional populations, and continue to do so today (Dowie 2009; Hecht and Cockburn 1989; Jacoby 2001; Neumann 1998). The long history of expulsion associated with parks have made them controversial in many developing countries, and it is for this reason that inhabited reserves in Brazil have been such a departure from business as usual.

Conservation set-asides involve transfers of land into state or, increasingly, privately owned conservation holdings. During most of the twentieth century, these programs followed the US model, which focused on demarcation and protection of state lands, but increasingly the model has been neoliberalized and is dependent on private property and exclusionary holdings, often regardless of the traditional holdings already in place. While lauded in some circles (Fraser 2010), the politics of such privatizations have become very contentious as grazing lands and agrarian and mixed farming holdings have gone into succession, or conservation easements are placed over inhabited landscapes and traditional holdings. The magnitude of the total areas shifted from occupied into dehumanized set-asides has not been quantified, but is likely to involve many million hectares. The Nature Conservancy, for example, owns more than 54 million hectares globally. These conservation exercises, often called re-wilding, may involve actual or functional privatization of state lands that are then usually managed by large conservation NGOs.

Socio-Environmentalism and Another Type of Transition. The term socio-environmentalism is most associated with Brazilian political movements that refer to the silvo-agrarian populations or forest peoples. Unlike the North American Muirist models, socio-ambientalismo or socio-environmentalism takes as its premise the possibility of positive interactions between peoples and environments based on historical settlement, local knowledge, and modern sciences. Socio-ambientalismo had its origins in the resistance movements against authoritarian regimes whose development model triggered massive forest destruction in the 1970s and 1980s. These movements were able to successfully articulate an independent model of forest livelihoods and resilience far different from "all or nothing"—the options of either complete conservation or complete destruction on offer by conventional conservationists or the military government. These movements halted large-scale deforestation through political action, better demarcation, and much better monitoring of deforestation. In Latin America, the politics surrounding these landscapes remains complex because the resources are not uniquely agricultural and form part of forest-based livelihoods. The claims invoke territory: identity, history, alternative knowledge systems inscribed in landscape, and stewardship advanced by natives, traditional peoples, extractors (like rubber tappers), quilombos (autonomous black heritage communities), and mapped out with participatory geomatics. Peasant economies that may have similar embedded knowledge remain politically invisible in this approach, although movements such as the international Via Campesina and Brazil's Landless People's Movement increasingly emphasize agroforestry and agroecological landscapes. Other peasantries have identified themselves with indigenous or black histories, rather than the politics of the agricultural plot, in a process of claiming a range of livelihood resources.

Forest peoples' movements present alternatives to set-asides and represent a range of different epistemes about nature, politics, and development. These approaches can be understood as post-modern in the sense that they reject universalizing practices of understanding nature, economic organization, and cultural life, and they insist on the validity of their understandings and socio-environments (Bray et al. 2003; Dove 2006; Greene 2006; Harris and Nugent 2004; Mathews 2009; Schmidt and Peterson 2009; Yashar 2005). This issue of the cultural turn in rural development raises many questions, especially as these societies develop formal institutional transitions toward rules and processes for inhabited forests. In this sense, these movements can be seen as part of a forest-agrarian reform based on heterogeneous property regimes and collective rules as well as new forest-based governmentalities. Because land management may reflect autonomous decision making, they have come up for criticism amongst those mainly interested in animal conservation (Figueroa et al. 2009; Redford and Sanderson 2000; Redford and Stearman 1993; Sanderson et al. 2002).

Payment for Environmental Services

Conservation expenditures are about 10 billion dollars per year, while carbon markets currently deploy some 110 billion dollars. With REDD, markets for carbon offsets are likely to heat up considerably, although they still remain profoundly controversial. The rise of this new economy will further restructure woodlands because they will set up a new set of institutional arrangements, power dynamics, and political economies over forest mosaics and matrix landscapes, and extend dimensions of state and private control deeply into the historical commons of extensive areas of land and air. In general, three competing landscape models are at play: investing in populist re-wooded or resurgence landscapes (hence inhabited and successional landscapes); efficiency landscapes (plantations); or conservation set-asides (re-wilding and traditional parks). These also involve political and distributional issues of how these transfers unfold; whether through direct payments to nations (including those with high potentials for corruption like the Congo Republic); regional states (like Amazonas in Brazil), forest plantations, and extensions of private reserves; or whether these are NGO landscapes, state and/or communal forests. What is certainly clear is the "natures" of forest resurgences and forest transitions are likely to be intensely debated as they become more visible, and as new forms of enclosures and pressures for more forest control, more monitoring, and more politics play out. In the carbon universe, forest property (often acquired for biodiversity purposes) has an extremely valuable and tradable derivative, the benefits of which are planetary, but the profits of which accrue to their owners—the capacity of forests for greenhouse gas (GHG) absorption. In this context, forests and their embedded (even if invisible) social lives must be understood as part of political economy debates with distributional and regional consequences, not just as neutral bio-environmental entities. Such an understanding is key to any ecology of justice or even rural stability, since the practice of resistance to large-scale enclosures involves nothing more complicated than a lighter in the dry season.

Conclusions

Questions of resurgent forested landscapes will define much of the rural development politics in the next decades. What drives woodland recovery are complicated, usually multicausal processes that are increasingly framed within the dynamics of globalization. Much of this recovery is occurring in the agricultural matrix, shaped by the social histories, the economies, institutions, imaginations, knowledge systems, and practices of the people who shape them. As forests become increasingly pivotal in global climate politics, understanding the dynamics of forest transitions, successions, and their social underpinnings—the social lives of forests—is a critical step for whatever resilience we might hope for in the maelstrom of twenty-first century climate change.

I close with a paraphrase of an aphorism from much-reviled Mao Tse Tung's Little Red Book: "Where do forests come from? Do they fall from the sky? No, they come from social practice."

9 ✳ Paradigms Lost: Tropical Conservation under Late Capitalism

JOHN VANDERMEER AND IVETTE PERFECTO

Conservation, Original Sin, and Redemption

When Adam and Eve were expelled from the Garden of Eden, our paradise lost created our life's goal: to find it again and regain redemption. Parallel narratives about conservation are hardly subtle: the "pristine" rainforest with its "cathedral-like" tree trunks majestically reveals the beauty produced by God or Darwin, but we were expelled by avaricious loggers who covet money over aesthetics, irresponsible peasants who lack condoms, and corrupt governments in "underdeveloped" countries. Perhaps hyperbole, but not, in the end, far off the mark in our experience talking with conservationists concerned with the tropics—a romanticism astounding for those among them who are scientists. More importantly, however, is that this romanticism is dangerously wrong for the preservation of the very pristine nature they wish to preserve.

In the nineteenth century the Hudson Valley School of art flourished in the Eastern United States. One of the most notable painters of this school was Thomas Cole, whose work was typical of the genre. In one of his most important paintings, *The Garden of Eden*, Cole depicts not just sublime joy for the humans in the Garden of Eden, but far more importantly, he depicts all the romantic images of nature that one could imagine—a babbling brook, mountains in the distance, palm trees, meadows, and lots of charismatic and peaceful-looking animals. His intent is clear, as it was to his numerous admirers in nineteenth-century United States. Connecting nature with the Garden of Eden was common in art of the period—indeed, Victorian England's penchant for nature study was in part moved by the notion that in nature we could discover something about the nature of God (Parrish 2006). This was also the position of the American Transcendentalists, such as Emerson and Thoreau. A subsequent Cole painting, *Expulsion from the Garden of Eden*, shows provocative light rays—the voice of God—streaming from the garden as Adam and Eve shamefully leave. Our original sin, which resulted in exile from the Garden of Eden, is our very nature, which has been disconnected from Nature. How powerful the doctrine of original sin has become in Western thought.

Later works by Cole provide the key juxtaposition. In the series *The Voyage of Life*, the same theme is presented four times—childhood, youth, manhood, and old age. A person sits in a boat upon a river leading to the end of life. In the child's boat, a guardian angel is poised on the stern. In the young man's boat the guardian angel stands on the shore to guide the boat as it enters its "independent" phase. The river flows toward Heaven and redemption (one needs to carefully guide one's boat to get there). It could not be clearer: original sin is connected with the loss of Nature, and redemption with regaining of Nature.

An important segment of the modern conservation movement remains tacitly embedded in this underlying schema. There is a nature that we have destroyed, and we must (1) let the nature that remains lie undisturbed, and (2) restore the nature that we have destroyed to its more primal state. Much of popular writing about nature embraces this framework, as do many basic texts (Terborgh 1999). It is, we believe, a framework that distorts and trivializes the very real problems we currently face in our need to stem the tide of biodiversity loss. Just as there never really was a real Garden of Eden, so there never was a real Nature: The living world has been in constant transformation for the past 3.8 billion years, and there is simply no part of the world that is inherently more Garden-of-Edenish than any other part of the world. There is no absolute about what should be preserved, nor what should be restored. It is a matter of human choice. If we retain the notion that we must seek some idealistic Garden of Eden, we can lose the centuries of secular thought, scientific research, and practices that could be brought to bear on the question of transforming nature. It is likely that conservation of nature will be best served by jettisoning ideas of original sin and undefiled nature and asking what it is about nature that we, as a collection of political actors, wish to conserve and why. But if we can come to ask those questions without the baggage of original sin and redemption, we will open up windows of opportunities for understanding, first, how nature is despoiled, and second, how we can put a stop to it and even reverse the trend.

In this chapter we propose first that there are certain scientific and political realities about biodiversity that we will not be able to change except under the most extreme circumstances. Second, we summarize some recent results from the discipline of ecology that may not be as well-known to the general community of concerned citizens. Third, in light of political realities and results from ecological science, we conclude that the standard paradigm for understanding landscape dynamics is often counterproductive. We suggest a new paradigm.

Scientific and Political Realities

THE TAXONOMIC LOCATION OF BIODIVERSITY

When the popular media approach the issue of biodiversity, the subject matter is almost always about charismatic megafauna—tigers, elephants, pandas, and the like. We also lament the probable extinction of these lovely, evocative creatures. The world will surely be diminished as the last wild gorilla is shot by a local warlord, or even as a rare bird species has its habitat removed to make way for yet another desperately needed strip mall or short-lived tropical pasture. However, such concerns are a very small tip of a very large iceberg. If we simply take mammals as an estimate of the number of creatures likely to be thought of as charismatic by the general public, we are talking about approximately 4,500 known species. By comparison, there are currently about 500,000 known species of insects—almost certainly a gross underestimate of how many actually exist. We have no idea how many species there are, since estimates range from about a million to as high as 30 million. While the latter estimate is probably exaggerated, even if there are only a million species of insects, we see that a focus on the 4,500 species that happen to look more or less like us is limited to a rather small fraction of the Earth's biodiversity. And to make the point even more dramatic, consider the biodiversity of bacteria. Microbiologists define two bacterial cells to be in the same species if their DNA overlaps by 70 percent or more; if applied to mammals, this would put all primates (if not all mammals) in the same species. Simply from the point of view of numbers, the world of biodiversity is mainly in the small things, from bacteria to insects, leaving the charismatic megafauna as a rather trivial subplot to the main theme.

THE FRAGMENTED TROPICS

Last year we flew over areas that had been continuous tropical rain forest in eastern Nicaragua, southern Brazil, and southwestern China. In a very general way, all three areas look very similar—a patchwork of forest fragments in a matrix of agriculture. With the exception of a few areas of the world (mainly in the Amazon and Congo Basins), the terrestrial surface of the humid tropics looks like this. And we always remain mindful that the world's biodiversity is concentrated to a great extent in this particular ecosystem:

if we are concerned in general with the conservation of biodiversity in the world, we ought to be concerned with what is happening in the tropics.

But if most of the tropical world is a patchwork of fragments in a matrix of agriculture, and most of the world's biodiversity is located in tropical areas, should we not be concerned with understanding the reality of that patchwork? Yet the vast majority of conservation work concentrates on the fragments of natural vegetation that remain, ignoring the matrix in which they occur. It would seem that ignoring one component of what is obviously a highly interconnected system is not wise.

The bias that favors concentrating efforts on the fragments while ignoring the matrix is far more damaging to conservation efforts than first meets the eye. The matrix matters. It matters in a variety of social and political ways, but, more importantly, it matters in a strictly biological sense. There is now little doubt that isolating fragments of natural vegetation in a landscape of low-quality matrix, like a pesticide-drenched banana plantation, is a recipe for disaster from the point of view of preserving biodiversity. Whatever arguments exist in favor of constructing a high-quality matrix, strictly from the point of view of biodiversity conservation, the quality of the matrix is perhaps the most critical issue. The concept of "the quality of the matrix" must be related to the natural habitat that is being conserved, but most importantly, it involves the management of agroecosystems.

Conventional wisdom has it that the natural habitats of the tropics have been devastated. Who can be against conserving those that remain? Yet, loss of habitat is only a very small part of the problem. It is true that much of the original forest and savanna cover of the terrestrial parts of the tropics has been dramatically altered since European invasion, but there is considerable evidence (as we see elsewhere in this volume) that tropical habitats were considerably altered at the time of contact. In the vast majority of cases, alteration has not been absolutely complete. The impressive scenes of devastation so common on conservation websites are actually not as widespread as one might think. In fact, as agricultural frontiers open, forest and savanna are cleared as needed and as technology and topography allow. Most of the tropics can be characterized not as deforested, but rather as fragmented.

This pattern of fragmentation has created a landscape that resembles an ocean archipelago. But here the islands are fragments of forest or savanna that remain in the ocean of agriculture. A real conservation program cannot ignore this reality! Even as the world struggles to protect the few remaining large areas of tropical habitats from further exploitation, we must acknowledge that a large fraction has already been exploited, and most of the world's biodiversity is, perhaps, located not in those few remaining protected natural areas, but in the far more extensive landscapes in which thousands of islands of natural habitat exist in the matrix of agricultural activities. Our purpose in this chapter is to examine exactly what must be done if conservation of biodiversity is to be successful in the long run.

If habitats are generally like islands in an archipelago, we are naturally led to ask whether it would be possible to treat the ecology of those islands as we would treat other islands. And here there is a body of ecological theory (the theory of island biogeography; MacArthur and Wilson 1967) that can be brought to bear on the problem: the biodiversity on any given island is proportional to its rate of immigration of new species divided by its rate of extinction. We emphasize that extinction at a local level is a perfectly normal process, one that can be seen on many islands and in many fragments (Cardillo et al. 2008; Feeley and Silman 2008; Gillespie 2001; Stratford and Robinson 2005), and even as species become extinct on individual islands, migration among islands prevents extinction over the species' entire range.

This picture, accepted as prevailing wisdom among ecologists today, has important consequences for biodiversity conservation. In a landscape that has been already fragmented, we face an island-like situation. For biodiversity conservation we must be concerned with the extinction rate and the immigration rate. The prevailing wisdom amongst many conservationists is that the political agenda is uniquely determined by the need to protect remaining natural habitats, which is to say focus on the islands in the landscape. For the most part, the extinction rate on an island is determined by the size of the island, smaller islands having higher rates of extinction than larger ones. And the critical fact we wish to drive home is that there is not much we can do about extinction rates, given a fragmented landscape. Thus, protecting the remaining patches of natural habitat is certainly an important component of a conservation program, but it is not nearly enough. Fixing the landscape at a particular level of fragmentation means fixing the extinction rates at some particular level. Nothing beyond preserving the patches of natural habitat can help in lowering extinction rates. Extinction is a normal biological process and will go on no matter what we do.

But we need to recall that extinction is only one side of the biodiversity equation. The other side is immigration, which we can do something about, as it is set by what happens in the "ocean"—the agricultural matrix within which habitat fragments occur. This means that the type of agroecosystem we allow to exist in the matrix will determine the immigration rate to the islands of natural habitat. The consequence of this elementary fact about the ecology of biodiversity has enormous sociopolitical consequences: A serious conservation program should focus on the type of agriculture done in the matrix, rather than on what happens in the fragments of natural habitat.

Completing the general picture, we note that promoting a high immigration rate means focusing on the agroecosystems, which in turn means examining what is happening sociopolitically in the agricultural matrix. This apparently unavoidable conclusion brings us face to face with the political turmoil that today occurs in the agricultural sector of the tropical world. Ignoring that turmoil is ignoring the only part of the biodiversity question that matters in a practical sense, the immigration rate.

Recent Advances in Ecological Theory

METAPOPULATIONS

Classical population ecology viewed the population as a homogeneous unit in space. Despite the fact that no environment in the world is homogeneous—and thus we expect some sort of spatial structuring in almost all situations—classical approaches treated populations as if they were without such structure. The most important deviation from this orthodoxy was a very simple insight by Richard Levins. Noting that populations were frequently arranged in space in a patchy fashion, he suggested that the beginning variable of analysis is the number of occupied patches. Thus, if we have a habitat that is divided into discrete patches (think of them as habitat "islands"), we ask, "What is the proportion of those discrete patches that are occupied by at least some individuals in the population of concern?" Rather than beginning with N, the number of individuals in the population, we begin with p, the proportion of patches occupied. With that change of focus, the development of the idea is simple and elegant. The rate of change of occupied patches must be related to the rate at which previously empty ones become occupied (m = migration rate) minus the rate at which subpopulations within a patch disappear (e = extinction rate). So we expect a balance between the migration rate and the extinction rate. In fact, the theory says that the percentage of habitats occupied by a particular species will be the difference between the migration and extinction rates, expressed as a percentage of the migration rate: $(m - e)/m$.

We are now in a position to ask what happens when a natural habitat becomes fragmented. It is clear that whatever spatial migration occurred (i.e., however much organisms moved from place to place in the original habitat), that rate will be lower in the fragmented situation. However, it is equally clear that the rate of local extinction will also tend to increase, given that the fraction of potential habitats in the smaller fragments are less likely to include all the requisites for all potential species. Thus, with fragmentation, m is expected to decrease, and the parameter e will increase. The goal

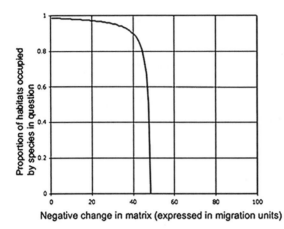

9.1. Hypothetical species migration rate decrease under agricultural landscape evolution

of conservation, obviously, is to reverse these trends. And we could clearly imagine planning production such that agricultural development is on a sustainable and biodiversity-friendly track, as much as preservation of remaining natural vegetation is a priority, especially in regions with significant disposable resources. But today we must realize that we live in a world where the major concentrations of biodiversity are in those areas with the least "developed" agriculture and the most poverty.

As the development of the agricultural landscape proceeds, the migration rate for any given species will change. We can ask how particular developmental patterns will affect particular species, and the elementary theory is not optimistic for particular developmental patterns. If, for example, agricultural landscapes evolve such that the migration rate of a particular species decreases, the theory says we should see a pattern something like that shown in figure 9.1. Even if migration rates go down slowly, the result is dramatic. With a decline of 20 percent, the occupancy of habitats goes from about 99 percent to about 97 percent, a decline that one could hardly measure. Another 20 to 40 percent results in a decline in occupancy only to 90 percent, but just another 10 percent decline causes extinction. It is a very sudden process, one that might not attract any attention at all, since the initial stages of the process suggest that nothing much is happening. Like many disequilibria processes, it is not incremental but reflects a change of state.

On the other hand, using a rational approach to designing the agricultural landscape (Nassauer and Opdam 2008), the graph in figure 9.1 could simply be turned around with improvement in the matrix increasing from right to left, encountering a threshold at about 48, where the species could once again survive, and then increasing to more than 90 percent of habitats occupied by the time the matrix value reaches 40. This is just a qualitative example of the kind of relationship we expect between the survival of species and the quality of the matrix, not a quantitative estimate of what to expect. The general theory says that sudden extinction of entire populations may be expected as the matrix declines in quality, no matter how many or how extensive the preserved natural areas. But it also says that improvements in matrix quality can restore landscapes where suddenly species that had been extinct from the landscape may be able to return. And we emphasize that this sort of "restoration" has nothing to do with the "natural" or "pristine" patches that remain in the area, but with action taken in the matrix in which those patches are embedded.

The Ubiquity of Extinctions

THE NATURE OF A LOW-QUALITY MATRIX

One thing is certain. The matrix within which fragments of native habitat are located is almost entirely made up of agroecosystems of various sorts. It is thus evident that any reflections on the quality of the matrix must deal with the nature of the agriculture practiced therein. There are two visions of agriculture contending for predominance in today's world: the industrial model, which does not seem to produce anything at all like a high-quality matrix, and agroecology (or ecological agriculture, or sustainable agriculture, or any of the other popular movements in the agricultural sector), generates much more landscape complexity and diversity.

While it is perhaps not necessary to review the problems that have been spawned by the industrial agricultural system, a few words about the intransigent nature of those problems (under business as usual) are warranted, and, perhaps more importantly, a reminder of where the industrial system comes from in the first place. We begin with a short description of the origin of the industrial system at the close of World War II, followed with a list of some of the problems it has created.

THE ORIGINS OF THE INDUSTRIAL SYSTEM

The characteristics of the industrial system most relevant for matrix quality are the massive spraying of biocides and the enormous scale of monocultural production. While other issues, such as fertilizers and genetically modified organisms, are part and parcel of the industrial system, we here emphasize the question of biocides and scale of production, since they are the particular features that affect the analysis of matrix quality from the point of view of biodiversity conservation.

With the outbreak of war in 1939, governments generally became interested in DNOC and similar chemicals as possible biological warfare agents. This spurred a great deal of government-sponsored research into chemical poisons. The United States was particularly active in biological and chemical warfare research and came up with a product that would see its application in the war against weeds: 2,4-D. (Later, as a component of the infamous agent orange, it would see war service in Vietnam.) In 1945, 917,000 lbs of 2,4-D were produced in the United States, and by 1950 the total had risen to 14 million lbs (Peet 1969).

But the real revolution was the synthesis of the chlorinated hydrocarbon dichloro-diphenyl-trichloro-ethane (DDT) in the late 1800s and the discovery of its insecticidal properties in 1939 from work done in the J. R. Geigy labs in Switzerland. It was quickly adopted by the armed forces in Britain and the United States. Tropical diseases, many of which were vectored by arthropods, were more important killers than the enemy in the practice of warfare. Finding effective solutions was a priority of war research, and DDT seemed to provide the final solution.

By 1945 other chlorinated hydrocarbons had been developed, and the arsenal of insect-fighting weapons was well established as part and parcel of war preparedness. Germany had made the same preparations, emphasizing organophosphates (e.g., parathion) rather than chlorinated hydrocarbons. World War II had thus created a high capacity for the production of biocides, with the general class of carbamates joining the chlorinated hydrocarbons and organophosphates to make the three major classes of insecticides we know today. Combine that with the herbicides that originally were developed as a byproduct of biological warfare research and it could be legitimately charged that World War II was the seed that germinated the agrochemical revolution.

While the war was a watershed for the chemical industry, allowing it to build up immense productive capacity, peace turned out to be a problem. Capacity suddenly turned into an overproduction crisis, and the industry had to scramble for new outlets. Agriculture was the obvious target, although household use was not insignificant. The industry developed some ingenious marketing strategies in postwar United States and Europe. War fever having reached a pitch, the public was especially susceptible to wartime rhetoric. What had originally been an argument that we needed the chemicals to defeat the enemy in war was easily translated into the need for these chemicals to defeat the new enemy in agriculture (Russell 2001). For example, the Industrial Management Corporation noted in an internal memo in 1946:

It's a sales story that's simple, effective, and true. It clears up the confusion in the average person's mind when you tell him—Yes, INSECTO-O-BLITZ is still exactly the same as supplied to the US Armed Forces.

The importance of this advertising blitz cannot be overemphasized. Throughout the times of the new husbandry, pest management had taken on the generalized goal of maintaining a healthy ecosystem through the five major categories of activities (biological control, mechanical control, cultural control, chemical control, and resistant varieties). Pests in agriculture were almost like germs in health, and the goal was to maintain a disease-free system. The dominant metaphor thus seems to have been a medical one. With wartime (and especially postwar) propaganda, the pests came to be seen as enemies to be vanquished, rather than germs to be controlled. The war metaphor replaced the medical metaphor.

The armaments in this new war were, of course, the new organic chemicals produced by the same corporations that produced them for the war effort. The wartime-induced productive capacity of the chemical industry was thus saved by the appropriation of pest control in agriculture. The attitude toward pests was altered, seemingly subtly, but in the end in such a way as to transform agriculture dramatically. The new metaphor meant that farmers changed from stewards who maintained the health of their farms to warriors who vanquished their enemies. The consequences were massive spraying of biocides in the years following World War II.

The attitude faced a major challenge with the publication of Rachel Carson's *Silent Spring* in 1962. Carson suggested that the massive use of pesticides was having a dramatic negative effect on the environment. Previously there had been much popular commentary about the human health effects of pesticides, a concern shared even by the pesticide manufacturers. But *Silent Spring* was the first popular account of the environmental consequences of pesticides, contributing not only to concern about environmental poisons, but perhaps providing the main springboard for the entire subsequent environmental movement.

What Carson said is now well known. Pesticides kill not only the targets, but also many species that are not targeted. Pesticides may concentrate in the higher trophic levels, thus making nonlethal doses at lower trophic levels quite dangerous at higher levels. Pests develop resistance to pesticides. The poisonous effects of pesticides and their residues may persist for a long time in the environment. These were the basic themes of her book (Carson 1962).

Silent Spring is an extremely well-documented book. Despite a Herculean effort at finding errors, mainly by representatives of pesticide manufacturers, only the most trivial errors were eventually encountered. Nevertheless, an immense and coordinated attack against the book was orchestrated by the pesticide industry, including an attempt by Velsicol to pressure Houghton-Mifflin not to publish the book in the first place. Book reviews were generally harsh and, as later discovered, frequently written by scientists receiving monetary rewards from the chemical industry. As carefully documented in subsequent works, most independent scientists received *Silent Spring* positively. Rereading the book even today suggests understatement was its actual problem.

Yet *Silent Spring*, despite its harsh yet substantiated warnings, was for the most part ignored, except for the growing numbers of people in the environmental movement. Carson's predictions have come true in many areas of the world, and further problems have emerged. Industrial accidents in pesticide manufacturing plants, plus worker exposure to toxins in such plants, have added to her concerns. The environment is now filled with chemicals that mimic estrogen, causing not only feminization in reptiles, but lowered sperm counts in humans, as documented in the remarkable book *Our Common Future* (World Commission on Environment and Development 1991). Unfortunately, the final chapter on pesticides has yet to be written. What began with the need to curb an underconsumption crisis ends with an environmental crisis of unprecedented proportions.

THE INDUSTRIAL SYSTEM PENETRATES THE GLOBAL SOUTH

While the initial birth of the industrial system can be seen as a response to a classical underconsumption crisis, its penetration into the Global South (which is where, as noted earlier, most of the world's biodiversity is located) followed a different pattern. As part of this same trend, industrial enthusiasts looked to the Global South for yet further market outlets. This need dovetailed quite well with the global political climate. Western capitalist countries legitimately feared the promise of communism, which provided, at least in theory, a seductive option for the downtrodden. Food and agriculture were key in the struggle for hearts and minds as the Cold War gained traction. Hungry people could be the victims of communist propaganda, as could farmers unable to support themselves. Projects to counter this possibility emerged as early as 1954 when the Eisenhower administration initiated Public Law 480 (PL 480). As US agriculture was undergoing the massive transformation initiated by the underconsumption crisis that emerged from the wars, it became evident that the oversupply of grains resulting from the expansion of the industrial model would lead rapidly to overproduction of grain itself. This clear politico-economic necessity fit well with the new fear of communist ideology, and PL 480 solved both problems. Grain would be shipped to "cooperative" countries. As insurrections emerged repeatedly in the Global South, export of food through PL 480 and military intervention (both overt and covert) became central features of US foreign policy. By 1956 food aid accounted for more than half of all US foreign assistance. By the early 1960s the full industrial model produced an enormous surplus of food in the Global North, much of which was exported under PL-480, solving the underconsumption crisis for companies like Cargill and Continental and serving the interests of the West in the global political struggle of the Cold War. This would all change with a new technological package. Recall the history of the tomato mechanization program undertaken by Jack Hanna and Toby Lorenzen of the University of California (Hightower 1973). Hanna was breeding a tomato to be harvested for a machine that did not exist, and Lorenzen was designing a machine to harvest a tomato that did not exist. The philosophy of technological development in agriculture is well-illustrated by this program, where complex and interconnected technologies are developed in consort with one another to create a "technological package."

Operating within this developmentalist framework, agronomist Norman Borlaug, working at the International Maize and Wheat Improvement Center outside of Mexico City, began a breeding program that took chemical fertilizers, pesticides, and plenty of irrigation water as underlying assumptions about the ecological background in which improved varieties of wheat needed to be developed. The degree to which any of the individual specific inputs contributed to the dramatic increases in production is debatable.[1] What is not debatable is that the ideas and practices of technological package were potent and yielded increased by as much as fivefold, compared with production technology that used none of the new inputs.

The trend of exporting the basic agro-industrial model to the Global South was a more or less continuous process as manufacturers of fertilizers, pesticides, machinery, and seeds independently moved their marketing operations there. However, as a political movement, the Green Revolution was born in July of 1965 when political tensions between the United States and India flared.[2] A new Indian prime minister, Lal Bahadur Shastri, took a critical view of the United States' bombing of Vietnam, to which the Johnson administration responded with a threat to cut off PL-480 food aid. This was a major threat, since India had already lost much of its food producing capacity due to the massive PL-480 distribution, which had effectively created an oversupply and dramatically lowered local prices, always a disaster for local farmers. PL-480 contracts were now to be renewed on a monthly (rather than yearly) basis, and it was made clear that India's attitude toward

US foreign policy would be a litmus test for continuation. This was the stick. But the United States also offered a seemingly unassailable carrot. If India would become a more pliable and cooperative political partner, the United States would provide not food but the new technology to produce it (the fertilizer/pesticide/machinery/irrigation/seed technologies—in short, the industrial system) to select farmers in India. This massive transfer of technology, first to India, then to the rest of the Global South, became known as the Green Revolution.

The actual performance of this new technological package remains a matter of considerable debate. Yields up to five times pre–Green Revolution technology have been reported. But what is being compared is often obscure. To be sure, if we begin with seed not necessarily adapted to conditions on the farm, with low nitrogen content in the soil, with low cation exchange capacity, with insect and disease pests having built up in the local environment, and with a lack of water in a drought-prone area, adding the fertilizer/pesticide/machinery/irrigation/seed technologies to the operation is almost certainly going to result in massive yield increases. And if those industrial inputs are subsidized to a great extent (which they were), economic yields will undoubtedly also increase dramatically. It is sort of like comparing advanced technology to no technology, and becomes a no-brainer. However, if the comparison were to be made to an alternative technological package where agroecological techniques were used to promote soil fertility and pest management, where small-scale irrigation projects were initiated at a local level, where traditional varieties of crops adapted to local conditions were used, and where machinery was adapted to local small-scale farming operations, it is not at all clear that biological yields would be greater with the industrial package, to say nothing of potential economic benefits. Organic production in fact provides biological yields equivalent to the industrial model[3]; thus, accepting the notion that the Green Revolution was a success remains, in our view, a controversial position.

On the other hand, from a strictly historical point of view, that revolution was one of the more successful in history. The industrial system did indeed penetrate the Global South. It is now the dominant ideological form of agriculture throughout the world, regardless of its actual performance. The average farmer in Latin America, Asia, and Africa does indeed want the fertilizer/pesticide/machinery/irrigation/seed technologies that constitute the industrial system. However, the new farmer's movements, discussed presently, have slowly but surely developed not only a political position toward land ownership and economic security, but a critical view of the Green Revolution/Industrial System, and they are ever more frequently openly talking about agroecological technologies.

Constructing the High-Quality Matrix

LANDSCAPE DESIGN IN THEORY: THE VIA CAMPESINA

The formation and continued evolution of the Via Campesina is a complicated issue (Desmarais 2007), beyond the scope of this chapter. However,
the position of the organization on two fronts is particularly important to
our analysis. First, the group has pioneered the idea of food sovereignty, an
all-encompassing view of rural development that is emerging as a logical
alternative to many of the crises of modern society, from food security to
urban violence. The second front on which Via Campesina is having a major global impact is in the expansion of the alternative, ecologically-based
model of agriculture, commonly referred to in the Via Campesina literature as agroecology. Almost all the member organizations of Via Campesina originated with popular struggles for land reform or other social justice concerns. However, within those struggles, environmental struggles
emerged, partly because of the participation of environmentally concerned
individuals and groups alongside the social justice groups, but partly a consequence of the concrete conditions created by the industrial model itself.

As has happened in the developed world, a growing realization of the
environmental and health problems created by the industrial system has
been penetrating the Global South and undoubtedly has an influence on the
members of the member organizations of Via Campesina. It is not difficult
to see a connection between pesticide poisoning and the conditions of social
justice. It is not the corporate board rooms of Monsanto that are sprayed
with pesticides, but rather those same small farmers and farmworkers who
are organizing the organizations that form the Via Campesina. Safe working conditions, long a fundamental demand of labor organizing the world
over, articulates naturally with ecological forces when the labor is agricultural. Individual farmers do not want themselves or their children exposed
to chemicals that the world now knows are tied to cancer and teratological
effects. The emphasis on sound environmental management is perhaps a
natural outgrowth of class consciousness when the class is rural.

Thus Via Campesina's environmental projection has evolved as a two-
pronged program: first, to oppose the most environmentally damaging aspects of the industrial system, and second, to promote an approach based
on ecological principles. Their opposition has taken the form of street protests, militant actions against the spread of transgenic crops, and a call to get
the World Trade Organization (WTO) out of agriculture altogether (Rosset
2006), arguing that food should be for the purpose of human nourishment,
not corporate profits. The promotion has taken on many of the same ideas

that emerged with the beginnings of the alternative agriculture movement. The Via Campesina approach has been a "farmer to farmer" approach, recuperating techniques that have been locally traditional and promoting their diffusion through interactions among farmers themselves. Articulating a technical agenda with this new approach to "bottom up" research is a challenge, but one that is actively pursued in various institutions and research networks.[4]

What is clear, however, is that a model of rural development that embraces ecological as well as social complexity within a matrix framework will, in the end, serve conservation and development purposes far better than a model of an imagined wild on one side and industrial agriculture on the other.

10 * Effects of Human Activities on Successional Pathways: Case Studies from Lowland Wet Forests of Northeastern Costa Rica

ROBIN L. CHAZDON, BRAULIO VILCHEZ ALVARADO, SUSAN G. LETCHER, AMANDA WENDT, AND U. UZAY SEZEN

Introduction: Succession and Natural Regeneration of Tropical Wet Forests

Succession is a process of vegetation change following a disturbance. Disturbances in tropical forests vary in frequency, intensity, and scale. They can be caused by natural events, such as windstorms, hurricanes, flooding, landslides, and fires (Whitmore and Burslem 1988), or by human activities, such as fragmentation, harvesting of timber and nontimber products, hunting, burning, and cultivation. Successional processes following large-scale forest disturbances tend to be highly stochastic and spatially variable, even in the absence of direct human intervention (Chazdon 2008b). Thus, human activities superimpose additional complexity onto an already spatially and temporally variable template. Biotic and anthropogenic legacies of landuse transitions and forest regrowth are intricately connected. For example, topography and soil fertility influence forest regeneration, both directly and indirectly, through effects on suitability for crop cultivation or pasture establishment, and thus can strongly affect rates and scale of land clearing for agriculture, the duration of land use, rates of agricultural abandonment, and seedling establishment following abandonment (Moran et al. 2000).

Studies in the Old World and New World tropics have documented pervasive, long-term human impacts on species composition and forest structure in tropical secondary forests (Zimmerman et al. 1995; van Gemerden et al. 2003). The rate, structure, and composition of forest regrowth are strongly affected by soil disturbance, residual vegetation, and proximity to seed sources (Chazdon 2003). Long-term effects emerge from cascading effects of the initial abundance, composition, and spatial patchiness of species that colonize abandoned agricultural areas (Chazdon et al. 2007). These characteristics make secondary forests particularly sensitive to human impacts. The composition of pioneer vegetation in Central Amazonia, for example, is strongly affected by pasture establishment and intensities of land use. In areas sharing the same soil type, elevation, and aspect, *Cecropia* species (*C. schiadophylla* and *C. purpurascens*) dominated early forest regrowth where forest was cleared and not used for pasture, whereas *Vismia* species (*V. guia-*

nensis, V. japurensis, and *V. cayennensis*) dominated regrowth in areas where pastures were established for four years and later abandoned (Mesquita et al. 2001). Differences in initial colonization affected the overall abundance and species richness of woody species within these stands (Mesquita et al. 2001).

In the Atlantic lowlands of northeastern Costa Rica, we have been studying secondary forest regrowth following abandonment of cattle pastures since 1992 (Guariguata et al. 1997; Capers et al. 2005; Chazdon, Redondo Brenes, and Alvarado 2005). We have also conducted studies on a large number of old-growth and secondary forest sites located throughout the region, spanning different ages since abandonment, as well as varying in the intensity and duration of prior land use (Chazdon et al. 2007; Letcher and Chazdon 2009). In addition, we investigated parentage and genetic diversity of an abundant canopy palm species regenerating in secondary forest areas (Sezen, Chazdon, and Holsinger 2005, 2007). These studies provide insights into five major ways in which human activities influence the process of secondary forest regeneration in this region and, presumably, in other regions of the wet tropics. Here we address five main types of human impacts observed in our study area: (1) remnant trees in pastures; (2) hunting and the density of mammalian seed predators; (3) duration and intensity of agricultural land use; (4) landscape structure and distribution of forest patches; and (5) invasion of an exotic cultivated palm in a secondary forest. Knowledge of how human activities affect forest regeneration can be used in constructive ways to shape successional processes in tropical landscapes and to enhance and restore biodiversity in complex landscape mosaics and biological corridors (Janzen 1988). Furthermore, this understanding can provide insights into how past land use may have molded features of forest structure and composition that we observe today (White and Oates 1999; Gillson and Willis 2004; Brncic et al. 2007; Heckenberger et al. 2007; Lentz and Lane, chapter 13, this volume; Heckenberger et al., chapter 24, this volume).

The Study Region

Our studies were conducted in the Atlantic lowlands of Northeastern Costa Rica in the vicinity of La Selva Biological Station, lat 10°26'N, long 84°00'W (McDade and Hartshorn 1994). Mean annual rainfall in the region is about 4,000 mm/year. La Selva is a 1600 ha biological reserve adjoining Braulio Carrillo National Park, forming the largest intact forest in the region. Since 1950, the region has been converted from nearly continuous forest cover to a mosaic of agricultural land use (predominantly pasture and cash crops) and forest fragments under natural forest management.

Since 1997, we have monitored vegetation dynamics annually in four, 1 ha plots (TIR, CR, LEPS, and LSUR), with another four plots added in 2005 (LEPP,

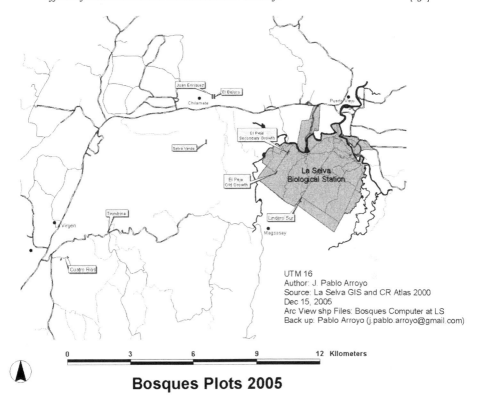

Bosques Plots 2005

10.1. Map of the study plots and La Selva Biological Station in Northeastern Costa Rica. Cuatro Rios (CR), Tirimbina (TIR), Juan Enriquez (JE), El Bejuco (EB), Lindero Sur (LSUR), and El Peje Second Growth (LEPS). Selva Verde (SV) and El Peje Old Growth (LEPP) are in old-growth forests.

EB, JE, and SV), including two in old-growth forests (fig. 10.1). These surveys include measurements of trees ≥ 5 cm diameter at breast height (DBH), saplings between 1 and 5 cm DBH, and seedlings < 1 cm DBH and > 20 cm height. In 2005 and 2006, seedlings were sampled in 1000 1 x 2 m² quadrats (0.2 ha), whereas saplings were sampled in 200 5 x 5 m² quadrats (0.5 ha).

Effects of Remnant Trees on Tree Regeneration

Remnant trees are commonplace on cattle ranches throughout Latin America (Guevara, Purata, and Maarel 1986; Guevara, Laborde, and Sanchez-Rios 2005; Harvey and Haber 1998). In northeastern Costa Rica, it is not unusual for 15 percent of the basal area of young secondary forests to be contributed by remnant trees (Guariguata et al. 1997). Trees and canopy palms are spared clearing for a variety of reasons, including shade for cattle, thatch,

Table 10.1. Remnant tree cover and vegetation characteristics in two 1-ha plots
in an eleven-year-old secondary forest

	El Bejuco (EB)	Juan Enriquez (JE)
Density of remnant trees (#/ha)	22	1
Species density of remnants (#/ha)	15	1
Basal area remnants (m²/ha)	5.78	0.04
Basal area nonremnants (m²/ha)	14.52	11.29
Species density of nonremnant trees (#/ha)	56	48
Density of trees ≥ 10 cm (#/ha)	339	357
Density of tree saplings (#/ha)	2,224	1,568
Density of tree seedlings (#/ha)	18,985	11,290

Note: The two plots are adjacent to each other and share similar soil types and
topography.

edible fruit, or difficulty in cutting. Areas beneath remnant trees receive
increased seed rain following pasture abandonment (Galindo-González,
Guevara, and Sosa 2000; Slocum and Horvitz 2000; Guevara, Laborde,
and Sanchez-Rios 2004), but few studies have quantified effects of remnant
trees during later stages of succession (Schlawin and Zahawi 2008).

We identified remnant trees as the largest trees in the plot (usually above
40 to 50 cm DBH), which are mature-forest species normally absent from
secondary forest canopies. Two secondary forest plots (EB and JE) in Chilamate, similar in age since abandonment (11 to 12 years in 2006), provided a
unique opportunity to compare the effects of remnant tree cover and composition on the abundance of regenerating seedlings and saplings of canopy
tree, based on data from the 2005 tree inventory and the 2006 seedling
and sapling inventory (fig. 10.1). The plots are within 50 m of each other
and share the same soils and topography. The EB plot had twenty-two remnant trees from fifteen species, composing a substantial fraction of the plot
basal area (28.4 percent) for trees ≥ 5 cm DBH. In contrast, the JE plot had 1
remnant palm tree, *Welfia regia* (Arecaceae), which did not contribute appreciably to plot basal area (table 10.1). When the basal area of remnant trees
was subtracted, the two sites showed similar total basal area (11 to 15 m²/ha;
table 10.1). The sites did not differ significantly in either tree or seedling
species richness (data not shown).

The forest plots differed substantially in the abundance of canopy tree
seedlings and saplings. Tree sapling and seedling densities were 1.4 and 1.7-
fold higher in the plot with remnant trees, EB (table 10.1). Out of seventy-

eight species of canopy tree and palm species with seedlings and saplings occurring in either or both sites, thirty-one species had more abundant regeneration in EB than JE. Nine of these more abundant species had at least one remnant tree present in EB. Tree regeneration was also enhanced by the presence of remnants during initial colonization. For some species, such as *Carapa guianensis* and *Pentaclethra macroloba*, the presence of several large remnant trees was associated with an increased number of canopy trees that colonized shortly after pasture abandonment in EB.

Of the thirty-one tree species with higher seedling abundance in the plot with remnant trees, twenty-eight (90.3) were dispersed by animals (birds or mammals), two by wind (6.4 percent), and one by explosive dehiscence. In contrast, of the twenty-one tree species with higher seedling abundance in the plot lacking remnant trees, only fourteen (66.7 percent) were dispersed by animals, and seven (33.3 percent) were dispersed by wind. The frequency of biotic versus abiotic dispersal between these two sets of species differed significantly ($X^2 = 4.51$; $P = 0.033$). Thus, biotic dispersal was associated with species showing *higher* abundance in the plot with remnant trees, whereas abiotic dispersal was significantly associated with canopy species showing *higher* abundance of seedlings in the plot lacking remnant trees. Remnant trees augmented the abundance of large-seeded, bird- and bat-dispersed canopy species, including species not represented by remnant individuals.

These results emphasize the importance of remnant trees as biological legacies following large-scale disturbances such as fire, hurricanes, or agricultural land use (Elmqvist et al. 2001; Slocum and Horvitz 2000; Guevara, Laborde, and Sanchez-Rios 2004). Remnant trees play multiple roles in enhancing regeneration, including production of their own seed, attracting diverse frugivorous birds and bats, providing perch and roosting sites for seed-dispersing animals, and ameliorating harsh microenvironmental conditions in abandoned pastures.

Effects of Hunting on Canopy Tree Regeneration

Plots located within La Selva Biological Station have been actively protected from hunting over the past twenty to thirty years by routine patrols by game wardens. In contrast, plots in the surrounding landscape have been exposed to continuous hunting pressure during past decades and have experienced less thorough patrolling (Hanson, Brunsfeld, and Finegan 2006). In response to protection from hunting, populations of terrestrial game mammals, such as the agouti (*Agouti paca*) and collared peccary (*Tayassu tajacu*) are noticeably higher at La Selva than in surrounding regions (Guariguata, Adame, and Finegan 2000; Hanson, Brunsfeld, and Finegan 2006). Based

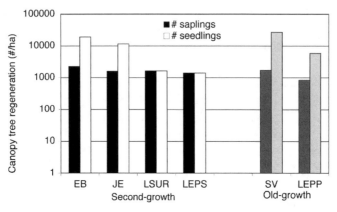

10.2. Seedlings and sapling density in four second-growth and two old-growth plots based on 2006 census data. FEB, JE, and SV sites are not protected from hunting, whereas LEPS, LSUR, and LEPP are located within La Selva Biological Station and have been actively protected from hunting for the past twenty to thirty years.

on data collected in 2006, we compared densities of canopy tree seedlings and saplings in six 1 ha plots. Two secondary forest plots (LSUR, LEPS) and one old-growth forest (LEPP) were located within La Selva Biological Station, and two secondary forest plots (FEB, JE) and one old-growth forest (SV) were located in Chilamate, 7 km west of La Selva (fig. 10.1). All secondary forests areas had previously been used for cattle pasture. Seedling and sapling data were used to compare effects of hunting between two second-growth sites at La Selva and two second-growth sites located in Chilamate.

Secondary forest plots protected from hunting within La Selva showed reduced densities of woody seedlings of canopy tree and palm species compared with sites off La Selva. Peccaries are frequently observed while visiting our study plots at La Selva, whereas we have never observed them in TIR or CR (also see Guariguata, Adame, and Finegan 2000). In secondary forest plots, canopy tree and palm seedling and sapling density were 10.1 and 1.3 times higher, respectively, in the two nonprotected plots compared with two protected plots (fig. 10.2). Due to the small number of plots, these trends are not statistically significant. These effects were not limited to secondary forests; however, seedling and sapling densities were 4.7 and 2.0 times higher, respectively, in the SV plot in Chilamate as compared with the LEPP plot at La Selva (fig. 10.2). In both old-growth forests, the palm *Welfia regia* was the most abundant species in seedling quadrats. At La Selva, *Welfia* seedlings composed 44 percent of the tree seedling community, compared with 72 percent at SV, where we observed high-density seedling patches beneath canopy trees. The highest density of *Welfia* seedlings at SV was

416 seedlings/m^2, compared with 48 seedlings/m^2 in old-growth forest at La Selva.

Reduced populations of seed predators in nonprotected areas has led to increased abundance, but not composition, of seedling and sapling recruitment, particularly of large-seeded species. Palms are particularly favored by peccaries (Beck 2005), so it is not surprising that *W. regia* seedlings show such pronounced effects. In the same region, the large-seeded legume *Dipteryx panamensis* also showed higher abundance of seeds in forest fragments exposed to hunting compared to protected continuous forest (Hanson, Brunsfeld, and Finegan 2006). Effects of hunting extend beyond changes in abundance, as large mammals and birds also play a critical role in seed dispersal and genetic structure of tree populations (Wright et al. 2007).

Effects of Land-Use History on Successional Pathways

We assessed effects of the duration and intensity of land use on successional pathways using data from a chronosequence study comprising thirty sites across the region, including La Selva (Letcher and Chazdon 2009). Seven sites were in old-growth forest, whereas twenty-three sites were in secondary forest regrowth from ten to forty-four years post-abandonment. Among the secondary forests, two sites were previously used for rice cultivation, three sites were cut, burned, and immediately left fallow, and eighteen sites were used for pasture. A modified Gentry transect approach was used to survey vegetation in five parallel strips of 2 x 100 m, each separated by 10 m. We recorded diameter at 1.3 m height and species identification for all woody stems ≥ 2.5 cm DBH for tree and shrub species and ≥ 0.5 cm DBH for lianas. To assess effects of duration and intensity of land use, we divided the twenty-three secondary forest sites into older (> 20 yr) versus younger (≤ 20 yr) and light use (0 to 10 yr in pasture) versus heavy use (> 10 yr pasture).

Both secondary forest age and land-use history (time in pasture) significantly affected woody species richness in secondary forests. Older forests and stands with lighter land use (< 10 yr in pasture) had higher species richness of woody species (Letcher and Chazdon 2009). Both factors explained 61 percent of the variance in species richness, with no significant interaction. The consistent effects of land-use history on woody species richness were not due to differences in stocking density, as density effects were taken into account in this analysis. Because woody species richness is strongly determined by initial patterns of tree colonization (Chazdon et al. 2007), these results suggest that initial colonization of woody species was richer in areas that regenerated shortly after logging or were used for pasture for less than ten years as compared with areas intensively used for pasture. Some of this enhanced richness may be due to resprouting of mature forest trees that were cut during forest clearance (Kammesheidt 1999).

Forest age, but not land-use history, had a significant effect on biomass accumulation, however, explaining 49 percent of the variance, with older forests showing significantly higher biomass than younger forests (Letcher and Chazdon 2009). Although other studies have demonstrated a significant effect of land-use history on biomass accumulation in tropical secondary forests (Aide et al. 1995; Hughes, Kauffman, and Jaramillo 1999; Steininger 2000), these effects can potentially be confounded with the effects of soil fertility. Zarin, Ducey et al. (2001) found no significant difference in biomass accumulation between secondary forests regenerating on former pastures and former slash-and-burn fallows in Amazonia.

Effects of Seed Sources on Genetic Diversity of Forest Colonists

Few studies have examined the effects of parental sources on the genetic composition of tree colonization during secondary succession. Sezen, Chazdon, and Holsinger (2005, 2007) conducted studies of parentage and spatial genetic structure of the founding population and second generation of seedlings of the canopy palm *Iriartea deltoidea* in 20 ha of secondary forest at La Selva Biological Station. The canopy palm *Iriartea deltoidea* colonized the LEP secondary forest area at La Selva Biological Station (fig. 10.1) during the first five years of regrowth, and many initial colonists became reproductively mature within twenty years. The major seed dispersal agents for *Iriartea* are toucans (*Ramphastos* species), although seeds are also dispersed by frugivorous tent-making bats (*Artibeus watsoni*) and seed-caching rodents. Based on AFLP analysis of DNA extracted from leaf tissue (DNA fingerprinting), parentage analysis was used to identify and locate the parent palms in 20 ha of adjacent old-growth forest (Sezen et al. 2005, 2007).

Reproductively mature trees in the secondary forest declined in abundance with increasing distance from the old-growth forest edge. Parentage analysis revealed extreme reproductive dominance by only 2 out of the 157 trees present in the adjacent old-growth forest; up to 48 percent of the genes in the founding population were contributed by these two parents. This initial founder effect led to a reduction of genetic diversity of trees in the second-growth area compared with trees in the neighboring old-growth population (Sezen et al. 2005, 2007).

In the area of secondary forest close to the old-growth edge, the higher density of founders led to depressed genetic diversity among their offspring; close to 30 percent of these seedlings were offspring of parents originating in the second-growth plot (Sezen et al. 2007). Trees from the neighboring old-growth forest continued to disperse seeds into both near and far second-growth areas. Seedlings sampled more than 500 m away from the old-growth border, however, showed a completely different pattern of parentage. Here, nearly 40 percent of the seedlings were offspring with both parents located outside of the sample area; only 10 percent were offspring of parents in the second-growth area (Sezen et al. 2007). The high fraction of seedlings originating from trees outside of the plot suggests that parental trees were located in secondary and old-growth forest patches in the surrounding landscape. The genetic contribution from these diverse sources resulted in a recovery of genetic diversity in the seedlings in the far end of the secondary forest plot.

These results have implications for the effects of landscape structure on recovery of genetic diversity for species dispersed by toucans and bats with large home ranges. In a landscape where founding populations originate from a small number of parent trees, founder effects can be dramatic, as illustrated here. Yet the presence of small fragments with reproductive trees can ensure more diverse genetic contributions in later generations, particularly where founding populations are less dense. Toucans and bats disperse a large number of old-growth forest tree species that colonize secondary forests, so these trends could well apply to other tree species.

Invasion of an Exotic Palm into Secondary Forest

When we established the monitoring plot in a fifteen-year-old secondary forest at the Tirimbina site in 1997 (fig. 10.1), we recorded numerous seedlings and saplings of a palm resembling the native canopy species *Euterpe precatoria* var *longevaginata*, the only *Euterpe* species in Costa Rica (Henderson and Galeano 1996). But the individuals in this plot appeared distinct, having multiple stems from the earliest seedling stages. We marked all stems ≥ 5 cm DBH each year from 1997 through 2006 along with all of the other trees. During the summer of 2007 we observed the first fruiting of several palms in the understory and identified them as *Euterpe oleracea*, a widely cultivated species known as *açai* in Brazil. Within 500 m of the TIR plot, a plantation of *E. oleracea* had been established in a nearby farm about twenty to twenty-five years ago, prior to the secondary forest establishment.

Our monitoring data showed that the *Euterpe oleracea* population showed exponential growth in the TIR plot from 1997 to 2007 (fig. 10.3), increasing from 1 to 42 stems. Following further observations of seedlings of *E. oleracea* in other secondary forest areas in the vicinity, in 2007 and 2008 we removed all stems ≥ 5 cm from within the study plot to prevent further spread by frugivorous birds and bats. The plantation, which now belongs to a pri-

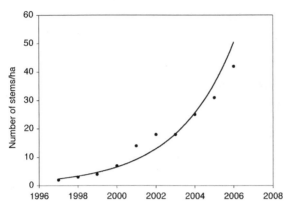

10.3. Exponential growth of *Euterpe oleracea* over ten years in a fifteen-year-old, second-growth forest plot near Timbina Research Center. The number of stems ≥ 5 cm DBH is fitted to an exponential function (R^2 = 0.959). A plantation of *E. oleraceae* is located about 500 m from the plot.

vate nature reserve, is also being cleared. *Euterpe oleracea* is considered to be an invasive tree species by the Food and Agriculture Organization of the United Nations (http://www.fao.org/forestry/27179/en/bra/). This widely cultivated species has also escaped from plantings in the Summit Botanical Garden in the Panama Canal Zone (Svenning 2002).

In secondary forests in Soberania National Park near Gamboa, Panama, eight species of exotic palms have invaded secondary forest areas, where they outnumber the native palm species (Svenning 2002). At least four of these species exhibit abundant regeneration and are probably naturalized. Meyer, Lavergne, and Hodel (2008) found that twelve palm species are now recognized as invasive in tropical mainland and island regions, although *E. oleracea* was not on their list. Another palm that can potentially become invasive in tropical secondary forests throughout the world is the African oil palm (*Elaeis guineensis*), which is now grown in forty-three countries and accounts for nearly 10 percent of the world's permanent cropland (FAO 2007; World Resources Institute [WRI] 2007). Exotic invasive tree species are highly abundant in second-growth forests of Puerto Rico; many have become naturalized and are altering successional pathways in major ways (Lugo and Helmer 2004).

Conclusions: What Is "Natural" about Natural Regeneration?

Human activities in tropical landscapes generate myriad effects on regeneration processes in secondary forests as well as in old-growth forests. In addition to the effects described here, climate change is affecting tropical forests worldwide (Clark 2007; Malhi et al. 2008). Tropical secondary forests are highly sensitive indicators of human impact. Therefore, the question is not *whether* humans are affecting tropical forest regeneration, but *how* we can understand and manage these effects to ensure forest integrity and long-term survival (Chazdon 2008a). Ranchers and farmers can be granted incentives to retain remnant trees in their pastures (Pagiola et al. 2007). Hunting and bush-meat harvesting can be reduced through programs that produce domestically raised animals or fish for protein sources. Buffer zones, forest fragments, and corridors can be managed to provide diverse genetic sources for secondary forest colonization. Invasions of exotic species can be curtailed by preventative measures or by removal programs.

Human activities have influenced tropical forest regeneration for millennia. Archaeological evidence suggests that white-tailed deer (*Odocoileus virginianus*) were overhunted by ancient Mayans during the late Classic Period (Emery 2007). The pejibaye palm (*Bactris gasipaes*) has been introduced and cultivated throughout the Neotropics for so long that its native origins cannot even be resolved (Clement 1988). But these historical realities do not placate today's concerns regarding potentially disastrous effects of hunting or invasive species. Throughout the world's tropics, secondary forests are following new successional trajectories under the influence of invasive species (Hobbs et al. 2006).

The distinction between natural regeneration and feral forests is fuzzy at best. The "natural" aspects of natural regeneration are precisely those elements that we do not completely control, modify, or domesticate, such as which fruits get dispersed by which animals, or which seedlings successfully grow up to become canopy trees. But regardless of how much we allow these natural processes to prevail, forests in early stages of succession will always be highly susceptible to human impacts and their long-term legacies.

PART II
Historical Ecologies

Human-Forest Relationships and the Erasure of History

KATHLEEN D. MORRISON

Just as paleoecology, once it finally caught the attention of ecology as a whole, radically altered interpretive frameworks away from notions of stability and balance and toward nonequilibrium models of ecological process, so too can integrated socionatural histories radically reshape existing notions about human-environment relationships. In this section, we present five papers that challenge long-held conventions in the natural and social sciences that carefully separate the natural from the cultural. These papers, furthermore, undermine myths about pristine forests, whether in the Amazon, island Southeast Asia, Mesoamerica, or southern India, forests ironically often seen as having recently come under attack by their human inhabitants. Across this tropical swath, new forms of historical research that integrate evidence and approaches from archaeology, history, paleobotany, and a host of chemical and geological analyses of past landforms, lifeforms, and artifacts are changing the way in which we view the past. The substantive accounts produced by such multidisciplinary projects not only point to long histories of human engagement with forests and other landscapes, but also challenge us to rethink the very categories we employ to discuss these places and their histories.

Given that the Amazon, as both Erickson and Heckenberger et al. discuss in this volume, actually has a long and rich history of human occupation, including urban centers involving significant landscape modification, what might this suggest about our understanding not only of these forests, but also of the vegetation dynamics that produced them? What happens to the common distinction between primary and secondary forests in the case of anthropogenic formations such as the forest islands discussed by Erickson for Mesoamerica and Fairhead and Leach for West Africa? Clearly, "secondary" need not mean "lesser," since in this framework much of the Amazon would need to be reclassified as secondary forest. Both cultural and natural histories tend to be relatively secure within existing interpretive frameworks, but new historical ecologies that give equal weight to historical data from both human and nonhuman records of the past seem to demand fresh consideration of inherited frameworks for understanding the modern world and for elucidating historical and ecological process. As a group, these papers point to the recognition of much more complex roles for humans in the creation, restructuring, and even renewal of forests—roles that have not just begun, but that have had long, if often unrecognized, histories.

In chapter 12, Janowski and colleagues discuss the complex forest worlds of the Kelabit and Eastern Penan of Borneo, two groups seen both by themselves and by anthropologists as highly distinct. The Kelabit practice a mosaic of food production and extraction activities from paddy rice to upland swidden, grazing, and collection of forest products. The Penan, in contrast, are (or were until recently) famously hunter-gatherers, nonagriculturalists dependent upon gathered forest foods, especially sago. Both groups draw a sharp distinction between wet rice, which requires significant labor to clear land and cultivate, and which grows in monocropped fields, and other vegetation forms, including swidden plots, which mimic the structure of the tropical forest, and plants that "grow on their own" and do not require the active tending and care rice paddies do. While the Kelabit valorize their work in the rice fields, the Penan shun it, preferring the much easier strategy of using wild sago and other forest products. Those who assume that strict nature-culture distinctions belong only to the industrialized west should take note here. What is even more striking, however, is the way in which this indigenous categorization elides major differences that agronomists, archaeologists, and others would recognize as clear distinctions between wild and domesticated plants, and between foraging and agriculture. While rice alone is associated with clear-cut property rights among the Kelabit, rights to forest plants that have been planted, tended, and otherwise encouraged are not treated separately from truly wild produce, nor indeed are the products of swidden plots. Janowski, Barton, and Jones point out that rice production may be only a few hundred years old in this region, as it is in some other parts of island Southeast Asia, but however long it has been practiced, the logic of rice stands in sharp distinction to what otherwise can be seen as a radical blurring between wild and domesticated, forest and arable, mine and yours, nature and culture.

Like other chapters in this volume, Janowski and her colleagues use a diverse array of sources—history, archaeology, paleoecology, and ethnography, among others—to highlight the more than 50,000 year history of human occupation in their study area. While more remains to be done to elucidate that history, it is clear that the tropical forest landscapes they consider are deeply entangled with human histories, and that the work of "culturing the forest" has a long history indeed. Though this is less a focus of the chapters in this volume, engagement between humans and forests also raise the issue of what are sometimes called "forest peoples." Both the Penan of Borneo and the so-called tribal peoples of Southwest India discussed by Morrison and Lycett (chapter 11) have long been regarded as human equivalents of pristine forests—untouched, ancient, or remnant peoples living an archaic lifestyle. These anthropological tropes, too, have begun to fade in the face of growing empirical evidence and conceptual challenges (Morrison 2002a), a development that must be seen in light of reconceptualizations of forest histories overall.

In our chapter, Lycett and I consider some of the ways in which social and political power and desire may help to structure human land use and consumption, and hence forest and landscape histories. In southern India, the valorization of rice increasingly led to differentiated physical landscapes in which water and labor were concentrated in areas of potential rice production while vast regions suitable only to the farming of dry grains such as millets became the domain of less valorized people as well. Dry tropical forests, already greatly reduced by the thirteenth century, were pushed back during eras of urban expansion and rebounded during periods of urban decline. However, the development of elite cuisines in southern India also sustained a trade in upland tropical forest products, prompting commercial foraging by ostensibly hunting-and-gathering groups, as well as by upland agriculturalists (Morrison 2002b). These upland forests in the Western Ghat mountains are often seen not only as the home of distinctive forest peoples, but also increasingly in terms of their significant biodiversity. In coming to terms with challenges to contemporary conservation, ecologists, development specialists, and local governments often cast upland peoples as forest destroyers when, in fact, multidisciplinary historical data show that humans may have played an important role in the creation of the forest itself. This fundamental misunderstanding, also evident in the work in this volume by Fairhead and Leach and Erickson, points to the power of existing prejudices and of entrenched narratives of decline and loss.

Lentz and Lane, in a systematic botanical analysis of two forest areas in Belize, present striking evidence for the long-term salience of human occupational history on species composition in tropical forests. While other papers use evidence of past vegetation, landforms, and land use, Lentz and Lane examine trees in contemporary forests to track the echoes of human activity on Maya landscapes. Comparing two sections within the Rio Bravo Reserve, one located adjacent to Dos Hombres, a small city occupied during the Late Classic period Maya civilization, and a control transect through an area not intensively settled in Late Classic times, in chapter 13 Lentz and Lane show convincingly that economically important tree taxa occur in greater numbers in areas once used by ancient urbanites, notwithstanding the hundreds of years that have passed since the Maya collapse. These "feral forests," to all appearances identical with those of the control transect, nevertheless stand as part of the human as well as the natural history of this region. In addition, they challenge easy understandings of forest dynamics; not only have these forests recovered, but they are also following divergent pathways, pathways whose route is a product of mixed socionatural histories rather than a simple consequence of either succession or human impact alone.

The significance of forests is not, however, simply a matter of concern for ecologists and social scientists. In chapter 14, Rangarajan points to the ways in which both forests and forest denizens have consistently been politicized in South Asia. While Lycett and I pointed to the entanglement of forests as well as lowland peoples in regimes of value surrounding forest products, Rangarajan focuses on the charismatic megafauna of the subcontinent, the lions, tigers, rhinoceroses, and other signature species of the region. Here a long colonial history of forest reservation, a policy dictated both by strategic interests of empire as well as nascent conservation concerns, led to a situation in which vast tracts of land came under the control of a government forest department. While this is also true of the western United States, the Indian context differs significantly in the density of human occupation around, and often inside, forest lands. Rangarajan makes the important observation that "the study of nature often treats humans as a uniform category while those who study society rarely see nature as a dynamic entity," a simplification all too common in a world of disciplinary specialization.

Rejecting such flat perspectives, Rangarajan draws on the rich historical record of South Asia to show the mutable and fluctuating boundary between forest and arable, and to illustrate the ways in which social and political conceptions of forests, animals, and people shaped these in demonstrable ways well before the industrial revolution. In many ways, our modern conceptions of the forest date to the British colonial period, when large-scale changes in management, manipulation, and control of woodlands was imposed. Mobile peoples and wild animals alike were seen as dangerous; to be hunted, reduced, and settled. Rangarajan's discussion of post-independence forest policy, including the role of "vertebrate nationalism" since the 1960s, brings to mind Neumann's work on wildlife reserves in Africa, a similar model of human exclusion and wildlife isolation being at work in both cases. In both cases, too, enforcing the radical segregation of humans and wildlife has encountered significant obstacles. Especially since those living with forests and large vertebrates are often the already-marginal of society, their exclusion from "wild places" tends to produce serious economic hardship.

The fact that these very groups may have helped create the landscapes now constructed as wild not only adds salt to their wounds, but ironically may also imperil the success of conservation efforts that seek to protect animals and forests. Rangarajan points to the evolving viewpoints of conservationists in India, which grapple with the multiple and sometimes incompatible rights of humans, tigers, grasses, and trees within a densely-populated, diverse, and politically dynamic landmass.

In chapter 15, Erickson takes on the notion of environment as a static backdrop to human adaptation rather than as dynamically created, an historical product formed both by human and nonhuman action. The latter perspective, which he associates with historical ecology, better accounts for the trajectory of human-landscape development in the Amazon. Here, as noted, archaeological and paleoecological evidence shows that the region played host to unexpectedly large and complex human societies prior to colonization. Far from being a "counterfeit paradise" with thin, nutrient-deficient soils unfriendly to agriculture, parts of Amazonia actually have organic and nutrient-rich soils, the famous "dark earths" that Fairhead and Leach note may also be present in tropical Africa. These soils are artifacts of human occupation, full of the debris of past occupants, the result of large-scale composting and active intervention by past residents. While earlier scholars (Meggers 1991) had posited that these areas were natural creations left by winding river courses, in fact these fertile soils were made in the course of human intervention in local environments, interventions that left a legacy of lush tropical forest stands. These forests, of course, stand atop another human legacy, this one less salubrious, since the reexpansion of forests in this region owes a great deal to colonial era depopulation, both the unintentional expansion of virgin-soil epidemics and also the active exploitation of indigenous peoples.

Together, these chapters suggest the necessity for developing new forms of scholarly work that integrate evidence and perspectives from both human and natural histories. While never simple to produce, such combined socionatural understandings can form the basis for entirely new attitudes toward both past and present. Even more important, they can also provide empirical support for new understandings of ecological and cultural processes, the processes that created—and continue to create—the landscapes we occupy today.

11 ✳ *Constructing Nature: Socio-Natural Histories of an Indian Forest*

KATHLEEN D. MORRISON AND MARK T. LYCETT

In 1988, biologist Norman Myers coined the term "biodiversity hotspot" to describe biogeographic regions characterized by both high levels of plant endemism and serious levels of habitat loss. The international nongovernmental organization (NGO) Conservation International adopted this conceptual scheme almost immediately, undertaking a global review in 1996 that identified twenty-five such hotspots and added quantitative thresholds for their designation (Conservation International).[1] This organization's definition specified that biodiversity hotspots must contain at least 1,500 species of endemic vascular plants *and* have lost at least 70 percent of their original habitat. Among these hotspots is the Western Ghat region of southern India, a chain of elevated lands running along the southwest coast of the peninsula. The hotspot label, which has been widely used, is thus not a simple marker of biodiversity, as might first be imagined; it also encodes arguments both about the "original" state of specific environments and about ongoing threats to them. While we would certainly not argue against the need for environmental protection and conservation, it is worth looking a little more closely at the problematic way in which humans figure into a concept like that of the biodiversity hotspot, a concept that resonates powerfully with both scholarly and popular traditions concerning both forests themselves as well as the forest peoples of southern India.

By most estimates, the "pristine" or primary forests of peninsular India are severely restricted in extent. The Greenpeace World Intact Forest Landscapes study (Potapov et al. 2008) classifies India's forest zone as 19 percent of the national land area, a figure similar to the Forest Survey of India's 19.4 percent estimate. Of that, however, Greenpeace labels only 5.8 percent as intact. Chaturvedi (1992), similarly, defines 45 percent of Indian forests as secondary, a definition that would include *all* of the dry deciduous forests of the interior. Conversely, much of the forest considered to be "intact" lies within the Ghats where, as we will see, some woodlands may have appeared only recently. In this chapter, we make the case that the upland forests of the Ghats, no less than their lowland brethren, are themselves products of dynamic historical phenomena rather than fragile remnants, survivors of a primitive condition.

We link the socio-natural history of the humid tropical forests of the Ghats with those of their drier counterparts across the peninsula, viewing this larger region as produced dynamically through the ongoing engagement of human agents with the social and natural world. In this sense, nature, or the environment, is always anthropogenic, and anthropogenesis is perpetual. The environment incorporates, shapes, and constrains processes and practices, even while it is produced by them. It is both the medium and outcome of society and the product of and template for social and cultural production. That is, it is a *social space* (*sensu* Lefevbre 1991) that is necessarily historical, and necessarily relational, not a simple thing with natural boundaries and stable properties. Humans have thus made the natural world both conceptually—through the creation of various ideas about nature, ecosystem, organism, and ecology—and materially—through millions of years of direct action in and on the landscape. It is this history of socionatural production that we take up here. We are less interested in the project of specifying an authentic, originary landscape or "state of nature," as either a baseline against which to measure loss or as an ideal, preanthropogenic condition.

In this volume, Rod Neumann cogently points out some of the internal inconsistencies in scholarly and activist accounts of long-term human-environment interactions and their consequences. Echoing a theme familiar from South Asia, Neumann notes that the dominant story of Africa's forests, where he has spent most of his career, are those of deforestation and destruction, with expanding populations threatening habitats and reducing biodiversity. In contrast, he notes, with some incredulity:

> According to the European Environmental Agency, Europe's current biodiversity is the product of centuries of human interaction with nature. . . . Land abandonment resulting from demographic and socio-economic shifts in rural Europe is thus "considered detrimental to biodiversity." . . . The EU's Forestry Strategy emphasizes "restoring traditional management" to minimize biodiversity loss. Hence, if a single dominant story could be written for Europe's forest it might read as follows: forest cover is the product of millennia of human occupation and use. Ergo, demographic and socio-economic changes leading to rural land abandonment and the collapse of traditional management systems threaten to reduce the biodiversity contained within many habitat types. (chapter 3, this volume)

With Neumann, we ask: Is it possible that humans in Africa—and South Asia—are so different from Europeans, whose historical actions have led to specific recognized and valued habitat types and whose traditional management systems are represented as beneficial to, rather than destructive of, biodiversity? Clearly, there is a double standard at work here. In Europe, land *abandonment* is seen to pose a threat to biodiversity, while in the colonized world land *occupation* is seen as the threat. While Neumann explores this contrast in terms of the rhetorics of contemporary conservation, here we move in a slightly different direction to consider the role of humans—and nonhumans—in the long-term history of south Indian forests, from the dry tropical and subtropical scrub forests of the interior to the mesic forest of the Western Ghats.

While anthropogenesis may be clear for areas long under intensive agriculture, such as the dry interior, it may be less obvious that the Western Ghats, too, are places in which nature and culture converge in significant ways, and where politics, social life, and human economies have played important roles in local histories. In South Asia, human occupation long predates the establishment of modern climatic conditions at the onset of the Holocene, so that it can fairly be argued that forests and human societies developed in tandem, linking forest history to human history in significant ways. In spite of this shared history, the stark divide between text-based environmental history and historical research based on material remains (e.g., paleoecology and archaeology) has led to such a sharp separation that each field continues to repeat truisms long discredited in the others. Both historians and development specialists, for example, have clung to a narrative of once-vast expanses of primeval peninsular forests, only recently degraded and deforested by human activity (Collins, Sayer, and Whitmore 1991; Bhat, Murali, and Ravindranath 2001; Gadgil and Guha 1992; Robbins 2001). Modern forests are almost uniformly seen as *remnants* of this history, a simple trajectory of change in which forests stand in neatly for nature and human action, especially agriculture, plays the role of culture.

Perhaps as a consequence of this dominant narrative, the significant regional and temporal diversity of forest history on the subcontinent is often not appreciated, and an image of a uniform, stable, precolonial landscape only moderately affected by human land use is common. Thus, a recent surge of research on colonial forestry (e.g., Guha 2000; Rangarajan 1999; Saravanan 2008; Sivaramakrishnan 1999) has spawned surprisingly little corresponding work on the precolonial history of South Asian forests (see discussion by Rangarajan 2002, 2012).

In this chapter we trace both agrarian and forest histories using archaeological, historical, and paleoenvironmental data. In particular, we follow the development of both elite and nonelite South Indian cuisines and track the implications of changing diets on the landscape, from field to forest.

The increasing demand for irrigated rice, vegetables, tree crops, and spices placed specific demands on agricultural production and the extraction of forest products. The elite South Indian meal includes elements from both upland (humid tropical) and lowland contexts; as such, its demands came to resonate across the peninsula (Morrison 2001). Changing agrarian practices and patterns of intermittent urbanism led to episodes of both deforestation and forest resurgence in the dry interior, while the demands of lowland consumers and states helped to transform Ghat forests as well, ultimately leading to significant patterns of woodland expansion. Coming to terms with these histories will be essential if we are to have an accurate understanding of modern forests, not as ancient remnants now under threat, but as complex outcomes of a long socionatural history.

Locations of Value: Cultural Representations of Forest and Field

We begin with a contradiction. For at least the last thousand years, South Indian textual representations have celebrated places of cultural value on the landscape, of which the two most prominent are irrigated fields and forests. Forests, in Indian literature, are often literally places apart: clearly distinguished from ordinary places of human residence and use, forests are often home to religious figures and may serve as locations for spiritual or even tactical retreat (Sivaramakrishnan 2009; Thapar 2012). As literary devices, forests serve as liminal places, providing transition or refuge, as for example in the great epic of the Ramayana. Forests, too, were (and are) seen as locations of alterity, containing differently organized peoples, even today distinguished as "tribals," a term used in contradistinction to caste society (Zimmerman 1987; Morrison 2002a). While many of these forest denizens served in precolonial armies and also worked as trackers, hunters, and traders collecting or growing various forest commodities, forests were also, as in European traditions, seen as potential lairs of bandits and other criminals (Guha 1999; Murthy 1994). Later British rulers would even classify entire ethnic groups as so-called "criminal tribes," codifying this prejudice into census categories. So danger and difference are the inverse, perhaps, of forests' association with holiness and separation from the corruptions of civilization.

Though tribals never had a high status, forests also served as reserved locations for elite sport, especially royal hunting, and the strategic value of dense forests was also widely recognized. The Amuktamalyada, for example, a poem attributed to the early sixteenth-century king of Vijayanagara, Krishnadevaraya, argues for the importance both of clearing forests within one's kingdom and of maintaining dense forests around its perimeter as a defensive measure (Sastri 1975). Later rulers protected forests as royal hunting reserves, institutional arrangements that codified the simultaneous valuation of forests as locations for activities of both the lowest and highest social strata.

Similarly, irrigated fields and especially rice paddies, feature prominently in both literary works and in the more sober, workaday texts describing donations, tax arrangements, and other transactions inscribed into stone or copper plates. Textual accounts of agrarian landscapes often represent them in terms laden with political, social, and gendered values stressing fertility, productivity, and just rule (Kotraiah and Dallapiccola 2003; Morrison 2009)—landscapes that necessarily contained specific kinds of crops such as rice, coconuts, and fruit trees.

Thus, it would be tempting to conclude from the historical literature that South Indian landscapes between about CE 1000 and 1700 consisted primarily of lush green paddy fields set within dense forests of abundant game. The contradiction is, of course, that the actual landscapes of southern India did not consistently (or even mostly) reflect these cultural valorized images. In the driest parts of the peninsula, forests were already relatively minor components of the landscape by the first millennium CE, notwithstanding their cultural importance; and in the dry interior, irrigated land was highly restricted despite tremendous efforts expended on the construction of canals, aqueducts, and reservoirs (Morrison 1995, 2009).

Despite their limited extent, however, these pockets of irrigated land were extremely important in facilitating urbanism, an accomplishment resulting in a sustained, though not precisely unidirectional, decline in the tropical dry deciduous forests of the interior for the last two thousand years. In the more humid tropical forests of the southern uplands, however, both natural and cultural factors led to both declines and expansion of woody vegetation. Here forest expansion is clearly reflected in paleoenvironmental records, but is generally ignored by both historians and development specialists, while the empirical evidence for new forests is almost uniformly attributed by ecologists to climate change, despite strong evidence for a human role in vegetation change (e.g., Gunnell, Anupama, and Sultan 2007).

Urbanization, Markets, and Exchange: Landscapes of Elite Consumption in the Dry Interior

Influential early syntheses of South Asian environmental history, such as that of Gadgil and Guha (1992), as well as others, helped to codify several truisms about forest history that have proven difficult to dislodge. Among these are the notion that the precolonial use of forests had minimal impact, major episodes of deforestation began only in the colonial period, precolonial forest use was tied exclusively to subsistence, and prior to colonial intervention, forests (and irrigation facilities) were managed by local communities, without exploitation, while following spiritual injunctions. Considering the received view of Indian environmental history in light of the textual valorization of forests and irrigated fields, we can perhaps see where the idea of precolonial sustainability emerged. The historical association of forests with religious ascetics, for example, moves easily into the idea of indigenous relationships with nature as essentially spiritual. To move beyond this, we suggest that at least two kinds of intellectual barriers need to fall. First, the established separation of agrarian from forest studies in South Asia will need to be eroded, recognizing both the ecological contexts of farming as well as the social contexts of forests. Second, it is critical to be able to assess the histories of vegetation and of human land use *independently* of past cultural representations. To this end, we and our colleagues have been conducting a program of archaeological, paleoecological, and historical research in the interior, tracking changes in land use, vegetation, and geomorphology over the last 5,000 years. While fully comparable data from other regions are lacking, we have also been able to draw out some larger regional patterns using existing research, especially from the west coast and Western Ghats.

In a nutshell, this work has shown that the dry deciduous forests of the interior developed in concert with human land use for more than 70,000 years, with the last 5,000 years seeing the most significant human impact. Here the expansion of agriculture, and then after approximately 400 BCE, of rice agriculture and of irrigation more generally, has had a major effect on soils and vegetation (Morrison forthcoming; Morrison 2009; Bauer and Morrison 2013). Agricultural practices have been responsive not only to changes in population, but also to evolving food preferences, as discussed later.

Upland and coastal forests in southern India have a shorter history of human land use than the interior—only about 2,000 years (reviewed in Morrison 2002b). Upland forests, while supporting diverse subsistence regimes, have also always been part of exchange networks, with culinary and ritual demand for forest products transforming local environments and social organization. Across the peninsula, both urbanism and emerging social differentiation that manifested in divergent forms of consumption have had a significant impact on regional environments.

During the first few centuries CE, and perhaps earlier, South India saw the development of a sharp differentiation in consumption practices, a difference still salient today. During these centuries there was a slow codification of several regionally differentiated but related elite cuisines based on irrigated produce, especially rice. Except in very moist environments, virtually all of the critical components of this cuisine require artificial irrigation, making the raw ingredients of meals offered to gods, as well as those consumed by elites, extremely capital and labor intensive—out of reach for many or even most of the population. Elite meals and expanded temple-based rituals also required varied spices and aromatics, many of which are derived from the more humid upland forests. Even the texture and presentation of the elite meal was moist, if not soupy, and was often served on disposable plates of banana or even lotus leaves. In addition to plant foods, dairy products such as yoghurt and ghee were critical elements, both associated with grazing animals, especially cows.

Foods of the poor, conversely, were (and are) very dry, an adjective that can also describe the major form of cropping, which is solely rain-fed. Consumption was built around a variety of hardy millets and legumes. Oils, too, were much more commonly derived from oilseeds such as sesame and castor, rather than coconut or animal-based products.

These diverse consumption practices were reflected in a mosaic of forms of production, from rainfed farming, to grazing, to intensive irrigated agriculture. Diverse forms of production also had varied environmental effects; permanently irrigated fields were associated with changes in hydrological regimes, the formation of paddy soils, the extension of disease vectors such as mosquitoes, and of course, transformed flora and fauna. Dry farming and grazing, too, were consequential, with erosion one of the most visible historical consequences of the expansion of extensive forms of production.

How did the development of elite cuisines affect south Indian landscapes? While rice and other highly productive crops themselves created the possibility for large, aggregated settlements in dry regions, the elite meal relied on extensive grazing land almost as much as canals and reservoirs. Thus, large cities not only implied zones of irrigation for the consumption of the well-to-do, but also dry farming zones for the support of poorer citizens and extensive grazing areas for all. If the presence of cities in the interior sounds like a recipe for ecological degradation, in many respects it was. But one surprising and interesting part of the history of human-environment interaction in this region is what happened when deurbanization took place in the context of an already-established irrigation network, something that would never have existed without the previous history of investment aimed at supporting the urban elite. This conjuncture allowed, however briefly, a resurgence in the dry deciduous forest cover, a small reversal in the long history of forest loss across the dry interior of the peninsula.

Much of our work comes from the area around the city of Vijayanagara, that great imperial capital whose rather improbable three-hundred-year history in the middle of one of India's driest regions represents an extreme manifestation of human ingenuity. We have studied the dramatic impact of this city on the regional environment, tracking changing strategies of agricultural production across the three centuries of urban occupation and beyond. By the early sixteenth century, when the city was home to several hundred thousand people, problems of erosion and reservoir sedimentation were already very severe. The pollen record of the fourteenth-century Kamalapuram reservoir reveals the regional vegetation of the time, including material derived from the iron-rich Sandur Hills (Morrison 1995). This record, as well as the Kade-bakele reservoir record studied by Bauer (2010) shows, most notably, a loss of woody vegetation during the fourteenth century as the city grew, and a rebound of trees and shrubs following the abandonment of the city.

This resurgent woody vegetation is at least in part created by the expansion of previously cultivated taxa, like *Phoenix* (date palms), while some appears to be a rebound in the woody vegetation of the nearby Sandur Hills. The Sandur Hills are a small "island" of more mesic vegetation similar to the western Ghat foothills, just south of the city of Vijayanagara. The Sandurs, the record shows clearly, were actually more sparsely forested in the sixteenth century than they are today. Woody vegetation on the Sandur Hills rebounded after the fall of the city, and by the early nineteenth century, documents suggest that the hills supported large animals such as tigers. Importantly, this was the period when textual records of the region became more detailed and abundant and in which most environmental histories of South Asia begin. By the late nineteenth century, however, mining and other human impacts were already prompting deforestation again, while very recent strip-mining for iron and magnesium has made further rebound unlikely.

Post-urban forest regeneration appears to be quite localized, restricted perhaps to the moister Sandur Hills. This is evident from the thousand-year Kadebakele reservoir sequence. Here the advent of dense urban occupation led not only to a never-reversed decline in tree cover, but also to a near-total loss of woody climbers, evidence of the consistent pressure of burning, grazing, and wood collecting. Thus, the forest record of the dry interior is complex, showing periods of forest loss and regeneration, a pattern connected not only with the tempo of urbanism and land use, but also linked to the specifics of soil and rainfall as well as the irrigation infrastructure made by previous generations. In this region, with its long record of agriculture, forest histories have mostly been ignored by scholars. Instead, ecologists and environmental historians have looked to the humid tropical and semitropical forests of the South Indian uplands, areas long seen as remote from the effects of human land use. This distinction is, however, overdrawn. Not only, as discussed later, were upland forests inhabited places for at least the last two thousand years, but they were also places directly linked to the agrarian worlds of the lowlands, links forged in part by consumption and exchange and enmeshed in relations of power.

Forest Products, Colonization, and the Entanglements of Upland Tropical Forests

Moving west to the more humid upland forests, there is at present no clear evidence for significant human use of these forests until the first century CE, a period archaeologists refer to as the Early Historic. This period also, and perhaps not coincidentally, saw the large-scale expansion of a far-flung trade in forest products across the Indian Ocean, into the Mediterranean and East Asia (Tomber 2008; Morrison 2002b; Morrison and Lycett 2013). Of the many forest products involved in this trade, a few were cultivated, but many, including pepper, were also gathered wild. The wealth of South India's tropical and semitropical forests, chiefly in the form of spices, resins, dyes, and other nontimber forest products (NTFP), fueled imperial and colonial expansion in South Asia and beyond. Medieval European demand or, as Freedman (2008, 1) puts it, *craving* for spices, the legendary "pepper forests" of the equally legendary Prester John drew the Portuguese and later European powers to the South Indian coasts, a search for "Christians and spices" that would change the world. During the sixteenth and seventeenth centuries, we have clear evidence for the emergence of occupationally specialized foragers, hunter-gatherers who collected forest products for world and local markets. This pattern of lowland-upland integration—and sometimes exploitation—was, however, significantly older. The consistent demand for spices, aromatics, and other forest products for elite cuisine and temple worship across the interior of peninsular India worked to ensure the integration of upland forests and forest-dwellers into lowland economies and ecologies. Lowland products, including rice and manufactured goods, moved into the uplands in exchange, relationships that, by the eighteenth century, had become marked by indebtedness and inequality (Morrison 2002c, 2006).

The expansion of both the regional and international trade in forest products led to an increased demand for rice, both to feed growing coastal cities and merchants and to send up into higher-elevation forests in exchange for gathered spices and other NTFPs. The concomitant expansion of rice growing into higher elevations (e.g., Hockings 1980) added to the pressures on upland forests, reducing subsistence options in what had previously apparently been flexible mixed economies involving both farming and gathering. Forest-dwellers, as best we can reconstruct this history, thus became increasingly enclaved and disadvantaged as they were brought closer into global economies (Morrison 2002c, 2006).

By the time British rule was established in this region, bringing not only a bonanza of historical documents and the beginnings of systematic ecological observation, Ghat forests had already been integrated into regional and international networks of power and exchange for more than a thousand years. Despite this, southern India's upland forests quickly established a reputation as pristine forests home to unchanging aboriginal populations (Morrison 2002c). In this view, still present to some extent in development and activist circles, both Ghat forests and Ghat peoples are seen as *remnants*—fragments displaced in time and space. Quite to the contrary, however, the historical development of specialized forager-traders out of more generalized foragers and agriculturalists led not only to their ethnographic *invention* as timeless yet noble savages, but also to a view of the upland forests *themselves* as somehow apart from the human history that had otherwise despoiled the peninsula. These kinds of perspectives resonated powerfully with existing Indian views of forests as places apart, a convergence of views that makes some of the misconceptions of contemporary environmental history in India more explicable.

Consistently represented as both remote and exotic, even within southern India, Ghat tribals are generally either regarded as essentially *part of* nature or, like the land-hungry Africans in the European Union's classifications, as nature's despoilers. Almost never are the Ghats' human residents represented as a part of mainstream history. They appear in exotic disciplines like anthropology, and in contemporary development discourse, but figure little, if at all, in historical and archaeological narratives about state formation, imperialism, or colonialism. And this is curious, given both the consistent participation of Ghat peoples in and the importance of Ghat products for these processes.

Perhaps it is this invisibility that has allowed upland peoples to be represented *both* as ahistorical forest beings, exotic from and irrelevant to the main currents of history, *and* as agentive threats to contemporary biodiversity. Indeed, the historical constitution of the Ghats as a place has tended to set aside consideration of politics, power, and cultural production, constructing instead both people and places there as natural and fragile. It is worth asking why the human contribution, both to the constitution of Ghat environments and to the spice trade, has been so consistently overlooked. This may be explained, in part, by the western perception that forest products did not, as manufactures clearly did, contain congealed human labor. South Asians, of course, knew better, but in spite of this, both indigenous and western traditions emphasize forms of production more fully "domesticated" by the state—intensive agriculture and craft production. Foraging, swidden, and other economic practices less easily contained by and legible to the state have tended to escape analysis as *cultural*, both by archaeologists and historians. Thus, in India, agrarian history is starkly set off from forest history, even when forests themselves yield produce. When plants become commodities, especially forest plants, it seems that their hybrid nature as socio-natural objects is easily forgotten.

While the forest products of the Ghats served international demand, they also circulated within South India as well. From the development of elite South Indian cuisines involving labor-intensive, irrigated products such as rice and forest products such as spices, to the establishment and elaboration of temple worship requiring aromatics such as camphor and sandalwood and later, elaborate meals, there operated within South India a political ecology of production and circulation that linked lowland and upland.

By the late precolonial and into the Early Modern period, as discussed in detail elsewhere (Morrison 2002b, 2002c, 2006; Morrison and Lycett 2013), Ghat forest products constituted major commodities in both peninsular and international trade, with products such as pepper, ginger, and cardamom grown in swidden plots, gathered wild, or both. Many upland peoples, later to be represented by anthropologists as untouched primitives, became specialist foragers, hunter-gatherers of commodities. Perhaps conceptual categories had so little space for commercial foragers that they had to be transformed into either the timeless "natural people" of early twentieth-century anthropology, or to the threats to nature perceived by colonial administrators and contemporary NGOs alike.

If material records from archaeology and paleoecology have exposed literary representations of South Indian landscapes as highly selective and not altogether reliable, it is important to note that physical records, too, require contextualization. For example, one higher elevation temperate area within the Ghats, the Nilgiri Hills, now contain numerous upland savannas, open grasslands often argued to be degraded evergreen Shola forests, newer culturalized places formed as products of degraded natural forests (e.g., Mohandass and Davidar 2009, 20). Savannas and forests coexist in these hills, prompting much discussion by ecologists. Grasslands, it has been argued, are recent human creations, caused by grazing, burning, and cultivation. Paleoenvironmental studies show, however, that this is *almost exactly the reverse of the actual progression of vegetation change.* Savannas are very old here and although the number of sampling sites is still insufficient to fully describe regional vegetation history and vegetation dynamics are clearly complex (Caner et al. 2007), moist savanna formations have characterized much of the vegetation of the area above about 1,500 m elevation since the Late Pleistocene. Several recent studies of pollen, soils, and carbon isotopes show definitively that much of the Nilgiri Shola forest is of quite recent

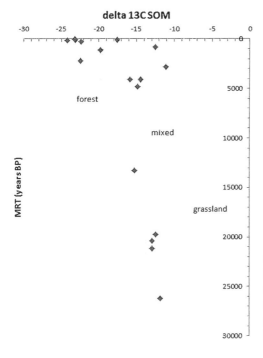

11.1. Carbon isotope values from Nilgiri soil organic matter. MRT = mean residence time, in years before present. Source: Data from Caner et al. (2003).

origin, formed within the last one to two hundred years (figure 11.1; Caner et al. 2003; Caner at al. 2007; Mariotti and Peterschmitt 1994; Rajagopalan et al. 1997; Sukumar et al. 1993; Vasanthy 1988). Most sampling locations record natural upland wet savanna vegetation in the early Holocene, open formations that were slowly afforested starting either at about 2000 BCE or, in other cases, more quickly transformed from grassland to forest during the last several hundred years. Here paleoenvironmental data fundamentally challenge previous ahistorical accounts of Niligiri ecology, leading the historically-oriented biologists conducting these studies to conclude that, in fact, the patchiness of the Shola forests is *not* a consequence of recent forest degradation. Even in these newer accounts, however, humans are only ever mentioned as agents of forest destruction, and evidence for recent forest expansions, although not found across all sampling locations, is never attributed to human action but to climate change instead. While the details of this change remain to be studied, it is critical to note that expansion of the Shola forest took place within a cultural landscape, in a region with a complex and long-standing history of human land use, and it is possible that the asynchrony of change within this small area may be better explained by causes other than regional or global climate change. Although the kinds of analyses of forest composition carried out in the New World tropics such as that reported by Lentz and Lane (chapter 13, this volume) have yet to be done in South Asia, it might be expected that over the centuries agriculturalists, foragers, and others have also modified tropical vegetation in this hotspot in ways that can be easily misrecognized as natural.

While vegetation change is almost always interpreted within the scientific community as evidence for changes in climate, independent forms of climate assessment, as well as a judicious reading of the archaeological and historical records, combine to show the poverty of this approach. Vegetation constitutes a complex outcome of both climatic and human forces, but in this case it seems clear—as we also see in parts of Africa, for example—that human activity, associated with the vital trade in forest products, has been instrumental in *creating* the evergreen forests of the Niligiris. Far from being mere agents of deforestation, then, Ghat residents actually *constructed* many of their forests over generations, helping create the very biodiversity hotspots they are now seen as threatening.

Discussion

In southern India, the diet of gods and elite humans that emerged during the last thousand years represents, if not a complete ecological anomaly as it does in the very dry interior, then at the least a very highly socialized product requiring upland forest products, labor-intensive irrigated produce, and a network of connections linking them. These social worlds and cultural desires led to very real material outcomes. To a large extent, the history of southern India's tropical dry deciduous forests is submerged within what is usually conceived as agrarian history, and hence tends to be viewed as a story of unmitigated forest loss and the expansion of "culture" at the expense of "nature." This account, clearly, is incomplete. Woody vegetation in the area around the sixteenth-century city of Vijayanagara was almost completely denuded, but it managed to rebound, at least on the moister slopes of the Sandur Hills, before being removed again by mining.

While it is clear that the landscapes of the dry interior are fundamentally anthropogenic, the Western Ghats forests, too, are dynamic products of a socio-natural history. The rhetorical construction of Ghat places, peoples, and/or products as either purely part of nature or as nature's despoilers, is not helpful. Forest history is human history and the conceptually separate spheres of environmental history, agrarian history, and political history are not merely connected, but mutually constitutive. Our tasks now are both to develop better frameworks for understanding these hybrid histories and to begin the hard work of making them.

12 ∗ Culturing the Rainforest: The Kelabit Highlands of Sarawak

MONICA JANOWSKI, HUW BARTON, AND
SAMANTHA JONES

The Kelabit Highlands and Its Inhabitants

The Kelabit Highlands in the Malaysian state of Sarawak is part of a tableland about 3,500 feet above sea level broken by lower ranges of mountains and hills, extending into East Kalimantan in the Indonesian part of Borneo (figure 12.1). This tableland is distinctive for the presence of numerous megalithic monuments, for the practice of wet as well as dry rice cultivation, and for the feasts of merit at which, until the 1950s, megaliths were still erected. It is inhabited by a group of peoples speaking what Hudson (1977) describes as Apo Duat languages, after the mountain range between the Kelabit Highlands and the Indonesian part of the tableland; this should more properly be described as Apad Uat, the local term for the range, meaning "root mountains" in the local languages.

Today the Kelabit Highlands are inhabited by a people who are known as, and call themselves, Kelabit. Eastern Penan, who are not Apad Uat-speaking people, have also lived in the Kelabit Highlands in the past and still live in areas immediately to the south and west of the Kelabit Highlands.[1] While the Kelabit are rice growers, the Penan were, until recently, dependant on wild sago palms as their main source of starch food. Since World War II, most Penan have become settled or semisettled (figure 12.2) and have begun to grow rice, although some groups continue to be primarily reliant on wild sago.

This chapter is based on data gathered in the Kelabit Highlands since 1986 by Janowski, chiefly in the community of Pa' Dalih in the Kelapang valley, and data gathered in the Kelabit Highlands and surrounding areas by a team including the three authors, as part of the three-year research project (from April 2007 to April 2010), *The Cultured Rainforest* (http://www.arch.cam.ac .uk/cultured-rainforest/), funded by the UK Arts and Humanities Research Council. This draws on anthropology, archaeology, environmental science, botany, and the use of GIS to investigate, compare, and store information about the present and past human uses of and relationship with the landscape and the natural environment in the Kelabit Highlands.

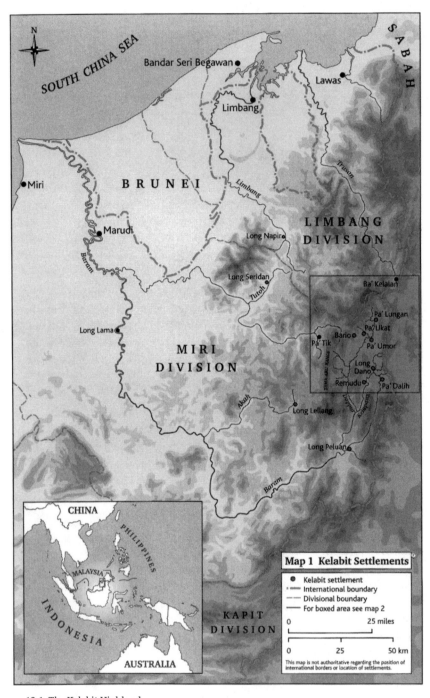

12.1. The Kelabit Highlands

12.2. The Penan camp at Ba Puak to the west of the Kelabit Highlands, July 2008. Source: Photo, Monica Janowski.

Before the project began, no significant archaeological data[2] or data on environmental history was available for the highlands, and there is minimal written information available. According to the Kelabit, their ancestors have always lived in the highlands. They tell of a race of superhuman giants, the Rabada people, living in the area in the ancient past, who were, they say, their ancestors. There is a story relating that all the peoples of the world originate in the Apad Uat Mountains, and that there was a flood in ancient times that carried everyone downstream except the Kelabit and related peoples.[3] The Penan say that their ancestors have roamed the area, which includes the highlands, for as far back as they know.

Culturing the Forest

The rainforest of the highlands is predominantly oak, with wet peaty areas, some scrubby *kerangas* vegetation (Browne 1952), and, on the mountains, cloud forest. The impact of human use is most clearly visible through the presence of wet rice fields, secondary growth deriving from recent dry swidden cultivation of rice and other crops, and buffalo pastures. Around the northern, western, and southern sides of the highlands on the Malaysian side, logging is taking place. A national park, Pulong Tau, was gazetted on

March 24, 2005, which focuses on the Tama Abu range of mountains extending along the western side of the Kelabit Highlands.

The forest of the highlands is central to the livelihoods of its inhabitants. The Eastern Penan, until recently, depended entirely on the resources of the forest (*tana* in Penan, using the same word that refers to the earth itself). They relied on sago starch as their staple starch, which they harvest from a variety of palms, but mainly *Eugeissonia utilis*. The Kelabit also rely heavily on the forest (*polong* in Kelabit) for much of their subsistence (Janowski 2004). All meat eaten on a daily basis by both Kelabit and Penan is from wild animals. Although in Pa' Dalih an increasing proportion of vegetable food is from plants in new-style *kebun* gardens, which are quite highly managed, much is still gathered in areas of young secondary growth after rice-farming or is from other crops planted in dry rice fields and the *ira* gardens, which are made on previously cultivated rice fields; these plants are in many ways treated as though they were wild (Janowski 1995, 2004).

The status of the highland forest as a truly wild place, independent of human intervention, is debatable. Much of the highland forest, except on high ridges, is likely to be at least to some extent anthropic and possibly anthropogenic.[4] This is both through rice cultivation and through past and present movement, manipulation, and cultivation of many other plants, both indigenous and exotic.

Despite their heavy dependence on the forest, the Apad Uat peoples of the highlands, including the Kelabit, organize their lives around the cultivation of rice. At present we have no idea how long rice has been grown in the highlands; it has been suggested that the Kelabit may belong to an originally "horticultural" complex of peoples (Sellato 1994), relying until a few hundred years ago on roots and tree crops, rather than grains. It is possible taro may have been grown in areas that are now wet rice fields, and that rice at some point, gradually or suddenly, supplanted taro as the crop of choice. Even if this is true, however, rice cultivation may be very old in the highlands; it may well have been grown in swampy areas in Borneo, including those in the highlands, before it was grown using dry methods (Janowski 2004). Nowadays, the Kelabit grow taro at the edges of wet rice fields, as well as many other crops together with rice in dry fields.

Rice cultivation, the cultivation and management of other plants, and the use of wild resources may be seen as "entangled" with each other. There is a patchy mosaic of forest types from dry rice fields to forest that may never have been farmed, intermingled with areas of secondary growth, full of species that have been planted, transplanted, or have become feral. The Kelabit, like the Penan, encourage favored wild resources. They do this in situ, leaving plants where they are and clearing undergrowth around them; they also plant and transplant wild plants into their dry rice fields, into *ira* gardens, and nowadays into *kebun*. Old rice fields are not clearly delineated from the forest, which is simply described as big or little, with no term equating to the English "virgin" forest, and no focus on the idea of forest that has not been affected by human activity. After the first year, sometimes two, in which rice is planted, a rice field becomes a space that partakes of both the domesticated and the wild. Transplanted plants and planted seeds generate a second growth forest on old rice fields. Fruit trees and bamboo groves are the longest lasting of this anthropogenic growth. They are usually planted during the first year of use and are harvested over many years once they are mature. The distribution of fruit trees and bamboo in the highlands is therefore due at least partly, and perhaps very greatly, to past human activity.

Sago palms are an important wild resource that is encouraged and managed, nowadays only by Penan, but quite possibly much more widely in the past. There are indications that the Penan care for, even manage, sago palms; the term *molong* is the Penan word for the harvesting of the palm with care, to ensure that it will regenerate (Brosius 1991; Langub 1989). As a result of management practices like these, the density of sago palms may have been affected by human activity, and the large groves of sago palm that are present in the highlands may be the result of human management. In the historical period, only the Penan, of the peoples living in and around the highlands, are known to have relied on sago starch; the Kelabit eat sago shoots but do not make starch from the trunk. However, it is likely that at some point or points in the past a large proportion of the population in the highland area, perhaps even the whole population of the highlands, may have depended partly or wholly on sago starch. Sago is a very ancient human food in Borneo; there is evidence at Niah on the coast of Borneo of sago consumption by humans before 40,000 BP (Barton 2005). There are indications that the Kelabit have relied on wild sago on long hunting trips (Harrisson 1959, 66) and that some Apad Uat peoples have relied on sago at times of rice shortages (Jayl Langub pers. comm.).

Rice-growing is also entangled with the use of wild resources in another way: it generates hunting and gathering areas by opening up areas for edible plants and by offering food to wild animals. A good deal of hunting and gathering by Pa' Dalih residents takes place in areas containing old-growth rice fields and, since the late 1800s when buffaloes were introduced to the highlands, in pastures created by their grazing. Thus, many hunting and gathering spaces are anthropic or anthropogenic, further blurring the distinction between wild and cultivated spaces.

Through *The Cultured Rainforest* project we are beginning to piece together a picture of the history of human manipulation of the environment

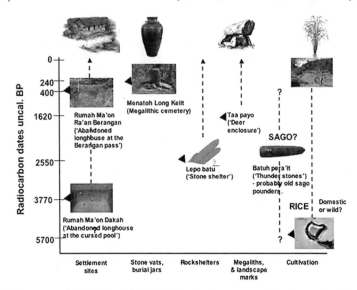

12.3. Human activity in the Kelabit Highlands: Summary of the key archaeological sites and associated radiocarbon dates from the first season of fieldwork of the *Cultured Rainforest* project. Boxes with triangles indicate radiocarbon date associated with that site. Question marks denote possible earlier, undated phases of human occupation. See Barker et al. (2008) for detailed description of all archaeological sites. Image prepared by Huw Barton.

in the highland area (see fig. 12.3). We have evidence for probable anthropogenic clearance for cultivation as early as 6,450 yrs BP[5] from an earth core at Pa' Buda,' an ancient river meander in the upper Kelapang now being used for rice cultivation. From archaeological excavation we have definite evidence of human presence and disturbance of the landscape soon after 3,770 yrs BP (Barker et al. 2008). However, no evidence of plants known to be cultivated or managed by humans has been found in the pollen or phytolith record as yet. Rice phytoliths have been found from before 6,450 BP, but these are likely to be of wild origin, as five species of wild rice are known to grow in Borneo (Gilliland 1971; Vaughan et al. 2008). Definite evidence of rice cultivation over the past 300 years, through phytolith, sediment, and microfossil identification, has been found in an earth core taken from a disused rice field at the edge of the village of Pa' Dalih (Barker et al. 2008).

That people in the highland area may have been engaged in altering the natural environment for a very long time is not a surprise. Humans have had a dynamic relationship with the forest and the landscape in Southeast Asia ever since our species entered the region. There are suggestions of anthropogenic burning in Borneo in the late Pleistocene (Anshari et al. 2004) and strong evidence of it from the mid-to-late Holocene (Anshari, Kershaw, and Van Der Kaars 2001; Hope et al. 2004). At the Niah Caves near the coast in Sarawak, hunter-gatherers appear to have been capable of identifying and removing poisonous compounds from plants such as the fruit *Pangium edule* and the yam *Dioscorea hispida* by 20,000 BP (Barker 2005; Barton 2005). This accords with the evidence for long-term manipulation in South American forests; but while this goes back perhaps 15,000 years, in Southeast Asia it may go back 50,000 years. The possibility of manipulating the environment for human gain was not invented with the arrival of Austronesian speakers in the region from about 4,000 BP, as the "Express Train" model suggests (Bellwood 2005; Diamond 1988).

This dynamic relationship points to the importance of understanding human relationships with the landscape and the environment in the region in a subtle, nuanced way. In this context, the use of the term "agriculture" in scholarship has arguably been unhelpful because it implies both radical breaks and unilinear trajectories. Both in the past and in the present, any sharp distinction between agricultural and nonagricultural peoples in Southeast Asia is arguably an artifact of human perception. The term agriculture implies a "eureka" moment when humans discovered that they could make things grow, but such a moment probably never happened. Rather, it seems likely that humans have always realized that it is possible to manipulate the natural world.

In this context, we should expect to find complex and multidirectional trajectories of change, rather than linear or evolutionary trajectories moving from a hunting and gathering way of life to an agricultural way of life. Such trajectories are likely to involve reliance at any one time on many different sources of livelihood and to be characterized by a combination of resilience and flexibility. This is indeed what we find in the Kelabit Highlands, where both Kelabit and Penan show a clear realization of the possibilities of manipulating both plants and animals.

Trajectories of change, and choices, are informed not only by physical restrictions placed by the environment. There are also cultural and cosmological reasons for choosing different portfolios of relationships with the environment. These reasons still remain to be fully elucidated, but we have some strong indications of what they may be, and that rice has played a major role.

Rice and Forest: An Imagined Divide

Despite the actual entanglement of rice and the forest, in the minds of Kelabit and Penan, there appears to be a clear divide between two broad ways of life:

one perceived as dependent on rice and one dependent on the forest. In the forest in and around the Kelabit highlands, humans are not at present, and probably were not in the past, under "food stress"; there is an abundance of food and of choice of livelihood. Hunger means not having access to the "right" food; it does not mean actually being without any food, and starvation is almost inconceivable. In this context, choices of livelihood are likely to be affected by preferences that develop for certain modes of interaction with the environment, and by the potential for "saying" certain things— socially, culturally, and cosmologically—through choice of livelihood and food. This may lead to an emphasis on certain activities and a deemphasis on others, which may not always be grounded in economic necessity or even convenience.

The growing of rice exemplifies this well. Rice in Borneo is sacred, its growing is highly ritualized, and growing it and eating it are associated with status and prestige. It appears to be associated with stratification among Borneo tribal peoples, as Sellato (1994, 212) has pointed out. This may also have been the case in the past and may explain the adoption of rice growing in the first place (Hayden 2003).

For the Kelabit, the distinction between a rice-growing way of life and a way of life that does not involve rice growing is very meaningful. Rice growing in the tropical forest is not easy, and they are quite clear about this; indeed, it is the point. If they only wanted to survive, the Kelabit are clear that they could make sago or grow root crops. Although this may not reflect reality, they believe that they and their ancestors have never relied on any starch other than rice at meals, although they grow taro, cassava, sweet potatoes, maize, Job's tears, sorghum, and millet for snack foods and, in the past, for making beer. For the Kelabit, to grow rice and to feed rice to others is equivalent to being a person of standing and status; I have suggested elsewhere that rice-growing in this context provides the basis for both kinship and hierarchy (Janowski 2007). Rice is seen as different from all other plants: while other plants grow on their own, *mulun sebulang*, rice can only grow if humans care for it. They see the cultivation of rice as initiating a particular way of living in the landscape and in the cosmos.

The Eastern Penan have been reluctant to take on rice growing and the way of "living in the landscape" that goes with it. The cultivation of rice generates a different relationship with the natural world. For the Kelabit, their relationship with plants that *mulun sebulang* is one dimension of their relationship with the natural world; it is complementary to—and arguably in tension with—rice agriculture. The cultivation of rice dominates and shapes the rhythm of Kelabit lives, and hunting and gathering are fitted into spaces left once the demands of rice are met. A rice field is carved out of the natural vegetation and represents a statement about separation from the forest. This is particularly true with the making of a wet field, since a wet field contains nothing but rice. For the Penan, until recently, it is rather their relationship with plants and animals of the forest that dominate and shape the rhythm of life, and rice cultivation is fitted into spaces that remain. For Penan, it is only the small circle around the cooking fire itself that separates them from the forest.

The special position of rice is expressed in attitudes toward rice-growing activities themselves. Not only among Penan, but also among Kelabit, rice growing is seen as a burden—as generating a kind of "world of work." Kelabit describe it as *lema'ud*, a term that has the same connotations as the Malay word *kerja*. These two words have a broadly similar meaning to the English word "work" and connote something that is opposed to pleasure or fun. Engaging in *lema'ud* is the source of status among Kelabit, simply in itself and because it makes possible rice harvests that are the foundation of *irau* feasts of merit, the "marking" of the landscape (see later) and the accumulation of heirlooms. It means being dependant on someone else for rice, which is equivalent to being a child (Janowski 1996). It is not physical necessity but cultural necessity that has driven the development of rice growing (Janowski 1988, 2004).

The effects of changes to landscape or vegetation, ranging from the large-scale changes involved in making a dry or wet rice field to the management of wild resources, are seen as creating the rights to the benefits of those effects among both Penan and Kelabit. However, where these relate to the use of resources other than rice—including planted resources—such rights are lightly enforced. Among both groups others are allowed to use a wild resource that has been marked or "assisted" by someone else, and among the Kelabit planted resources other than rice are freely shared with others (Janowski 1995). Where a rice field is made, however, rights are created that are more definite, more strictly enforced, and longer lasting.

A distinction then needs to be made between the actual physical effects that humans have on the environment and the way in which they categorize and value these effects. The effective entanglement of rice-growing and the forest discussed earlier is not recognized by the people of Pa' Dalih in attitudes to plants or animals; even though many plants are assisted, managed, or actually planted, only rice is marked as special. This is expressed in the sharing of foods: while most plants, even cultivated ones, are freely shared with others without the creation of a debt, the sharing of rice creates indebtedness and a deep sense of shame, leading eventually to dependence and ultimately enslavement.

12.4. The stone etuu near the village of Pa' Mada said to be the culture hero Tukad Rini's sharpening stone. Source: Photo, Monica Janowski.

The success associated with rice growing is commemorated on the landscape through the placing of *etuu*, or marks (figure 12.4). These include megaliths as part of cemeteries (*menatoh*) or placed to commemorate important individuals: upright stones (*batuh senupid*), carved stones (*batuh narit*), stone burial jars (*longon batuh*), stone tables (*batuh nangan*), ditches (*nabang* and *abang*), and mounds of stones (*perupun*), as well as wet rice fields themselves. The erection of some of these types of *etuu* continued until the 1950s, and took place at great feasts of merit called *irau*, involving huge expenditure of rice. The Kelabit and other Apad Uat peoples say that these *etuu* were all made by themselves, their direct ancestors or culture heroes said to be their ancestors. *Etuu* are seen as evidence of their rights over the land (Janowski and Langub, forthcoming).

For the Kelabit, rice growing places humans on one side of a kind of imaginary Great Divide, on the other side of which is the rest of nature—the forest, the mountains, all that lives "on its own." Until they adopt rice growing, the Penan are on the other side of the Great Divide (Janowski 1996). They are living in and of the forest and have made no clear and explicit division between the forest and themselves. Sago processing does not separate the Penan from the forest; it is done in the forest, and it is of the forest. There is no creation of a separate, humanized space through the harvesting of sago as there is with the growing and processing of rice.

The way in which the Penan view the Kelabit perception of a Great Divide remains to be fully investigated. They do appear to see themselves as living in a different relationship with the forest than the Kelabit, and nomadic Penan emphasize their attachment to the forest *as it is*, and do not appear to wish to alter the environment as do the Kelabit (Janowski and Langub, 2011). It may not be going too far to say that they have made a conscious choice to live on the other side of a divide they conceive of in a similar way to the Kelabit, and which they show reluctance to cross (Langub 1993).

Conclusion

We argue here that both Kelabit agriculturalists and Penan hunter-gatherers interact with the environment in a way that is manipulative, altering and managing the natural vegetation to different degrees, while also relying on purely wild resources. There is a continuum of ways in which this occurs, and use of wild resources is intertwined and entangled with the management and manipulation of other resources. However, despite this continuum, a sharp divide is imagined between rice cultivation on the one hand and all other forms of interaction with the natural environment on the other. For both the Kelabit and the Penan, rice growing is emblematic of a transition to a different way of living in the landscape. We have described these two ways of life, which coexist in close contact with each other in the area in and around the Kelabit Highlands, as separated by a conceptual Great Divide.

We postulate that the Great Divide exists more in the mind than in reality, as Kelabit, like Penan, rely on the forest for much of their subsistence. Despite this, the two groups appear to conceive of themselves as living in very different cosmological spaces, although within the same or adjacent physical space. The different choices of relationship with the natural world made by Kelabit and Penan imply different social structures and different statements about the ways in which they, as humans, choose to embed themselves in the cosmos (Janowski 2007; Janowski and Langub, 2011).

The complexity of use of the landscape in the Kelabit Highlands, the extent of reliance on wild resources, and the anthropogenic nature of the landscape have been little recognized in intellectual or public discourse. The government of Sarawak, which, like the Brooke rajas up to World War II and the colonial government until 1963, appears to share the rice orientation of its citizens, and does not legally recognize uses of the land apart from actual cultivation of rice in the establishment of Native Customary Rights. Although the Kelabit mourn, in the context of logging, for the loss of forest, this does not destroy the center of their lives: rice growing. The still-nomadic Penan, in contrast, lose their entire way of life with the loss of the forests, and have been active in mounting blockades and protests. They are gradually, however, being forced by circumstances across the Great Divide, and are settling and taking up rice cultivation. This is not historically a one-way street; there is evidence that many groups have crossed back and forth into and out of the rice-growing way of life all over Borneo (Sellato 1994). However, it is now a street up which it is hard to reverse.

13 ⁕ Residual Effects of Agroforestry Activities at Dos Hombres, a Classic Period Maya Site in Belize

DAVID L. LENTZ AND BRIAN LANE

Introduction

Scholars have long been intrigued by the ability of the ancient Maya to build vast social networks and extensive civic-ceremonial centers in the midst of the Central American tropical forest. Notwithstanding this fascination, today the interaction between the ancient Maya and the forests that surrounded them is poorly understood. The purpose of this study is to provide some insight into ancient Maya forest management strategies and observe what impact the Maya may have had on the modern forests of the Petén region. Because trees can live for many decades, even centuries, the denizens of a mature forest that has been undisturbed for a millennium may be populated by descendants from only a handful of generations. Accordingly, a fundamental hypothesis of this study is that past forest influences of the ancient Maya can be observed in the forests of today. The fieldwork for this study took place in the vicinity of Dos Hombres, an ancient Maya archaeological site located in the Rio Bravo Reserve of the Orange Walk District in northwestern Belize (figure 13.1).

The objective of this research was to evaluate phytosociological data collected from vegetation surveys in two areas of the Rio Bravo Reserve: one area located adjacent to Dos Hombres, a medium-sized Maya civic-ceremonial center, and a second located in an area that revealed no signs of human occupation at any time (the control). Dos Hombres was large enough to include plazas, temples, and palaces, all of cut stone architecture. The control area was extensively surveyed by archaeologists, but no evidence of any settlement was found there. Both survey areas were in zones labeled the same type of forest by a team of ecologists who carefully mapped the vegetation of the entire reserve (Brokaw and Mallory 1990). Thus the two study areas were located in ecologically similar zones, but the main variable was the dense occupation of one area (the land adjacent to Dos Hombres), while the control area lacked any indication of human occupation.

The dominant vegetation of northwestern Belize has been variously described as tropical rainforest (Wagner 1964), tropical semideciduous forest

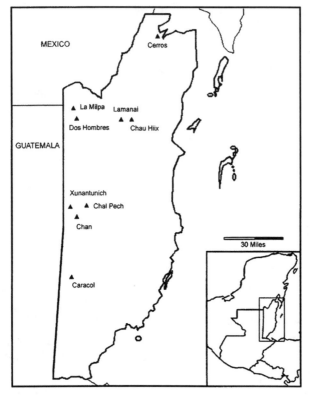

13.1. Map of Belize

(Rzedowski 1981; Greller 2000) or as quasi-rainforest (Lundell 1945). The climate of the area is decidedly tropical with an average rainfall of 1500 mm (Waddell and Royal Institute of International Affairs 1961) and temperatures sometimes as high as 38° C in the dry season, which lasts from November to May (Bolland 1977). The region has been studied botanically by various scholars beginning in the nineteenth century (Millspaugh 1895, 1903), by others in the early twentieth century (Standley and Dahlgren 1930; Lundell 1934), and more recently by Balick, Nee, and Atha (2001). In one of the first studies that took an ecosystem approach to the examination of the floristics of the region, Brokaw and Mallory (1990) created a vegetation map of the Rio Bravo Reserve area. The three major vegetation zones of the Brokaw and Mallory study were the upland forests, the lacustrine or swamp forests, and the extensive transition forests in between. The upland forests occupy the low, rolling hills with good drainage and deep, black, fertile soils. This is a common type of vegetation in the Petén, and the soils associated with these forests were highly prized by the ancient Maya for agricultural purposes. Swamp forests, often referred to as "bajos," were rooted in what many regard as less desirable agricultural land. Nevertheless, trees, especially logwood (*Haematoxylon campechianum*), from these areas were esteemed by the ancient Maya as construction material (Lentz and Hockaday 2009) and the swamp forest margins probably were of productive use for agricultural purposes. The transition zone between the swamp forests and the upland forests contain many of the forest species of both of the adjacent zones with soils that are reasonably fertile. Because it is in this zone that the occupants of Dos Hombres established their settlement, the transition zone is the focus of our study.

The human land-use history of the study area began sometime around 900 BC during the Middle Classic period (Dunning and Beach 2000). At that time, the region was sparsely populated by Maya subsistence farmers living in small, isolated communities. The first indication of urbanization and major landscape modification occurred just prior to the Classic period, from 300 BC to AD 250 (Adams 1995). During this time, much of the region's natural forest cover seems to have been extensively altered by anthropogenic processes (Campbell et al. 2006) or removed to make room for annual crop agriculture. As a result of these extensive human activities, by the Early Classic period (AD 250 to 600) much of the landscape was severely eroded (Dunning et al. 1999). Dynastic Early Classic stelae have been described from the nearby site of La Milpa, just 12 km north of Dos Hombres, but at that time, monumental architecture was still absent from Dos Hombres. During the Late Classic period (AD 700 to 900), however, Dos Hombres underwent a massive monumental building program at the site center, indicating not only an increase in social complexity, but high settlement densities as well. This population density likely resulted in extensive forest clearance (Dunning and Beach 2000), particularly in the area adjacent to the site core. Pollen records provide evidence of widespread forest removal in the region surrounding Dos Hombres during the Late Classic period. Despite indications of forest clearance on a broad scale, though, the presence of arboreal species, such as zapote (*Pouteria sapota*) and related economic species in the Sapotaceae (Dunning et al. 1999) were still observable in the pollen rain. Apparently, even in the face of widespread deforestation, at least some tree cover remained on the landscape. Following the Maya "collapse" in the ninth century, northwestern Belize was largely abandoned. The forests that once dominated the landscape gradually returned, although the imprint of the Maya occupants appears to have remained in subtle ways.

Models of Maya forest use and agricultural activities have evolved from a suggestion that the Maya used simple slash-and-burn techniques (Thompson 1954), to a reliance on orchards (Folan, Fletcher, and Kintz 1979; Lentz and Ramírez-Sosa 2002; McKillop 1994) or "artificial rainforests" (Wiseman 1978), kitchen gardens (Ball and Kelsay 1992; McAnany 1992b; Robin 1999), house-lot gardens maintained with lithic mulch (Lohse and Findlay 2000), a "managed mosiac (Fedick 1996), forest gardens (Ford and Nigh 2009), and forest resource management practiced in concentric zones of intensity from a residential core outward (Killion 1990; Kunen 2004; Lentz et al. 2002). Any of these attempts to maintain trees in the Rio Bravo area could have contributed to the continuing arboreal pollen rain that endured throughout the Classic period. Another possible explanation for the consistent presence of arboreal species in the Precolumbian pollen rain is that there may have been inherited forests or ancestor estates that were part of a land tenure system controlled by kinship as practiced by the Mam (McAnany 1998). Quite possibly, the rulers of Dos Hombres had clearly defined kinship ties and long-standing connections that may have given them control of inherited forests (McAnany 1995). The Itza maintained tended stands of forests called päk'-al, where useful trees were protected and sometimes planted (Atran 1993). Some authors (Burman 1992; Gómez Pompa, Flores, and Fernández 1990) described the phenomenon of "sacred groves," which were patches of vegetation of special species associated with a deity or, for various reasons, existed on sacred ground. The Yucatec Maya maintained forest gardens known as pet kot that were carefully managed forests where useful plants were encouraged and others were eliminated (Gómez Pompa, Flores, and Sosa 1987). Whatever the mechanism, somehow the arboreal species evident in the pollen profiles recovered from northwestern Belize were from tree species being conserved or cultivated in the face of strong demographic pressure.

After the abandonment of the Rio Bravo area by the Late Classic Maya, the next major wave of settlement in northwestern Belize arrived with the advent of Europeans in Central America. The Spaniards made infrequent forays into the region in the sixteenth century and created strongholds in Honduras to the south and Yucatan to the north, but were largely uninterested in Belize. The void eventually was filled by British woodcutters in the seventeenth century, who set up logging camps to extract the highly sought after logwood (*Haematoxylon campechianum*), which served as an essential source of textile dye in the woolen industry at the time. After decades of conflict with the Spanish, the British, using the logging camps as a justification, established a claim on the land. Belize, then known as British Honduras, eventually became a Crown Colony and retained a close association with the United Kingdom until independence in 1981.

The trade in logwood precipitously declined in the nineteenth century as the use of newly developed chemical dyes obviated the need for dyes from logwood (Thomson 2004). By this time, many of the European forests had

been denuded, and the demand for wood for shipbuilding and furniture man-
ufacture spawned a new demand for tropical woods, mahogany (*Swietenia
macrocarpa*) in particular. Other tree species, such as Santa Maria (*Calophyllum brasiliense*), banak (*Virola koschnyi*), salmwood (*Cordia alliodora*), ziricote
(*C. dodecandra*), waika (*Symphonia globulifera*), Spanish cedar (*Cedrela odorata*), cypress (*Podocarpus guatemalensis*), balsa (*Ochoma lagopus*), ceiba (*Ceiba
pentandra*), palumulato (*Astronium graveolens*), rosewood (*Dalbergia stevenso-
nii*), mayflower (*Tabebuia pentaphylla*), quamwood (*Schizolobium parahybum*),
and yemeri (*Vochysia hondurensis*), collectively referred to in the timber trade
as "lesser known species," have been only of minor commercial interest (Benya 1979; Record 1930). Initially, British lumbering operations were carried
out largely by slaves working with oxen. Individual trees of value would be
located in the forests, cut with hand axes, trimmed with handsaws, then
dragged to the river banks for eventual shipment overseas. In the twentieth
century, this laborious process was modernized with power saws and bull-
dozers, but the same extractive strategies remained in place. Lumbermen
would go into the forest, locate trees of interest, cut them down, then haul
the trimmed logs to the nearest roadway. Tropical hardwood forests were, for
the most part, only selectively cut, rather than clear-cut, so that even though
the loggers could inflict extensive damage with their trails and logging roads,
they nevertheless left the overall structure of the forest intact, with the exception that the numbers of sought after species were reduced. By the 1950s the
mahogany trade went into a permanent decline, largely because the supply
of large timbers had been exhausted (Thomson 2004) and the British government made a general decision (Downie 1959) not to replant the forests
with native hardwoods because of the long wait for the trees to mature. For
example, it takes mahogany trees forty to one hundred years to reach maturity, too long for even the most patient investor. The exploitation of pine
(*Pinus caribaea*) as a commercial forest product began in the 1940s (Dobson
1973), but this was in the Maya Mountains and the lowland coastal areas of
Belize, and did not impact the tropical forest areas of northwestern Belize.

The tropical forests of the Rio Bravo Reserve, where our research project took place, were last selectively logged in the 1950s. In the 1980s, the
Massachusetts Audubon Society and the Nature Conservancy purchased
the land from the Coca-Cola Company, the previous owner. Currently, the
Program for Belize manages the land as a nature preserve. In short, what
we know about the past land use in the study area was that it was used intensively for agriculture and forest product extraction by the Maya during
the Late Classic period, followed by logging efforts in post-Contact times.
The forests regrew during the Post Classic period after the Maya "collapse,"
and the forests were then selectively cut during historic times, probably

multiple times, with a focus on a few commercially desirable hardwoods. This provides the backdrop for the condition of the forests described in the following sections.

Materials and Methods

Forest inventories were conducted in the Rio Bravo Reserve in areas labeled transitional tropical forest in the Brokaw and Mallory study. Plots of 500 m² (50 m x 10 m) were laid out in forested areas adjacent to the Dos Hombres site and in the control area. We knew the control area was unoccupied in Maya times because it had been intensively surveyed by archaeologist Jon Lohse just prior to our study, and he found no evidence of any structures or monuments from any time period. Vegetation surveys at Dos Hombres and the control area were carried out using an identical methodology. Survey lines were established with the aid of a portable transit mounted on a tripod. The latitude and longitude of each plot were recorded with a GPS unit. All trees greater than 10 cm in diameter within each plot were measured for diameter at breast height (DBH). Total height also was calculated using a handheld clinometer. In this way, we defined 10 m wide rectangular plots that were laid end-to-end through the forest. The species of each tree was recorded along with its diameter and height. Each time we encountered a new tree species, we made herbarium voucher specimens of the plant. Because most of the trees were quite tall, vouchers were obtained using long pole pruners or by ascending the trees with the aid of climbing griffs. More than 100 collections were prepared and more than 686 individual trees were identified. Voucher specimens have been stored at the New York Botanical Garden herbarium.

Forest inventory data were analyzed using SPSS statistical software. Basic descriptive statistics for mean height and mean DBH were calculated for the control and Dos Hombres plots. Also calculated were indices of evenness and species richness, as well as the Shannon Weiner index of general diversity. We were able to determine if the tree species were of economic interest to the ancient Maya by references to published accounts of paleoethnobotanical data from Maya sites in the region as summarized in Lentz (1999). The economic species content of inventories from Dos Hombres and the control plots were compared using the Pearson Chi Square statistic with a significance level set at 0.05.

Data Analysis

Results of the forest surveys through the Dos Hombres area and the control area showed some similarities between the two, but many differences. In the

Dos Hombres survey we measured 280 individual trees from thirty-eight species and seventeen families (table 13.1). The most common tree was *Pouteria reticulata* in the Sapotaceae family, with forty-five individual trees. Other common tree species in the Dos Hombres forest were wild lime (*Trichilia minutiflora*), monkey apple (*Drypetes brownii*), copál (*Protium copal*), and sabal (*Sabal mauritiiformis*), an exceptionally useful palm. The Dos Hombres survey contained a number of trees of economic importance to the Maya, for example, breadnut or ramón (*Brosimum alicastrum*), the rubber tree (*Castilla elastica*), ceiba (*Ceiba pentandra*), baal che (*Lonchocarpus guatemalensis*), chicle (*Manilkara zapota*), zapote (*Pouteria sapota*), hogplum (*Spondias mombin*), copal, and many others. The most diverse families of trees in the Dos Hombres area were Sapotaceae, Fabaceae, Moraceae, Apocynaceae, and Sapindaceae.

The most common species in the control survey, identified ninety-three times, was boyjob (*Matayba apetala*), a tree in the Sapindaceae. Other common tree species in the control included the pi tree (*Gymnanthes lucida*), Santa Maria (*Calophyllum brasiliense*), chicle (*Manilkara sapota*), nargusta (*Terminalia amazonia*), negrito (*Simarouba glauca*), and sapotilla (*Pouteria reticulata*). Useful fruit-producing trees in the control plots included siricote (*Cordia dodecandra*), hog plum (*Spondias mombin*), wild standing fig (*Ficus americana*), wild tamarind (*Cojoba arborea*), chicle (*Manilkara sapota*), wild grape (*Coccoloba* spp.), and negrito (*Simarouba glauca*). Other economically important trees in the control survey included copál, mahogany (*Swietenia macrophylla*), and madre de cacao (*Gliricidia sepium*). Overall, the control plots contained 406 trees from thirty families and fifty-four genera. The most common families were Anacardiaceae, Clusiaceae, Euphorbiaceae, Polygonaceae, and Sapindaceae (table 13.2).

Comparative statistics, including species richness, evenness, and general diversity for the tree inventories at Dos Hombres and the control area can be found in table 13.3. At Dos Hombres, species evenness was 0.82, but only 0.76 in the control area. Species richness, however, was greater in the control plots (21.08) than in the Dos Hombres area (15.12). The Shannon Wiener Index of Biodiversity was also higher in the control area (1.325 versus 1.293 for the Dos Hombres area), indicating that the control area was slightly more diverse than the Dos Hombres area. Most significantly, as shown in table 13.4, of the individual trees found at Dos Hombres, a higher percentage (36.4 percent) was recognized to be of economic value to the ancient Maya than at the control site (28.3 percent) (Chi Square = 5.032; p = 0.025). Also, a greater percentage in the number of tree species found at Dos Hombres (55.3 percent) was used by the ancient Maya, as compared with 33.9 percent in the control area (table 13.5; Chi Square = 4.215; p = 0.04).

Table 13.1. Tree inventory from Dos Hombres vegetation survey

Taxon	Common Name	Maya Name	No. of individuals	Density: number/ha
Acacia cookie Saff.*	bull horn acacia	zu'ioin	6	10.00
Acacia dolichostachya S.F. Blake*	wild tamarind, black tamarind	jiesino	2	3.33
Acacia glomerosa Benth.*	white tamarind	kixche	1	1.67
Alseis yucatanensis Standl.	wild mammee	ha'as che	1	1.67
Aspidosperma cruentum Woodson*	red mylady, white mylady		10	16.67
Aspidosperma megalocarpon Muell. Arg *	white mylady	pech max	8	13.33
Bourreria oxyphylla Standl.	wild crabboo	bake	1	1.67
Brosimum alicastrum Sw.(Pittier) C.C. Berg*	wild breadnut, ramón	sac ox	10	16.67
Castilla elastica (Liebm.) C.C. Berg	rubber tree	hule	1	1.67
Cecropia peltata L.*	trumpet tree	ix-coch	2	3.33
Cedrela odorata L.*	Spanish cedar	culche	2	3.33
Ceiba pentandra (L.) Gaertn.*	ceiba	yaxche	1	1.67
Cojoba arborea (L.) Britton &* Rose	wild tamarind, barba-jelate, barba jalote, cola de mico, quebracha		1	1.67
Cupania belizensis Standl.*	grandy betty	chacpom	6	10.00
Diospyros sp.*			1	1.67
Drypetes brownii Standl.	male bullhoof, bullhoof macho, wild monkey apple	succoutz	30	50.00
Eugenia axillaris (Sw.) Willd.	white stopper	ich-huh	14	23.33
Ficus insipid Willd.*	red fig		1	1.67
Hirtella americana L.	pigeon plum, wild coco-plum, limoncillo		1	1.67

(continued)

Table 13.1. *continued*

Taxon	Common Name	Maya Name	No. of individuals	Density: number/ha
Lonchocarpus castilloi Standl.	black cabbage bark	man chich	1	1.67
Lonchocarpus guatemalensis Benth.	dogwood	baal che ke	4	6.67
Manilkara zapota (L.) P. Royen*	chicle, sapodillo	ya	1	1.67
Nectandra salicifolia (HBK) Nees.*	timber sweet	tzununte	3	5.00
Pouteria amygdalina (Standl.) Baehni	silly young	pacecen pacece	4	6.67
Pouteria durlandii (Standl.) Baehni	wild mammee, plantain stick, mamey cedera, zapotillo	kaniste	13	21.67
Pouteria reticulate (Engl.) Eyma	sapotilla	tzitz'ya	45	75.00
Pouteria sapota (Jacq.) H.E. Moore & Stearn *	mamey apple, mame, salute, zapote	xchaca ha'as	1	16.67
Pouteria sp.			1	16.67
Protium copal (Schltdl. & Cham.) Engl. *	copál	pom che	18	30.00
Sabal mauritiiformis (H. Karst.) Griseb. & H. Wendl.	sabal, bayleaf	xiiri	18	30.00
Sideroxylon floribundum Griseb.	cream tree	subul	3	5.00
Samira salvadorensis Standl.	johncrow redwood, high ridge redwood, palo colorado	sac te m'ooch	8	13.33
Spondias mombin L.*	hog plum	jujub	13	21.67
Stemmadenia donnell-smithii (Rose) Woodson*	horse's balls, cojeton, cojeton large, chaklakin, comulyote, huevo de caballo		10	16.67

(*continued*)

Table 13.1. *continued*

Taxon	Common Name	Maya Name	No. of individuals	Density: number/ha
Swartzia cubensis (Britton & Wilson) Standl.	bastard rosewood	buluche	1	1.67
Talisia oliviformis (Kunth) Radlk. *	kianep	va'um	1	1.67
Trichilia minutiflora Standl.	wild lime	chaltecoc	32	53.33
Trophis racemosa (L.) Urb.*	red ramon	catalox	4	6.67
			Total: 280	

* Used by the ancient Maya.

Discussion and Summary

A careful analysis of the data observed in the forest surveys reflects a broad pattern of past environmental impact. A particularly interesting set of statistics relates to the reduced levels of biodiversity and species abundance in the Dos Hombres area when compared with the control plots. Notwithstanding the lowered diversity and tree abundance in the Dos Hombres transect, this same area showed a greater evenness and significantly higher percentage of trees known to have been economically important to the ancient Maya. One possible interpretation of this observation is that the Maya were practicing some kind of agroforestry or arboriculture in the immediate area around Dos Hombres, ostensibly attempting to manage their forests to optimize the presence of economic species. They could have done this in any of the ways described earlier, or perhaps they employed a method involving selective cutting—protecting naturally occurring trees, combined with arboriculture. Most of the economically important species listed in table 13.1 are trees that are not known from the ethnographic record to have been cultivated by modern Maya. Notable exceptions are *Spondias mombin*, *Manilkara zapota*, and *Pouteria sapota*, which have all been cultivated. Thus, there may have been the cultivation of some trees combined with the intentional removal of less desirable tree species and the active protection of useful ones. The data presented suggest a model of land management whereby some form of arboriculture, or at least a vigorous form of agroforestry, was being practiced near Dos Hombres. The residential core zone in the vicinity of Dos Hombres may have been managed intensively with an array of

Table 13.2. Tree inventory from uninhabited area (control)

Taxon	Common Name	Maya Name	No. of individuals	Density: number/ha
Acacia sp.*		tzukzuk	3	4.00
Annona glabra L.*	alligator apple, bobwood, corkwood, yobapple	xmac	3	4.00
Aspidosperma cruentum Woodson*	mylady, red mylady, white mylady, milady		6	8.00
Astronium graveolens Jacq.*	glassy wood	culinzis	1	1.33
Bucida buceras L.*	bullet tree	pocte	12	16.00
Bursera simaruba (L.) Sarg. *	gumbolimbo		1	1.33
Byrsonima bucidifolia Standl.	craboo		1	1.33
Calophyllum brasiliense Cambess.	Santa Maria		28	37.33
Calyptranthes sp.			1	1.33
Cassipourea guianensis Aublet	waterwood		1	1.33
Cecropia peltata L.*	trumpet tree	ix-coch	1	1.33
Clusia lundellii Standl.	chunup, hubu, matapalo		1	1.33
Clusia rosea Jacq.	matapalo		1	1.33
Coccoloba acapulcensis Standl.			10	13.33
Coccoloba belizensis Standl.	berry tree, wild grape, bob, papa turro, uva montes, uva silvestre		3	4.00
Coccoloba cozumelensis Hemsl.	wild grape, manzanilla		18	24.00
Cojoba arborea (L.) Britton & Rose*	wild tamarind, barba-jelate, barba jalote, cola de mico, quebracha		1	1.33
Cordia dodecandra A. DC.*	chack opte, Siricote, zericote		1	1.33

Table 13.2. *continued*

Taxon	Common Name	Maya Name	No. of individuals	Density: number/ha
Cupania rufescens Triana & Planch.*	white grandy Betty, xucuroi		2	2.66
Diospyros salicifolia Willd.*	guayabillo, silion		1	1.33
Erythroxylum guatemalense Lundell	redwood, swamp redwood, wild pepper	ka-ka-ti	5	6.67
Ficus americana Aubl.*	fig, small fig, matapalo		1	1.33
Gliricidia sepium (Jacq.) Kunth ex Walp.	hotz, madre de cacao		5	6.67
Gymnanthes lucida (Sw.) Rothm.	pi	pij	48	64.00
Haematoxylon campechianum L.*	inkwood	ec	6	8.00
Hirtella americana L.	pigeon plum, wild coco-plum, limoncillo		3	4.00
Hyperbaena winzerlingii Standl.	pinch me back	ix-coch	2	2.67
Laetia thamnia L.	bullyhob, night perfume, perfume de la noche	bakelac	1	1.33
Lonchocarpus guatemalensis Benth.	dogwood	baal che ke	2	2.67
Lonchocarpus rugosus Benth.	dogwood	canasín	2	2.67
Manilkara zapota (L.) P. Royen*	chicle	ya	26	34.67
Matayba apetala (Macfad.) Radlk.	boyjob, bastard willow, mabehu, zacuayam		93	124.00
Metopium brownie (Jacq.) Urb.*	black poisonwood	chechem	23	30.67
Plumeria obtusa L. var. *sericifolia* (C. Wright ex Griseb.) Woodson	cojoton	nichte chom	2	2.67

Table 13.2. *continued*

Taxon	Common Name	Maya Name	No. of individuals	Density: number/ha
Pouteria amygdalina (Standl.) Baehni	silly young	pacecen pacece	1	1.33
Pouteria durlandii (Standl.) Baehni	wild mammee, mammy plantain stick, cedera, zapotillo	kaniste	3	4.00
Pouteria reticulate (Engl.) Eyma	sapotilla	tzitz'ya	10	13.33
Protium copal (Schltdl. & Cham.) Engl.*	copal	pom che	2	2.67
Pseudobombax ellipticum (Kunth) Dugand	mapola		1	1.33
Sabal mauritiiformis (H. Karst.) Griseb. & H. Wendl.	sabal, bayleaf	xiiri	5	6.67
Sebastiania confusa Lundell	white poison-wood, chechen blanco	iki-che	4	5.33
Sebastiania tuerckheimiana (Pax & K. Hoffm.) Lundell	poison-wood, ridge white poison wood, chemchem blanco, reventadillo		1	1.33
Senna sp.			1	1.33
Simarouba glauca DC.	negrito	pasac	14	18.67
Simira salvadorensis Standl.	high ridge redwood, johncrow redwood, palo colorado	sac te m'ooch	2	2.67
Spondias mombin L.*	hog plum	jujub	1	1.33
Swartzia cubensis (Britton & Wilson) Standl.	bastard rosewood	buluche	2	2.67
Swietenia macrophylla King	mahogany	chacalte	2	2.67
Terminalia amazonia (J.F. Gmel.) Exell *	nargusta	vanxan	23	30.67
Ternstroemia tepezapote Schltdl. & Cham.	river craboo	uixlilil-caax	1	1.33

Table 13.2. *continued*

Taxon	Common Name	Maya Name	No. of individuals	Density: number/ha
Trichilia hirta L.	red cedar	pay-huy	1	1.33
Trichospermum grewiifolium (A. Rich.) Kosterm.	balsa wood, lagroso, mahaua, moho	cha-hib	7	9.33
Vitex gaumeri Greenm.*	fiddlewood	yaxnix	1	1.33
Zuelania Guidonia (Sw.) Britton & Millsp.	waterwood	tamax	5	6.67
Zygia cognate (Schltdl.) Britton & Rose	turtle bone		1	1.33
			Total: 406	

* Used by the ancient Maya.

orchards and kitchen gardens so that a high percentage of economically valuable species were interspersed among domiciliary and other structures. A similar pattern is very much in evidence at the site of Ceren, where fruits and trees were preserved in situ as a result of a nearby volcanic eruption in AD 590 (Lentz and Ramírez-Sosa 2002). Drawing on the example of Ceren, we hypothesize that the modern forest studied around Dos Hombres is an anthropogenic relict of former land-use practices of the ancient Maya. The control area in the Rio Bravo drainage, although not a place where the Maya constructed substantial houses or other structures, was too close to densely populated areas to have been unused by the ancient Maya. In all likelihood, this land was exploited with a different management plan, perhaps with fields planted in annual crops produced in some type of intermittent fallow system or simply as a forest that was less intensively managed for the promotion of economic species.

Whatever conservation practices the Maya may have utilized in the Rio Bravo area, however, evidence from many sources (e.g., Beach et al. 2009) describes a rapid population increase during the Late Classic period coupled with widespread reduction in forest cover and concomitant land degradation, mostly in the form of erosion, that put a taxing strain on the resource base of the inhabitants. Probably working in concert with other factors, this forest degradation eventually led to a precipitous decline in the carrying capacity that may have contributed to the rapid depopulation of this area at the end of the Classic period. Although the forests eventually returned

Table 13.3. Species diversity in uninhabited forest and forest adjacent to Dos Hombres

Study Area Location	Area of Plots (ha)	No. of Individuals	No. of Species Represented	Species/ Area (ha)	Index of Species Richness 1 (d1)	Evenness Index (e)	Shannon Wiener Index of Biodiversity (H)
Uninhabited Forest	0.75	406	56	74.7	21.08	0.762	1.325
Dos Hombres	0.6	280	38	63.3	15.12	0.823	1.293

Table 13.4. Number of economic tree species found in uninhabited forest area and forest adjacent to Dos Hombres

	Location of vegetation survey	
	Uninhabited forest	Dos Hombres
No. of species used by ancient Maya	19	21
% within location of transect	33.9%	55.3%
No. of species not used by ancient Maya	37	17
% within location of transect	66.1%	44.7%
Total Number of Species	56	38

Table 13.5. Statistical analyses comparing the number of economic tree species found in each of the two study areas

	Value	df	Significance (2-sided)	Significance. (1-sided)
Pearson Chi-Square	4.215	1	.040	
Fisher's Exact Test			.056	.033
No. of cases	94			

during the Post Classic period, after the Maya "collapse," their composition and biodiversity were indelibly influenced by the land-use practices of the ancient Maya, as seen in the results presented herein. These findings support the concept that human activities, especially in areas of intensive occupation over long periods of time, tend to lower alpha diversity of surrounding habitats.

The Maya, however, were not the only people to have shaped the modern forests of northwestern Belize. Impacts of recent logging efforts also were apparent. For example, it was said that mahogany was once abundant in Belize, with four to five trees per acre in so-called virgin forests (Dobson 1973). We observed only two mahogany trees in our plots for a density of 0.6 trees per acre, a dramatic reduction. Many of the other tree species traditionally harvested in Belize were missing from our vegetation surveys, with exceptions of *Calophyllum brasiliense*, *Cordia dodecandra*, and *Ceiba pentandra*. Only *Calophyllum brasiliense*, one of the species less commonly targeted by the lumber industry, was recorded with anything resembling robust numbers. Apparently, the logging industry during the last several centuries has brought about a reduction in the abundance of desirable timber species, which has created considerable challenges for modern loggers. It is curious that the ancient Maya had the impact of increasing the number of species they found useful. This increase suggests that their attempts at conserving economic tree species were at least somewhat successful, even if their approaches to forest management were eventually overwhelmed by excessive resource demands by a burgeoning population. In contrast, the more modern approach to forest management has been largely extractive in nature, with a focus on immediate returns and less regard for long-term productivity. The overall impact on the modern forest in the Rio Bravo area, considering both ancient Maya and recent logging initiatives, has been the reduction in the number of commercial timber species, but an increase in the number of tree species considered valuable to the ancient Maya. Perhaps our modern society has something to learn from the ancient Maya in terms of sustainable agroforestry practices. Whatever the lessons we can learn from the mistakes and successes of the past, the human impact on the forests of northwestern Belize is unmistakable and serves as a reminder that even subtle landscape modifications may resonate for centuries.

14 ⁎ *Forest as Faunal Enclave: Endangerment, Ecology, and Exclusion in India*

MAHESH RANGARAJAN

Prelude to a Contest

May 2008 found extensive reportage in India about the fate of its forests and charismatic megafauna. The Sariska Tiger Reserve in the dry thorn forest of the Aravalli Hills, not so far from the capital of Delhi, was slated to get a pair of tigers air lifted in from another tiger reserve. Restoration would commence in the Sariska Tiger Reserve, the first such reserve where the species became extinct. Not to be outdone, its neighboring state, Madhya Pradesh, is readying plans to reintroduce lions to Kuno sanctuary, where they were exterminated in the nineteenth century. The ranges of mega carnivores and their forest homes have shrunk to enclaves where they are often in conflict with India's marginal peoples (Sethi 2008; Kaushik 2008; Johnsingh 2006).

What forests mean has perhaps never been as contentious. The new millennium saw significant initiatives in forest restoration. In 2005 a Tiger Task Force attempted a synthesis of science, conservation, and citizenship (Ministry of Environment and Forests 2005). The Forest Rights Act of 2006 sought to defuse insurgency fueled partly by stringent forest laws. The policy attempted to secure islands of forest in a sea of humanity, but while tempering the excesses of an overly bureaucratic past (Aiyar 2003).

Faunal fortunes and forest futures have never been far from politics. Since the 1870s, the Forest Department, then a creation of the Crown, has been a key player. Owner of more than a fifth of the land mass, it controls land and labor, deploys capital and cuts timber, auctions contracts and permits clearance, protects wildlife, and executes civil works. Even if sections of forest are to be faunal enclave, there is debate regarding who ought to demarcate them. It is difficult to agree on what it ought to be: nature reserve or living space, tourist space or grazing ground, mine or jungle. It is difficult to agree on which agency to trust: the Forest Department or the judiciary (Salve 2008).[1] Judicial pronouncements have even altered what is defined as forest (Lele 2007). Different sections view the forest in mutually contradictory terms. An articulate middle class constituency has been mobilized, and urban school students are urging Mogiya Scheduled Tribals to move out of tiger habitat (Sahgal and Thiyagarajan 2008; Singh 2008; Bindra 2008).[2] Armed forest guards with the status of the police will halt tiger poaching (Thapar 2006; Terborgh 1999). The Campaign for Survival and Dignity argued that wildlife preservationists:

> Use their class and institutional power against these communities rather than alongside them. Conservation has not succeeded and can never succeed if it is based on an authoritarian, repressive model inherited from an Empire. (Campaign for Survival and Dignity 2007)[3]

The divide is sharp: preserve nature or ensure rights. A wildlife enthusiast saw it as a "dangerous piece of legislation" (Thapar 2006; Guha 2006).

Seeking middle ground are groups that stop short of either control and command regimes over the forest or complete devolution (Madhusudan 2005). Others see a more cooperative model with devolution as ideal (Ramnath 2008). There are also signs of a dialogue across these divides.[4] The study of nature often treats humans as a uniform category, while those who study society rarely see nature as dynamic entity (Worster 1996). Only a dialogue across boundaries can provide a perspective on what ails the forest (Gadgil and Guha 2000; Madhusudan and Raman 2003).

Of Hunt and History

In order to move ahead, it helps to ask how we got here. The millennia of history of human interactions with forests in India do not suggest any equilibrium in the past. Successive rulers tried to push back the forest line to augment revenues and enhance control of the countryside. Sanskrit texts and Persian records often refer to the forests as untamed wild spaces with dangerous beasts and men. Yet relations were complex. Animals could be signifiers of the forest landscape, with Sanskrit pharmacopeias placing the lion at the apex in the *aranya* or forest and man of *grama* or village (Smith 1991). Marshes were labeled *anupa*, and the dry cereal-bearing land as *jangala*. The latter had black antelope; the former the geese and elephants (Zimmermann 1987). Still, the forest was not simply a polar opposite of cultivated space or of the city. The lion symbolized power and the elephant intelligence. In classical Tamil literature, landscapes could be represented by trees, flowers, or animals (Valmiki 1984; Ramanujan and Tolkappiyar 1985; Thapar 2001).

The immensity of the monsoon forest in premodern times is difficult to comprehend. Islands of cultivated arable stood out in a vast backdrop of woodlands.[5] Even four centuries ago forests were "appreciably more extensive," stretching via the Kaimur range close to the river Ganga (De-loche 1993). The forest line shifted due to war and disease, with land under cultivation reverting to jungle. Mughal portraiture depicts a north Indian landscape of antelope herds hunted by cheetahs and lions. Conflicts with cattle owners were common (Divyabhanusinh 2005; Koch 1998). Imperial power was asserted in large-scale hunts, a huge area being encircled with nets. "If anyone, whether native or traveler had the misfortune to wander into the corral, he could be taken into slavery or slain just like one of the wild animals" (Schimmel, Waghmar, and Atwood 2004, 202). Mature tree forest was important for elephants, and their extensive range four centuries ago in central India indicates a moist climate and dense vegetation (Suku-mar 2003; Trautmann 1982; Digby 1971). Mughal rulers controlled supply sources of this animal that was vital to warfare (Habib 1982a).

Power, production, and cultures had much to do with the forest as an en-tity. In dry western India, rulers controlled the use of scarce grass and browse for war horses. But the wild could also signify danger (Kumar 2005; Guha 2002; Eaton 1993; Richards 2003). From ancient times kingly hunts were a reaffirmation of dominance over the wild landscapes (Falk 1973). Rulers often entered into ritual arrangements with forest peoples. Spring hunts in western India were dedicated to the goddess Gowri: the hunter was a devotee ready to face danger (Pal 1978). In southern India, the rulers' power in Pudu-kottai was derived as much from Vedans (forest peoples), who helped obtain game meats and forest fruits, as by patronage of priestly scribes (Brahmans) (Waghorne 1994). The Pudukottai rules took a stern view of trespassing hunt-ers but their reach had its limits (Guha 1999).

There was no clear linear movement from forest to farm and city land-scapes. The continuum of jungle and arable was the norm not the excep-tion,[6] with lines not always sharply etched on the ground between the woods, pastures, and fields.

ADMIRABLE EDIFICE OR FESTERING SORE? (WILLIAMS 2003)

British rule in India unleashed changes with no precedent. Theirs was a more absolute notion of property, with a harsher view of itinerant groups on the agrarian fringe. Mobile peoples and wild animals were seen as marauders to be eliminated or bound down. Initially, the emphasis was on binding down the one and eliminating the other (Bayly 1990). As for-ests became more valued for timber there were concerns about the con-sequences of denudation. Prior to this, however, the creation of *cordon sanitaire* enabled a spread of merchant capital into the jungle. Timber for the railway was a more pressing factor by the 1860s and 1870s. By 1904, there was 600,000 square kilometers of government forest, the most extensive on Earth (Guha 1989; Grove 1995; Cederlöf and Sivaramakrishnan 2006).

The impositions designed to remake the forest from floor to canopy to suit imperial aims unfolded through Working Plans from the mid-1870s onwards. Rules were framed to ban fires and grazing, lopping and felling, to curb swidden cultivation and snare setting. The woods were mapped for their capacity to generate more stock per acre. British Indian forestry consisted of control over when, where, and how labor was used vis-à-vis wood, fish, and fauna (Prasad 2003). The wooded estate was too vast to permit plantations except in certain tracts. The regime aimed to hasten regeneration of *sal* and *shisham*, teak and deodar. As a leading social reformer wrote in the late nineteenth century in western India, "In the past those farmers who had very little land and could not survive on this produce went into the forest and would gather wood, fruits and leaves," but now, the Forest Department had taken control of "all mountains and hills and valleys" (Deshpande 2002, 132). The impact was uneven but it was doubtless a critical watershed in ecological and economic terms. For a sample of regional histories see Saikia (2005), Saberwal (1999), Sivaramakrishnan (1999), and Bhargava (1999).

Interestingly, the same writer also wanted farmers' crops protected from wild boars and damages compensated from salaries of officials (Deshpande 2002)! The protection of game and elimination of vermin ran as parallel programs. This reclassification of animals as benign (those that were eaten) and those that ate human flesh was to shape official policy. From the mid-1870s onwards, more than 1,500 tigers, 3,000 leopards, and 4,000 wolves were killed in British India and a few princely states *every* year for bounties (Rangarajan 2001a; Chakrabarti 2006). There was an inherent contradiction in such measures: carnivores helped to control herbivores, which raided crops. Further, the very groups blamed for depleting ungulates with traps were critical for the success of such measures. The impact of such killing was uneven. Even large hunts in the savanna of the *tarai* could not dent tiger numbers. Elsewhere, and especially where cattle raids were common, depletion was for real (Sankhala 1977).

In Assam's hills in India's tropical northeast, forest-dwelling villagers and officials exterminated the Sumatran rhino; the price of its horn being an incentive (Milroy 1934). Agrarian expansion could also obliterate habitat, though the boundary lines of government forest halted the plough, keeping the forest intact. The period of about a half century ending in the late 1920s was decisive for large taxa. By 1914, tigers had vanished from the Indus basin. Keystone species that vanished from much of their habitat included the wild buffalo, swamp deer, and Javan rhinos from the Sundarbans.

But it was also the start of a protective umbrella. Fears of extinction led to protection of fauna specified in Shooting Blocks, protecting them from overkill even by elite sportsmen; Kaziranga was set aside as a refuge for the nearly extinct rhino in the 1900s.

This shift was more than mere response to decline. Exploitation and conservation went hand in hand (Beinart and Hughes 2007). Elimination of carnivores led to huge spaces of empty forest where they were like "ghost species" (Meyer 2006). Changes in the fire regimes could bring about huge ecological shifts. Swidden cutters, seen as destructive of timber, were often evicted from forested zones. This eviction was despite the fact that in places like the Banjar Valley (now in the Kanha Tiger Reserve) such practices had often opened up land for wild ungulates. "Firing" was restored, though under the watchful eye of foresters (Rangarajan 1996). The forest was transformed into an intensively managed landscape, though the tug of war between rival users continued (Morris 1937).

PRINCES AND COERCIVE CONSERVATION

A third of the land mass and a fourth of the people in British India were distributed among approximately 500 princely states. The value of game species as trophies for princes led to their protection, with some princes even asserting exclusive rights over tiger hunting (Prater 1940). Yesterday's vermin became today's game.

The easing of pressure on carnivores went so far as to include near total protection of lions in their last forest home. The protection was not absolute: they were safe unless shot by a very privileged few. When lions lifted cattle in the Gir Forest, the ruler compensated losses in cash, perhaps the first scheme of its kind in the world (Rangarajan 2001a). A few princes even tried restoration: Gwalior imported lions from the Sudan. Colonel Kesri Singh later recalled how the 1,000 square mile forests (some 2,600 sq km) of Kuno contained a few tribal hamlets. The lions were shot once they turned on cattle and people (Singh 1969). Tiger reintroduction succeeded in Dungarpur, Rajasthan. But the warrior princes zoned the forests, with certain tracts for tiger hunts and others open to cattle (Ranjitsinh 1997; Divyabhanusinh 2006). Reintroduction was rare. More common was a renaturing of the wild to attract animals from adjacent territories. Bhavnagar's princes controlled grazing and dug water holes in the Mytiala hills, attracting lions from Junagadh's Gir Forest (Dharamkumarsinh 1978).

Large-scale princely hunts were still the norm (Hughes 2008). Exclusion could be brutally enforced, as commoners of that era would later describe. "It was a hated past," they said, "of conscripted labor and brutal suppression of human freedoms. With the former goes a dense wood where herbs, berries and wildlife flourishes; with the latter goes a dense wood that shelters rampaging pigs" (Gold and Gujar 2003, 71). The hierarchies in the world of men found expression in who could impose their vision on the natural world.[7]

Contesting India's Forests

After 1947 and independence, the imperative to modernize the Indian economy overcame interest in the forest. Even in the late nineteenth century, half of north India was covered with mature tree forest, savanna, and secondary growth (Richards, Hagen, and Haynes 2008). Given stagnant agricultural yields in the early twentieth century, expanding the area under the plough was given primacy. By the latter half of the twentieth century, demographic growth exerted pressures on forests on a subcontinental scale. Populations were slow growing, with the density in 1881 being about 80 to a square kilometer, more than twice that of 1600 (Guha 2001). A free-for-all prevailed until the end of the 1960s. Still, the consolidation of forests formerly controlled by the princes and aristocracy, now under an independent central government, created a unified jurisdiction of India's forests for the first time. (On wildlife, see Shahi 1977; on forests, see Chambers, Saxena, and Shah 1989, 198.) Industry, mining, and hydropower made major inroads into the forest. Crop protection guns helped cultivators to control deer and boars. The clearing of the *tarai* (Singh 1993; for the Nepal side, see Mishra and Ottaway 2008) landscape in north India was aided by new pesticides (Farmer 1974). As one senior official found, "Untamed nature disorderly, chaotic and wayward. Man has been constantly battling his environment. He constantly fights to produce a semblance of order" (Randhawa 1965, 3). To lay claim to being a Divisional Forest Officer, it was essential to prove one's worth as a good shot. Accuracy with the gun was a measure of a forester's skill (Sankhala 1977). But the forest was mainly seen as timber manufactory and as potential arable land. Multispecies forests of more than half million hectares made way for exotic monocultures to provide pulp (Savur 2003), though the personal interest of India's new rulers protected some small wildlife refuges.[8]

A sea change only came about at the end of the 1960s, as ecological patriotism came of age. Prime Minister Indira Gandhi (1966 to 1977 and 1980 to 1984) made assertive nationalism a major plank of her politics. Nature as emblem had a symbolic importance well beyond material gain. The tiger became a symbol of an Indian ecological nationalism. Saving it mattered as much to the rulers as wiping it out had been a priority for their imperial predecessors a century earlier. She even shelved a major hydel project to save a rainforest (for a similar case see Robin 1998, 134, 138; on Silent Valley, see Rangarajan

2006). The unfolding of conservation policy has to be placed in the larger political context. There were signs of change even earlier but it was only under Indira Gandhi that these ideas acquired force in government policy. Her son and successor Rajiv continued a similar set of policies from 1984 to 1989, after which India has had a succession of coalition or minority governments. The major legislative initiatives were undertaken when the ruling party had massive parliamentary majorities and controlled most, though not all, states.

Vertebrate nationalism was a common thread through the tiger crises of the late 1960s, the early 1990s, and the mid-2000s. The tiger was symbol of preservation, to be saved at all costs. India of the period from 1969 to 1989 was a far more centralized polity than it was before or since. At the same time, high rates of growth of cereal production helped ease up the pressure to extend agriculture.[9] Nature would renew itself once rival users were shut out of the forest (Greenough 2003). Tiger numbers were the index of success, and the tiger census of 1972 marked its transformation into conservation icon. As in princely hunting reserves, tiger reserves had core areas with no human settlement. Buffer zones permitted grazing and cultivation (Rangarajan 1996; Sanghal 2008). A similar plan was under way for the lions of Gir Forest (Choudhury 1970; Dharamkumarsinh 1959; Wynter-Blyth 1956). Pugmark counts were used to assess numbers as foresters imbibed the game crafts of princely India. The use of these princely game crafts was to be a leitmotif, even at a time when wildlife biology made impressive strides in India and elsewhere (Lewis 2003). Authoritarian solutions to the crisis worked in a limited sense, with the area under protection growing tenfold in just over three decades, to over 150,000 square kilometers (Rodgers and Panwar 1988). Even critics attested to the recovery of habitats and ecosystems (Karanth 2001; Karanth et al. 2004; Thapar et al. 2003). A veteran naturalist has recorded how deer had been more trusting of humans in a princely reserve free from logging (Krishnan 1959).[10] Once all hunts ended a few years later, even tigers became more diurnal (Thapar and Rathore 1999). Enclosure, the stricter the better, seemed the answer to endangerment (National Archives of India 1933; International Conference on the Protection of Fauna and Flora of Africa 1933; Neginhal 1974).

Conservation at a Crossroads

Seclusion was thus the lynchpin of strategies of ecological recovery. But by the end of the twentieth century, populist and radical critics of the official policies had gathered more force. While welcome, official policies still have a long way to go (TPCG and Kalpavkriksh 2005; Kothari et al. 2000). Forest conflicts center on two issues: eviction and exclusion.

Eviction is often the preferred first response of conservation planners (Johnsingh, Sankar, and Mukherjee 1997; Rangarajan and Shahabuddin 2006). It is not easy to agree when displacement is coercive and when it is voluntary (Padmanabhan 1998; Choudhary 2000). For foresters, displacement is a solution, but for those displaced it is the beginning of new challenges (Panwar 1978). In 1978, forcible eviction of refugees from the Sundarbans Tiger Reserve, West Bengal, led to massacre (Economic and Political Weekly Correspondents 1985; Vardhan 1978). Such violent displacement would be difficult to contemplate today (Jalais 2005; Mallick 1999).[11] The aftermath of displacement, however, can be highly variable. Displaced by a new wildlife sanctuary in Kuno, central India, Sahariya Scheduled Tribals grapple with debt (Kabra 2003). In contrast, evacuees from a southern tiger reserve, Bhadra, have gained the title to irrigated fertile lands. The conservation outcomes are equally diverse and contentious (Madhusudan and Arthur 2007). On the whole, the immediacy of conflict between carnivores and cattle does recede as human settlements move out (Karanth 2007; Madhusudan and Mishra 2003). In Bhadra, recovery rates of vegetation and wild animals have been monitored and show steady progress. The Tiger Task Force Report of 2005 suggested careful identification of vital wildlife habitat and more transparent offers to evacuees (Ghate and Beazley 2007).

Exclusion remains more widespread than eviction in Indian forests. Conservation does not always work as a consequence. Landed upper strata prey on the forest wealth, while laboring poor barely survive on what they glean from the jungle (Robbins 2000). Preservation can also align with dominant landed castes and service elites in towns and villages. The drive against tiger poaching often becomes a campaign against specific nomadic tribal groups (Radhakrishna 2008; Kabra 2006), whose space in the Denotified Tribes Campaign runs counter to middle-class ardor for endangered animals (Dutt, Koleta, and Hoshing 2007). Such peoples are not always ecologically benign, but threats to the vegetal complex from beyond park boundaries can be far more intense than those within. Yet, the overwhelming focus of conservation is on relocating villages from the forest's bounds or on shutting off access. Soft targets are easier to pick, and greater threats are more insulated from regulation (Shahabuddin, Kumar, and Shrivastava 2007; Nagendra, Pareeth, and Ghate 2006).

DOES FAUNA HAVE A FUTURE?

It has been held that local support is *not* essential, as conservation via *diktat* can flourish in highly nonegalitarian societies. Conservation by fiat has a long history beyond India (Brockington 2003; Schama 1995; Spence 1999).

But in recent years, a growing awareness of rights has emerged as a check on arbitrary authority in Indian parks (Chhatre and Saberwal 2006). Such trends are still nascent but hold promise.[12] Much of peninsular India has room for broad-based initiatives that combine protection with welfare. This is more intractable in central and western India with its deeper disparities. *Can* the forest survive as faunal enclave? Probably, but only if the wider circumstances change for the better (Madhusudan et al. 2006). Secluded zones are *essential* for species with fragile life cycles, but larger landscapes with coexistence are equally essential.[13] The challenge will be to reorder the forest separating and recombining conservation with livelihoods. This will enable the forest *with* its fauna to survive.

15 ✳ Amazonia: The Historical Ecology of a Domesticated Landscape

CLARK L. ERICKSON

Introduction

The classic literature on Amazonia presents the past and present cultures of the region as determined largely by the environment to which they adapted. What appears to be a lush, bountiful setting for human development is actually a counterfeit paradise, according to some scholars (e.g., Meggers 1971). Environmental limitations, such as poor soils and a lack of protein resources, combined with limited technologies, few domestic animals, and abundant, unoccupied land restricted social development. The simple societies of Amazonia did not evolve into what we recognize as civilization. In this traditional view, the environment as an immutable given or a fixed entity to which human societies adapt (or do not, and thus fail and disappear).

Historical ecology provides a radical alternative perspective for understanding human-environment interaction over the long term and the complex human histories of Amazonian environments. Historical ecology focuses on *landscape* as the medium created by human agents through their interaction with the environment. Although landscapes can be the result of unintentional activities, historical ecologists focus on the intentional actions of people and the logic of indigenous knowledge, particularly the understanding of resource creation and management. Historical ecologists, borrowing from the new ecology, argue that disturbance caused by human activities is a key factor in shaping biodiversity and environmental health. Because much of human-environmental history extends beyond written records, the archaeology of landscapes plays an important role. Through the physical signatures or footprints of human activities, technology, engineering, and knowledge embedded in the landscape, historical ecologists have a perspective of more than 11,000 years regarding human-environment interaction in Amazonia.

What Amazonian people did to their environment was a form of domestication of landscape (Erickson 2006). Domesticated landscapes are the result of careful resource creation and management with implications for the diversity, distribution, and availability of species. Through long-term historical transformation of the environment involving transplanting of plants and animals, selective culling of noneconomic species and encouragement of useful species, burning, settlement, farming, agroforestry (forest management), and other activities discussed in this chapter, humans created what we recognize and appreciate as *nature* in Amazonia. Through the perspective of historical ecology, we see that nature in Amazonia more closely resembles a garden than a pristine, natural wilderness. Rather than "adapt to" or be "limited by" the Amazonian environment, humans created, transformed, and managed cultural or anthropogenic (human-made) landscapes that suited their purposes. The cultural or anthropogenic landscapes range from the subtle (often confused with "natural" or "pristine") to completely engineered. In this chapter, employing the perspective of historical ecology, I survey examples of human activities that have created, transformed, and managed environments and their association to biodiversity.

Amazonia: Wilderness or Cultural Landscape?

Amazonia-as-wilderness is an example of the *myth of the pristine environment* (Denevan 1992a), the belief that the environments of the Americas were relatively untouched by humans prior to European conquest. Native people are believed to have been too few in number, technologically limited, or living harmoniously with the Earth to significantly change nature. The assumption also reflects the *myth of the noble savage* (or Ecological Indian)—that past and present native people lived in harmony with nature until Europeans and modern world systems arrived (Redford 1993).

Archaeologists, however, have demonstrated that prior to the arrival of Europeans, much of Amazonia was occupied by dense populations practicing intensive agriculture and urbanized societies that significantly contributed to creating the environment that is appreciated today (Denevan 1992a; Erickson 2006; Heckenberger 2005; Lehmann 2003; Stahl 1996). Scholars now argue that much of the tropical rainforest is the result of a "rebound effect" created by the removal of these people and their activities by European diseases, civil wars, ethnocide, slavery, and resource expropriation.

Contrary to popular notions, Amazonia is diverse in environments and was probably more so in the past. While rainforest covers approximately one third of the region, the majority of Amazonia is deciduous forest, palm forest, liana forest, forest island, savanna, and wetland (Goulding and Barthem 2003; Moran 1993; Smith 1999). In addition, historical ecologists argue that much of Amazonia's diverse ecological patchwork of diverse habitats is anthropogenic and historical (Posey and Balée 1989; Balée and Erickson 2006a). Archaeologists have shown that before the native population collapse after 1492, much of Amazonia was transformed by burning, settlement, roads, ag-

riculture, and agroforestry into forest clearings, savannas, parkland, countryside, and forest islands (Denevan 1992a, 2001; Erickson 2006; Heckenberger 2005; Heckenberger et al. 2003; Posey 2004; Stahl 2006). Amazonia
had fewer trees 500 years ago, and the existing forests were more similar to
gardens, orchards, and game preserves than wilderness.

Amazonia: Counterfeit Paradise or Anthropogenic Cornucopia?

Environmental determinism has a long history in anthropological studies
since the nineteenth century. In Amazonia, the main spokesperson of environmental determinism, Betty Meggers (1954, 1971, 2001), explained the
presence of simple societies and relatively nomadic lifeways of Amazonian
people in historical and ethnographic accounts as evidence of environmental limitations imposed on human cultural development. The poor quality of
tropical soils is said to have restricted agriculture to simple systems such as
slash-and-burn (swidden) (Carneiro and Wallace 1960; Meggers 1971). Adopt-
ing the idea from natural scientists and developers that the lush, rich vegetation of the tropical forests is actually a fragile ecosystem growing on poor soils,
Meggers (1971) coined the term *counterfeit paradise* to describe Amazonia.

Swidden, the most common traditional agriculture today, was assumed
to support low population densities. Without large populations, surplus to
support nonfarmers, class stratification, and cities, Amazonia could never
develop civilization. Environmental determinists also pointed to primitive
technology (the wooden digging stick, stone ax, and wooden machete) as a
reason for simple agriculture. Others examined the lack of animal protein as
an environmental limitation, proposing that the availability of protein determined settlement, population density, and inter- and intra-societal relationships in Amazonia (Gross 1975). Unlike societies in the Old World, Amazonian people had few domesticated animals to provide reliable protein;
thus, they were assumed to have relied on unpredictable hunting of easily
overexploited wild animals. Based on ethnographic cases, scholars argued
that settlement sizes, duration, and regional patterns could be explained by
the lack of protein. In more extreme interpretations, Amazonian patterns of
warfare, settlement spacing, and mobility were explained by fierce competition over limited hunting resources (Chagnon and Hames 1979).

Meggers (1979, 1995, 2001) proposed catastrophic climate change as another element of environmental determinism to explain periodic settlement
abandonment and changes in pottery styles in the archaeological record.
She hypothesized that cycles of mega-El Niño events throughout prehistory
caused severe and extended floods and droughts that caused frequent societal collapse, encouraged nomadic patterns of settlement, and limited social

development. Recent El Niño events have caused droughts and flooding in Amazonia, often resulting in large forest fires that have been exacerbated by uncontrolled development of the region. Pre-Columbian societies faced similar challenges and survived. However, the evidence presented for catastrophic climate change by mega-El Niños and its impact on humans has been challenged (e.g., DeBoer, Kintigh, and Rostoker 1996; Erickson and Balée 2006; Stahl 1991; Whitten 1979).

Few contemporary scholars support environmental determinism. In the 1960s, scholars documented intensive agriculture in pre-Columbian Amazonia including house gardens, river levee farming, raised fields, terraces, Amazonian Dark Earth (ADE), and anthropogenic forest islands (Denevan 2001; Denevan and Padoch 1987; Langstroth 1996; Lathrap 1970, 1987; Lathrap, Gebhart-Sayer, and Mester 1985; Posey 2004). Archaeologists and geographers highlighted the potential of farming river levees and banks when floods recede (Hiraoka 1985; Smith 1999). Raised fields, terracing, and ADE (discussed later) are capable of continuous high yields and are associated with dense populations, large permanent settlements, and complex society (Denevan 2001; Erickson 2006; Lehmann 2003; Neves and Petersen 2006; Valdez 2006; Walker 2004). These strategies take advantage of patches of naturally fertile soil and technologies of soil creation, transformation, and management and negate environmental determinism. Since swidden agriculture depends on metal axes and machetes to efficiently clear mature forest, pre-Columbian farmers, using digging sticks and stone axes, probably continuously cultivated fields and practiced agroforestry, rather than clearing mature forest. Research has also documented that swidden agriculture is often far more productive per area than has previously been assumed.

Scholars have also noted that most groups studied as examples of protein limitation live inland, far from major water bodies and fish. In fact, Amazonian people were primarily riverine and relied on fish and other aquatic resources rather than game animals as the main source of protein (Beckerman 1979). In addition to rivers and lakes, fish were systematically harvested in large numbers using networks of fish weirs (Erickson 2000a). Furthermore, sources of protein included maize (Lathrap 1987; Roosevelt 1991) nuts, fruits, and insects common in the humanized forests (Beckerman 1979; Clement 2006).

Elements of a Domesticated Landscape

Evidence of landscape creation, transformation, and management of domesticated, engineered, humanized landscapes in Amazonia includes anthropogenic burning, settlements and associated landscapes, mounds, anthropogenic forest islands, ring ditch sites, ADE, raised fields, transportation and communication networks, and water management, fisheries management, and agroforestry.

ANTHROPOGENIC BURNING

Fire is the oldest and most powerful technology of environmental creation, transformation, and management available to native people. For most natural scientists and conservationists, fires caused by humans are considered a threat to Amazonian rainforests and biodiversity. Complex fire histories documented in lake sediment cores, soil stratigraphy, and archaeological sites suggest that humans regularly burned Amazonia in the past (Oliveira and Marquis 2002; Lehmann 2003; Sanford et al. 1985). Anthropogenic fires are distinguished from natural fires by their regularity, context, timing, and patterns (Pyne 1998).

Hunters and gatherers burn landscapes to attract browsing game, clear the understory for easier movement and harvesting of wild plants, encourage economic species attracted to light gaps and disturbance, and hunt game through cooperative drives employing fire and smoke. Farmers employ burning to clear and prepare fields, gardens, orchards, and settlements, fertilize fields, incinerate garbage, and reduce bothersome insects (Pyne 1998). Burning and the production of charcoal is a key element in the formation of ADE (discussed later). Most scholars now agree that fire plays a key role in the creation and maintenance of Amazonian environments, in particular the savannas and dry deciduous forests that cover much of Amazonia (Langstroth 1996; Oliveira and Marquis 2002).

SETTLEMENT AND ASSOCIATED LANDSCAPE

Human settlements may be one of the most persistent and permanent transformations of the Amazonian environment. Scholars have recorded a wide variety of settlement types and regional settlement patterns for past and present Amazonian people (Denevan 1996; Durán and Bracco 2000; Erickson 2003; Heckenberger 2005; Neves and Petersen 2006; Roosevelt 1991; Wüst and Barreto 1999). While most settlements were small (less than 1 ha), the archaeological site under the present day city of Santarem in Brazil covers 4 km^2 and the Faldas de Sangay site in Ecuador is possibly 12 km^2 (Roosevelt 1999). Traditional communities included some that had large, open, clean central plazas and streets along which houses were arranged in linear, grid, radial, or ring patterns.

A typical house in an Amazonian indigenous community is a simple example of resource use and local landscape transformation. The foundation requires four to six upright wooden posts plus additional beams (each representing a tree). Earthen floors are often raised 10 to 20 cm for drainage during the wet season (1.5 – 3.0 m³ for a 3 × 5 m house). Thick layers of palm and grass thatch cover the roof. A typical Pumé community in Venezuela would require 13,498 fronds of palm, which are replaced every two to three years, and 750,000 fronds from 125,000 palms for a large communal house of the Bari who also live in Venezuela (Gragson 1992). Vegetation around the house is cleared to bare ground for protection against snakes and for aesthetic reasons. A small but densely packed house garden is established for production of a variety of plant species and is also a compost pile for kitchen waste. In humid tropical regions, houses last five to ten years. In summary, the environmental impact of a single house is profound: rearranging and altering soils, accumulation of organic matter through garbage and human wastes, deforestation and opening of forest canopy, cutting of construction and roofing materials, replacement of natural vegetation with economic garden, crop, and orchard species, and mixing of the soil horizons. Denevan (2001) estimated a pre-European conquest native population of 6.8 million for Amazonia. Assuming five people per household, some 1,360,000 houses were required at any time. The environmental impact described above for a single household is now multiplied by more than one million houses across the landscape.

House gardens were associated with individual residences and larger clearings for staple crops in the forest, with raised fields in savannas and wetlands, or on exposed river banks beyond the settlement. Stream channels and wetlands were criss-crossed with fish weirs (corrals for harvesting fish). Any standing forest within a 5 km radius was a managed forest. Pathways were hacked through the forest and roads within settlements were often raised or defined by earthen berms and other infrastructure. In the savannas, large earthen causeways with adjacent canals served as roads and canoe paths. In addition, each settlement required firewood, game, fish, and other wild resources in quantity.

A community's permanent transformation of the environment for these basic needs and infrastructure is staggering. As a result, the forested environments that are typical today were scarce in the past and of a much different character. Based on the archaeology, these communities were stable, long-lived, and sustainable despite this impact.

MOUNDS

Many Amazonian cultures were impressive mound builders (Denevan 1966; Durán and Bracco 2000; Erickson and Balée 2006). Farmers built mounds in the Llanos de Mojos of Bolivia, Marajo Island, and the lower and central

Amazon basin and Pantanal of Brazil, the Llanos de Venezuela, Mompos basin of Colombia, Sangay in the Upano Valley and Guayas Basin of Ecuador, and the coastal plains of Guyana, Brazil, Uruguay, and Ecuador. Mounds were constructed of earth, with the exception of the *sambaquis* of coastal Brazil, which are primarily of shell. Excavations show that many mounds served multiple functions, often simultaneously. Mounds generally contain fill or layers of domestic debris (bones, shell, and other organic food remains, pottery, and stone tools) typical of settlements. Some mounds have such a high percentage of broken pottery that scholars apply the term "potsherd soils" (Langstroth 1996). Mounds were formed over considerable time through the collapse and leveling of wattle and daub buildings, accumulation of refuse and construction debris, and the intentional addition of fill from adjacent large borrow pits, often filled with water. Mounds in the Llanos de Mojos and on Marajo Island contain hundreds of human burials in which a large pottery urn with lid was used for a coffin (Nordenskiöld 1913; Roosevelt 1991). Other mounds were used as chiefly residences or ceremonial centers (Rostain 1999; López 2001).

Although most are small, the Ibibate Mound Complex in the Bolivian Amazon covers 11 ha and is 18 m tall with more than 250,000 m³ of fill (Erickson and Balée 2006). Mounds are often found in groups of up to 40 for Marajo Island (Roosevelt 1991), and more than 50 mounds for the Huapula site (Rostain 1999). Mound construction required mass movement of soils, transformation of local topography, soil enrichment, and change in vegetation composition. Our study of the Ibibate Mound Complex in the Bolivian Amazon demonstrates that the biodiversity on the mounds was significantly richer than that of the surrounding landscape and consists primarily of economic species, some 400 years after abandonment as a settlement (Erickson and Balée 2006).

ANTHROPOGENIC FOREST ISLANDS

Forest islands, ranging in size from a few hectares to many square kilometers, are common throughout the savannas and wetlands of Amazonia (fig. 15.1). Most are raised less than one meter and often surrounded by ponds or a moat-like ditch. Excavations in forest islands in the Llanos de Mojos and Pantanal document their anthropogenic origins and use for settlement, farming, and agroforestry (Erickson 2000a, 2006; Walker 2004; Langstroth 1996). In Bolivia, archaeologists estimate the existence of 10,000 forest islands (Lee 1995; CEAM 2003). The Kayapó of Central Brazil create forest islands (*apêtê*) of improved soils through additions of organic matter from household middens and recycling of crop debris for intensive cultivation of crops (Posey 2004; Hecht 2003).

15.1. Forest island in the savanna,
Machupo River, in 2006. Source:
Photo, Clark Erickson.

RING DITCH SITES

Ring ditch sites are reported in the Bolivian Amazon Mato Grosso, Acre, and Upper Xingu River regions in Brazil (Erickson 2002; Heckenberger 2005; Pärssinen et al. 2003; Ranzi and Aguiar 2004). These sites consist of a closed or U-shaped ditched enclosure or multiple ditches. Heckenberger (2005) describes numerous sites with large open plazas and radial roads marked by earthen berms extending through residential sectors enclosed by deep semicircular, moat-like ditches and embankments. Early explorers described villages that were protected by wooden palisades and moats. If palisaded, typical ring ditch site would require hundreds or thousands of tree trunks, a considerable environmental impact.

Ring ditch sites in Acre and the Bolivian Amazon, described as geoglyphs because of their impressive patterns (circular, oval, octagon, square, rectangle, and D-shapes), appear to be more ceremonial than residential or defensive (fig. 15.2). Some ring ditch sites are associated with ADE. Modern farmers in the Bolivian Amazon intensively farm these sites and those covered with forest are good locations for hunting game and gathering fruit.

AMAZONIAN DARK EARTH (ADE)

As discussed earlier, soils have been central in debates about environmental potential and cultural development in Amazonia and play a major role in enhancing resource biodiversity and biomass. Rather than adapt to limited soils, we now recognize the ability of Amazonian farmers to improve and manage marginal tropical soils through creation of settlement mounds, forest islands, raised fields, and Amazonian Dark Earth (ADE).

Indian black earth (*terra preta do indio*) or ADE is an important subclass of anthrosols or anthropogenic soils and associated with archaeological

15.2. An octagon-shaped ring ditch site in the Bolivian Amazon. The ditch measures 108 m in diameter and 2 m deep. Santiago, Baures in 2006. Source: Photo, Clark Erickson.

sites (Smith 1980; Erickson 2003; Lehmann 2003; Glaser and Woods 2004; Neves and Petersen 2006). A lighter color ADE, *terra mulata*, often surrounds *terra preta*. Amazonian Dark Earth is estimated to cover between 0.1 percent and 10 percent or 6,000 to 600,000 km² of the Amazon basin. Amazonian Dark Earth sites range from less than 1 hectare to as large as 200 ha in size, and ADE was probably used for settlement, house gardens, and permanent fields, rather than for slash-and-burn agriculture, the common practice today. Scholars believe that these soils were created specifically for permanent farming. Today ADE is prized by farmers for cultivation and in some cases, mined as potting soil for markets in Brazilian cities.

Amazonian Dark Earth is rich in typical domestic debris found in archaeological sites including potsherds, bone, fish scales, shell, and charcoal. The extremely dark color and fertility are due to large quantities of charcoal and other organic remains that sharply contrast to the surrounding poor reddish tropical soils. In contrast to slash-and-burn agriculture, in which complete combustion of felled forest is the goal, ADE farmers practice "slash and char," a technique to produce biochar or charcoal through low temperature, incomplete combustion in a reduced atmosphere. Biochar has been shown to be a high-quality soil amendment for enhancing and maintaining soil fertility over hundreds of years. In addition, ADE is a rich habitat for beneficial microorganisms. Once established, ADE is a living entity that may sustain and reproduce itself (Woods and McCann 1999). The presence of intact ADE after 400 to 500 years is evidence of its permanence, sustainability, and resilience. Ethnobotanical studies document high biodiversity on ADE (Balée 1989; Smith 1980). The number of soil microorganisms in ADE alone may be quite large. Although understudied, potential contributions of microorganisms in ADE to overall biodiversity is substantial.

If ADE was formed as the simple unintentional byproduct of long-term residence in a locale, we would expect to find black earth sites at any location

15.3. Pre-Columbian raised
fields, canals, and causeways
in the Bolivian Amazon. The
clearing is now a ranch and the
causeways are used as paths. San
Ignacio in 2006. Source: Photo,
Clark Erickson.

where past human occupation was dense and of long duration. Archaeological sites fitting these criteria are common throughout Amazonia, but
do not have ADE. This suggests that ADE formation, which involves careful
production of biochar and management of soil microorganisms, is intentional soil engineering.

RAISED FIELDS

Raised fields are probably the most impressive example of landscape engineering at a regional scale in Amazonia (Denevan 1966, 2001; Erickson
1995, 2006; Walker 2004). Raised fields are large platforms of earth raised
in seasonally flooded savannas and permanent wetlands for cultivating
crops (fig. 15.3). Excavations and agricultural experiments suggest that
raised fields serve multiple functions, including drainage of waterlogged
soils, improvement of crop conditions (soil aeration, mixing of horizons,
and doubling of topsoil), water management (drainage and irrigation),
and nutrient production, capture, and recycling in canals alongside each
platform. Crop production in experimental raised fields is impressive and
up to double that of nonraised fields (Erickson 1995, 2006; Stab and Arce
2000; Saavedra 2006). Based on high productivity and substantial labor
costs to construct, raised fields were probably in continuous production. In
addition to traditional crop cultivation on the platforms, aquatic resources,
such as edible fish, snails, reptiles, and amphibians, could be raised in the
adjacent canals. Canals also trap organic sediments and produce organic
"green manure" and "muck" that can be periodically added to the platforms
for sustained cropping.

Raised field agriculture represents a massive landscape transformation
at a regional scale through rearranging soils, changing hydrology, and imposing a heterogeneous micro-topography of alternating terrestrial and

aquatic ecosystems on landscapes that originally were relatively flat and biologically homogeneous and of limited production. Landscape engineering of this magnitude substantially increased biodiversity and biomass in savannas and wetlands.

TRANSPORTATION AND COMMUNICATION NETWORKS AND WATER MANAGEMENT

Transportation and communication networks in the present and past have significant environmental impacts at the local and regional scale. Paths, trails, and roads connect settlements and people and, like modern roads, bring development and new settlements, expand farming, and cause environmental change. All Amazonian societies use elaborate networks of paths and trails and roads between settlements, gardens, fields, rivers, resource locations, and neighbors. The Kayapó maintain thousands of kilometers of paths (Posey 1983 cited in Denevan 1991). Posey (2004) documents subtle anthropogenic impact along Kayapó paths created by the discard of seeds from meals and snacks and transplanting of economic species along path clearings. These resources also attract game animals, making them easier to find and hunt. The long linear disturbance and light gap created by clearing and maintenance of paths produces distinct anthropogenic vegetation communities that penetrate deep into the forest.

Some advanced Amazonian societies built impressive formal roads, causeways, and canals of monumental scale (fig. 15.4). Large and small sites in the Tapajós and the Upper Xingú regions are connected by traces of networks of straight roads with earthen berms, suggesting hierarchical sociopolitical organization at a regional scale (Nimuendajú 1952; Heckenberger 2005). The late pre-Columbian inhabitants of the Llanos de Mojos and Baures regions in the Bolivian Amazon completely transformed the environment into a highly patterned landscape of complex networks of raised earthen causeways and canals (Denevan 1991; Erickson 2001, 2009; Erickson and Walker 2009). These earthworks had multiple functions, including transportation and communication, water management and production of aquatic resources, boundary and territorial markers, and as monumental ritual and political statements. Canals brought water for irrigation and provided drainage when necessary.

Transport and communication by water is a basic element of tropical forest culture (Lathrap 1970; Lowie 1948). Nordenskiöld (1916) pointed out that most of the major headwaters of Amazonian river drainages connect to the headwaters of adjacent river drainages. Some of these aquatic connections, such as the Casquiare Canal between the major Negro and Orinoco drain-

15.4. Four pre-Columbian causeways and canals connecting forest islands in the Bolivian Amazon. The palm-covered causeways are 3 to 4 m wide and 1 m tall with adjacent canals of 2 to 3 m wide and 1 m deep. Baures in 2006. Source: Photo, Clark Erickson.

ages and the Pantanal between the Guaporé and the Paraguay drainages, are partially anthropogenic. Artificial river meander shortcuts are common in the Llanos de Mojos of the Bolivian Amazon, Amapá Region of the Central Amazon basin, and the Ucayali River of Peru (Abizaid 2005; Denevan 1966; Nordenskiöld 1916; Raffles and Winkler-Prins 2003). The large meander loops of typical rivers often require hours or even days of paddling to move short distances. The problem is solved by cutting short canals between the neck of a large looping meander. In a number of cases, these anthropogenic canals created a new river course, dramatically and permanently changing the regional hydrology.

Inter-river canals are common in the Llanos de Mojos of Bolivia. Pinto (1987) describes a complex network of natural channels combined with artificial canals to allow canoe traffic over 120 km perpendicular to natural river flow. In other cases, artificial canals tapping the headwaters of two adjacent rivers diverted the flow of one into the other, permanently transforming the hydrology of two drainage basins (CEAM 2003).

FISHERIES MANAGEMENT

Fishing is now recognized as the major traditional source of protein in the Amazon basin (Chernela 1993; Beckerman 1979; Erickson 2000b). In contrast to other civilizations that domesticated fish, Amazonian people artificially enhanced the natural habitats of wild fish to increase availability through creation of artificial wetlands and expanding the capacity of existing wetlands through construction of raised field canals, causeways, and other water management techniques.

The Baures region of Bolivia is an excellent example of landscape domestication for the improvement of natural fisheries (Erickson 2000b). Low linear earthen ridges zigzag across the seasonally inundated savannas between forest islands with a funnel-like opening located where the earth-

15.5. A network of pre-Columbian fish weirs in the Bolivian Amazon. The brush-covered fish weirs measure 1 m wide and 50 cm tall. Straight features at the top and bottom of the image are causeways and canals, and circular features are artificial fish ponds. Baures in 1999. Source: Photo, Clark Erickson.

works changed direction (fig. 15.5). These features are identified as fish weirs based on descriptions in the ethnographic and historical literature. Fish weirs are fences made of wood, brush, basketry, or stones that extend across bodies of water. Baskets or nets are placed in openings to trap migrating fish. Most fish weirs are simple ephemeral structures on a river or shallow lake. In contrast, the fish weirs of Baures are permanent earthen features covering more than 550 km^2. Small artificial ponds associated with the weirs are filled with fish and other aquatic foods when the floodwaters recede. These were probably used to store live fish.

AGROFORESTRY

Countering the view of Amazonian forests as pristine and natural, historical ecologists have shown that these forests are, to a large degree, the cultural products of human activity (Balée 1989; Posey and Balée 1989; Denevan and Padoch 1987; Posey 2004). Amazonian people past and present practiced agroforestry: tree cultivation and forest management (Peters 2000).

Analysis of pollen, opal phytolith, and sediment from lakes document local and regional anthropogenic disturbances of Amazonia over thousands of years, including burning, clearing, farming, and agroforestry (Piperno and Pearsall 1998; Mora Camargo 2003; Piperno et al. 2000). Much of what was originally misinterpreted as natural change due to climate fluctuations is now considered anthropogenic. Records show a steady increase of "weeds" and secondary forest species, many of which are economic species, and later domesticated crops that thrive in open conditions and heterogeneous mosaic of forest and savanna and intermediate states created by human disturbance. At the same time, the frequency of species characteristic of closed canopy forests decreases until the demographic collapse after 1491. Fire histories are also documented in association with the formation of the anthropogenic forest. Evidence of fruit and nut tree use and human disturbance is documented by 10,500 years ago in the Central Amazon (and see discussion of dates in the Colombian Amazon in Roosevelt 1996; Mora Camargo 2003; see discussion of evidence for domesticated crops at some sites in Amazonia in Piperno and Pearsall 1998; Piperno et al. 2000).

The long-term strategy of forest management was to cull noneconomic species and replace them with economic species. Sometimes this involves simple thinning, planting, transplanting, fertilizing, coppicing, and weeding of valued species to enhance their productivity and availability. Many wild plants are often found outside their natural range due to transplanting, cultivation, and habitat improvements. In other cases, wild and domesticated trees are tended as orchards.

Slash-and-burn or swidden agriculture is typically characterized as involving low labor inputs, limited productivity per land unit, and a short period of cultivation followed by longer periods of fallow or rest. Researchers have pointed out, however, that swidden fields are rarely truly abandoned and unproductive during fallow. In Amazonia, agriculture is typically combined with agroforestry. In the initial cutting and burning to clear a field or garden, certain economic species are left to thrive while unwanted species are removed. In addition to basic food crops, useful fruit and palms are often transplanted to the clearing. As fields fall out of cultivation because of weeds and forest regrowth, the plots continue to produce useful products, long after "abandonment."

Anthropogenic forests are filled with fruit trees; eighty native fruit trees were domesticated or semidomesticated in Amazonia (Clement 2006). Fruit trees, originally requiring seed-dispersing frugivores attracted to the juicy and starchy fruits, became increasingly dependent on humans through genetic domestication and landscape domestication for survival and reproduction. In addition, humans improved fruit tree availability, productivity, protein content, sweetness, and storability through genetic selection. Forest islands of cacao trees are agroforestry resource legacies of the past inhabitants of the region (Erickson 2006). Agroforestry and farming also attract game animals that eat the abundant crops, fruits, and nuts. Farmers often grow more food than necessary to attract game. As a result, "garden hunting" is particularly efficient (Linares 1976). Many game animals of Amazonia would have a difficult time surviving without a cultural and historical landscape of human gardens, fields, orchards, and agroforestry. The biodiversity of animals can also be enhanced by domestication of landscape.

Even hunters and gatherers contribute to anthropogenic forests. The nomadic Nukak of the Colombian Amazon change campsites seventy to eight times a year (Politis 1996). When establishing a new location, a small number of trees are felled and hundreds of palm fronds are collected for construction of a simple lean-to structure. Wild fruits and nuts are collected and some end up discarded. After the camp is abandoned, palm seeds take root in the clearing and thrive. Repeated over hundreds of years, the selective cutting of trees for nomadic camps, creation of small light gaps or openings, and distribution of seeds can substantially change the forest composition to one rich economic species of plants and animals.

Conclusions: Lessons from the Past?

Amazonian Dark Earths, agroforestry, raised field agriculture, transportation and communication networks, urban settlements, mounds, artificial forest islands, river cut-offs, water control, and fisheries management are clear examples of landscape creation, transformation, and management by pre-Columbian native people in Amazonia. What they transformed was often less productive and biologically diverse than what resulted. In other cases, human activities reduced biodiversity. Most landscapes that are today appreciated for their high biodiversity have evidence of human use and management, even if those landscapes are relatively unoccupied today. Environments with high biodiversity are a result of, rather than in spite of, long human disturbance of the environment.

Bolivian informants state that the best hunting and farmland is on pre-Columbian earthworks deep in the forests. Recognized as having the highest biodiversity in Bolivia, the Tsimane Indigenous Territory is covered with raised fields, causeways, canals, and settlements under what is now continuous forest canopy. These cases of present-day biodiversity, treasured by scholars and the public alike, were ironically created under conditions of intensive farming, urbanized settlement, and dense populations. Were these native practices sustainable? Sustainability usually refers to rational continuous harvest of a resource without destroying the capacity of that resource to reproduce. The longevity of settlements, agriculture, and cultural traditions and the dense populations supported in what are now considered biologically diverse environments are evidence of sustainability.

Are the past strategies of environmental management defined by historical ecology applicable to the modern world? Many goals of pre-Columbian native people, modern inhabitants of Amazonia, scientists, planners, and the general public coincide: the management of environmental resources for a comfortable life and sustainable future in what most consider a fragile ecosystem. Increasingly, the reservoir of existing biodiversity is found in humanized landscapes. The failure of conventional solutions, such as fencing off nature and excluding native people, highlights the need for strategies that embrace the coexistence of nature and humans. Environmental management informed by time-tested strategies for specific landscapes may be more appropriate than existing solutions. Because humans played a role in the creation of present-day biodiversity, solutions will have to include people.

Amazonian Dark Earth as a means to mitigate global warming is an example of applied historical ecology. Low-temperature biochar or charcoal, the key ingredient of ADE, and ammonium bicarbonate produced from urban wastes are the byproducts of biofuel production. Burial of biochar treated with ammonium bicarbonate is an excellent nitrogen-based organic fertilizer *and* an ideal form of carbon sequestration (Marris 2006). Controlled burning, traditionally considered degrading to the environment, is being reintroduced as a management strategy. Once removed from their homelands in the establishment of parks, native people are now integral participants in the management of some ecological reserves and indigenous territories (Chapin 2004; Posey 2004). Many small farmers living along the Amazon River continue to practice sustainable strategies from the past within a modern urban context (Smith 1999).

Many conservationists consider the idea that humans as a keystone species created, transformed, and managed biodiversity through their activities as dangerous and detrimental to fund-raising to protect what they advertise as pristine wilderness (Chapin 2004). Native rights advocates worry that Amazonian people will be viewed as bad environmental stewards and lose claims and control of indigenous territories (Redford 1991; Chapin 2004; Conklin and Graham 1995). Others declare that those who argue against the ideas of the Amazon as a counterfeit paradise fan the flames of tropical rainforest destruction by encouraging reckless development of already transformed landscapes (Meggers 2001).

I believe, however, that ignoring the complex human history of environments in Amazonia would be unwise. A vast indigenous knowledge spanning hundreds of generations about the creation, transformation, and management of environments is physically embedded in the landscape, encoded in the distribution and availability of plant and animal species, documented in historical and ethnographic accounts, and in some cases, still practiced by native Amazonians.

PART III
Market Dynamics

Market Dynamics and Regional Change

NICHOLAS K. MENZIES

The chapters in this volume demonstrate that the theoretical purity of Adam Smith's assertion that timber is simply another marketable commodity (Albritton Jonsson, chapter 4, this volume) is muddied by the layers of social, economic, and symbolic meaning associated with forests. The five chapters in this section describe forest ecosystems whose structures and composition have been deliberately modified to favor the production of marketable goods. The authors examine how and where demand for forest products is generated, reminding us that the usual market metrics of the costs of labor, production, and transport are poor predictors of the monetary value paid by consumers for products as diverse as shea butter for cosmetics, fresh vegetables and fruit, or industrial rubber.

Branding is a time-honored device to differentiate a product to create added value in the market. Raymond Bryant's chapter on teak is a reminder that branding has also been a powerful mechanism through which states have laid claim to a resource and by extension to the land where it is found and the people who live there. Teak is not just a high-quality timber. It was a royal tree reserved for the rulers in Burma and Thailand well before the heyday of the British Empire. The discourse of science gave legitimacy to claims that the species is special. The colonial and post-colonial Burmese state had no compunction in employing violence against the land and people, while trade interests associated with the colonial enterprise and the post-independence ruling elite persuaded affluent customers in the metropole that teak furnishings conferred status to their lives. Today, science, violence, and seduction are still associated with teak as it makes a transition to being marketed as a "green" product, sustainably produced without harming the environment—even as conflict and violence continue to scar the land and people where it is produced.

Carney and Elias remind us that the keepers of specialized knowledge about trees and forest products are often those whose livelihoods have been intimately tied to them, not the more visible—usually male—power brokers in a community. In West Africa, colonial scientists had little success in establishing plantations of *Vitellaria paradoxa* to expand shea butter production into an industrial enterprise. *V. paradoxa* is a savanna species that has been tended over the centuries in agroforestry systems. In Burkina Faso, its distribution and range are the outcome of selection and protection by the women who collect and process the nuts. It is their gender-specific knowledge that guides the selection of trees that yield fruit with desired characteristics and makes it possible to meet the demands of new markets such as the the high-quality, high value shea butter that is now in demand for "natural" cosmetics.

Perhaps counterintuitively, the producers at the end of commodity chains that span continents may be more responsive than large state agencies or corporations to changes in market demand among consumers in distant places that are far beyond their own direct experience. My account of the rediscovery and marketing of Ancient Forest Tea in Xishuangbanna tells how high volume plantation production of Pu'er tea has been eclipsed in the last ten years by the tea that minority farmers continued to cultivate under the forest canopy even when their land had been collectivized and they themselves were labeled by the authorities as backward, unscientific, and of low quality. Ironically, these farmers have been quick to turn these negative labels to their advantage, marketing their tea as organic with a unique historic heritage and cultural authenticity—the "seduction" element of branding that Bryant refers to. The story of Ancient Forest Tea highlights, too, one of the caveats of branding—with the value-added from the niche market comes the need for credibility and quality control.

The first three contributions to this section trace the dynamics of markets for specific forest products: teak, shea butter, and tea. Fox and Fujita Lagerqvist reorient the frame of reference from the commodity to the region to trace the impact on landscapes and communities of the dramatically changing economic and political geography of the border region of Thailand, Laos, and Xishuangbanna (China).

The three countries Fox discusses have had different trajectories of development and change but share a recent history of a dramatic transformation of their markets with the implementation of bilateral and regional trade agreements. The introduction of rubber to Xishuangbanna was a government-directed initiative in response to China's exclusion from international markets by an economic blockade and then by self-imposed political isolation. Domestic industrial demand defined the market. The international border marked the boundary for production. In Thailand, the state's preoccupation with securing the forested lands on its periphery by excluding longtime forest users was tempered by the introduction of new crops under the Royal Project Foundation to meet new urban markets for vegetables, flowers, and fruit. With the dismantling of trade barriers, the old national center–periphery axes of demand and supply have given way to a much larger regional market where comparative advantage in the factors of production now drives land-use decisions. Most dramatically, the availability of land in northern Laos is setting the stage for a transformation of the landscape into what is in effect a continuation of the extensive rubber plantations across the border in China.

Shifting scale to view these developments from the perspective of Lao villages near the Chinese border, Fujita Lagerqvist shows farmers assessing the opportunities afforded by expanded markets and choosing to plant rubber, albeit on their own terms in smallholdings managed by households. Lao farmers have been quick to make use of informal trans-border kinship networks and small business connections to secure the capital and technical skills they need to manage the new crop. The state agencies, international nongovernmental organizations (NGOs), and corporations (from China in this case) that might be expected to be most familiar with regional and global markets have been less adept at responding to the new environment— although time will tell whether rubber production in northern Laos will continue to be dominated by smallholder producers or whether they will ultimately be bought out by the corporations.

In global and regional markets, forests and forest products are not undifferentiated commodities. Time, place, politics, the dynamics of international trade, and social practice all affect the desirability of a product in the marketplace. The very existence of a marketplace is a function of forces as large as regional wars and economic integration in Southeast Asia, or as individualistic as urban consumer preference for cultural authenticity and environmental health. The women who process shea butter for cosmetics in Burkina Faso, the forest tea producers from ethnic minority communities in the mountains of China's Xishuangbanna prefecture, and smallholder rubber producers in Laos and China are actors in a world of constantly changing social and economic forces. Physically distant as they may be from the final consumers of these products, it is often those who are closest to the forest and the trees, the more disadvantaged or marginal members of society, who are able to spot and to exploit new market opportunities that larger, more cumbersome entities such as states and big businesses have missed or ignored.

16 ✳ The Fate of the Branded Forest: Science, Violence, and Seduction in the World of Teak

RAYMOND L. BRYANT

A defining trait of modern capitalism is the role of the brand in shaping consumer choice. As brand-name goods deepen their colonization of the mind, they exert great influence on a wide array of consumer products and associated production processes (Klein 2000). However, this story is neither new nor confined to the world of manufactured goods. The branding of nature is a case in point. This process has long ruled the fate of forests home to sought-after trees (such as mahogany and teak), species whose biophysical "properties" underpin brand identity. Yet very little is known about the ways in which brands and branding relate to the social life of forests. This chapter provides initial insight into this phenomenon. It does so with reference to teak, a valuable hardwood found notably in Burma.

The branding of teak reflects a wider human attempt to manipulate nature. Yet the brand stands out inasmuch as it is closely associated with economic activity under capitalism. The word comes from the Old Norse *brandr* (to burn) whence it was incorporated into English as "an identifying mark burned on livestock etc with a hot iron" (Thompson 1996, 113). This phenomenon was linked to the development of trade insofar as it facilitated exchange: (1) it was a means to *mark ownership* of a product, thereby providing the seller some guarantee of possession; (2) it permitted *informed discrimination*, as buyers could readily link product to producer; and (3) it helped to *identify quality*, which might be due to skilled labor or superior conditions of production. As such, the brand joined such things as the circulation of money and the legal enforcement of contracts as an essential ingredient in market elaboration. The industrial revolution brought further change. Use of brands became widespread as rapid developments in manufacturing, communications, advertising, business organization, trade, consumption, and law transformed social relations (Blackett 2003). By the twentieth century, the preeminent meaning of brand was overtly commercial and targeted on consumers (Trentmann 2004).

In the process, the brand has criss-crossed nature wherever it is feasible and profitable to do so. It has been a means by which to romanticize the wealth of nature, drawing on discourses of beauty, purity, and innocence (Slater 2003). Indeed, it is the rich, discursive potential of nature's products—a mirror in which to reflect all manner of human desire and calculation—that presents brand makers with bountiful opportunities (yet something apparently missed by the guru of modern capitalism, Adam Smith: see Albritton Jonsson, this volume). Yet such opportunities require use of complex tools to manage brands from first production of a good to its ultimate consumption.

Three tools concern us here: science, violence, and marketing. *Science* has long played midwife to the birth of a brand. "Objective" truth in science has helped to build brand identities based on such things as "proof" of disease prevention, product quality, or product purity. Science thus provides "facts" about a product that render it desirable to consumers. *Violence* has long links to brands. It, too, is multifaceted: murder, injury, imprisonment, lawsuits, sweatshop labor, livestock branding and slaughter, tree felling, collateral ecological damage, and so on. Whether disciplining people or disrupting nature, the aim is to create production and consumption conditions fit for brand growth. *Marketing* is the best-known tool today. Here we enter a world of intangible ideals, consumer research, and brand placement—divining what consumers want or can even be taught to want. The choice of a logo, a color scheme, packaging, or self-help messages all help to communicate desire, fantasy, and aspiration.

The case of teak (mainly from Burma) illustrates some of the ways in which science, violence, and marketing come together in making a brand, a case to which this chapter now turns.

Science of "Superior" Teak

Scientists helped to crystallize perceptions of teak as the world's premier timber by providing a raft of "facts" about its legendary properties, even as they oversaw the systematic creation of the branded forest in which it grew. Yet knowledge was fragmentary. How extensive were the forests? Could teak be planted, and how did it grow best? What were its key properties, and how could these be matched to commerce?

The British imported scientific forestry to Burma in the nineteenth century to answer such questions. A bureaucracy was created in which foresters disseminated state-of-the-art knowledge about teak. Such knowledge facilitated central control even as it contributed to a mystique about that "most valuable of all known timbers" (Brandis 1888, 103).

Colonial foresters thus turned their scientific gaze on Burma's forests. One urgent activity, especially in a local context of chronic instability and

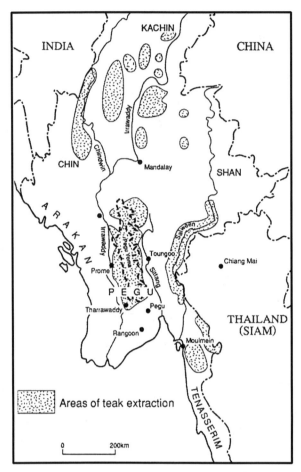

16.1. Teak forests of Burma

violence, was to map all forest in which teak was found. As such, considerable effort was expended in ascertaining precisely where those forests were.
The result was a detailed mapping of Burmese teak that pinpointed core
areas (fig. 16.1).

Working plans spelling out what timber was to be cut when and where,
as well as under what conditions, complemented such maps. A process of
"internal territorialization" was thus undertaken, with momentous ramifi-
cations for people and forests. Teak-bearing forests became reserved forests,
the production home of this high-value tree.

Another urgent activity was the need to assert direct control over the
trees themselves. Fortuitously, a traditional system of branding was in
place. A royal tree since the eighteenth century, it was forbidden to fell or

16.2. Girdling (ring-barking) of teak

damage teak except with royal approval. That approval was signified by the girdling (ring barking) of mature teak, a process that branded *and* killed the tree in preparation for commercial extraction that would follow after two years of further seasoning on the stock (fig. 16.2).

This system had its advantages (Nisbet 1901). First, branding was vital in regulating use in multispecies forests, as it served to mark teak trees as government property (but still did not prevent theft; see Baden-Powell 1874b, 17; Burma Forest Department 1899, 8). Second, girdled teak was easiest to extract. Timber needed to be hauled to streams by elephants and then floated to saw mills; because girdled teak was lighter than green teak, it was easier to haul and float.

Finally, scientists suggested *commercial* benefits inasmuch as girdling resulted in timber of superior quality to that derived from green teak. There was dissent (e.g., Government of India 1868). Yet, according to Brandis (1888, 103–104), the case was clear:

> The principal value of teak timber . . . is its extraordinary durability . . . once seasoned, teak timber does not split, crack, shrink, or alter its shape. In these qualities it is superior to most timbers. In contact with iron, neither the iron nor the teak suffers, and in this respect it is far superior to oak. It is not very hard, is easily worked, and takes a beautiful polish. It has great elasticity and strength, and is not very heavy . . . [yet] being exposed to the wind and to the action of the sun, the timber of a girdled tree seasons more rapidly and more completely than that of a tree felled green.

This argument was persuasive precisely because it interwove scientific knowledge, aesthetic sensibilities, and commercial preference. The argument about durability was vital, as this was prized by wood-workers.

Colonial science thus underlay teak's emerging brand status. Indeed, an army of scientists experimented on teak and other wood to ascertain their strength, density, yield, and texture. This process was designed to compare them with each other as well as to ascertain how different growing conditions and harvesting arrangements affected those properties (Simmonds 1885; Gardiner 1942). For some, the virtues of teak were overstated. Baden-Powell (1873, 5) criticized "indiscriminating demand" for teak, urging his colleagues to promote other trees instead. Time and effort was devoted to this task (e.g., Ribbentrop 1900; Rodger 1921). Timber lists were even sent to merchants with phrases such as "equally useful with teak" or "stands next to teak" (Simmonds 1885; Howard 1923).

This effort came to nought. As Brandis (1888, 103) put it: "Great efforts have been made to find substitutes, but no timber has been brought to market in sufficient quantities combining the many valuable qualities which teak possesses." Gardiner (1942, 653–54) added that, "of the 200/300 kinds of timber trees found in Burma, teak . . . holds pride of place . . . [and] the utility of the timber depends not on one outstanding quality but on a combination of several." Thus, despite fears over substitutes (such as concrete and plastic), the demand for teak held up well (Hopwood 1935). Exports in 1940 were at record levels—460,000 tons per annum—and worth a then-staggering sum of about £2.25 million (Morehead 1944, 58; Scott 1945, 83). Accounting for 75 percent of world trade, Burma teak was "the commercial product of outstanding importance and value in the pre-war years" (Burma Forest Department 1947, 10).

That Burma's forests were enormously valuable despite the spread of substitutes suggested that teak was somewhat insulated from the vagaries of the market precisely because it enjoyed brand status. Science played its part here, but so too did violence.

Violence in the Branded Forest

The branded forest is often a "violent environment" (Peluso and Watts 2001). Rather than settling conflicts, branding tends to sharpen differences and encourage violence against people and trees. Burma's teak forests are a case in point, as resource militarization and structural violence went hand in hand to blight life there (cf. Peluso and Vandergeest, chapter 5, this volume).

Those forests were home to many who opposed the nation's rulers in pre-colonial and colonial times. As "a traditional hiding-place for malcontents" (Foucar 1956, 72), they were rife with "bandits" and would-be revolutionaries (Maung 1976). The capture of teak logs was a favored practice providing a powerful shot in the arm to cash-poor insurgents. Forest officials were targeted and sometimes killed, and villagers were also dragged into the violence.

Yet such violence appeared modest when compared with the extreme conditions that existed after independence in 1948. Teak forests thus became the focus of strategizing by insurgent armies intent on capturing lucrative revenue. Notable among them were the Karen, an ethnic minority group that launched a rebellion against the Burman-controlled state shortly after independence. Indeed, their sixty-year struggle to establish a sovereign state was partly reliant on such revenue, especially after the 1960s when they were pushed back to the Thai border by the Burmese army (Bryant 1996).

But it was the Burmese state that turned the teak forests into veritable killing fields. Central control of the forests had always been a hit and miss affair, even under the British. The early post-colonial years were worse, however. Indeed, the beleaguered government mounted a large-scale military campaign, "Operation Teak," in which army units secured the Sittang River between Toungoo and Rangoon so that teak could be rafted to Rangoon (*The Nation*, November 10, 1955).

Violence increased further following the 1962 military coup, as livelihoods were disrupted and residents sometimes forced to bear arms. A counterinsurgency campaign was launched that targeted those who lived in or near the forests and was designed to deprive insurgents of local food, funds, intelligence, and recruits (Smith 1999). This campaign was effective: the Burmese army swept all before it. Teak was indeed an important part of the army's strategizing here (Global Witness 2003). Hence, a campaign of terror was accompanied by large-scale extraction that led to rapid forest depletion.

The systematic structural violence that was embedded in the elaboration of scientific forestry practices beginning in colonial times added to the mayhem. Indeed, in important respects, this process was even more transformative of the social life of the teak forest than the resource militarization just discussed. It was vital, too, to the cultivation of Burma's teak brand.

Much of what transpired under scientific forestry was designed to introduce "government" in the sense of the term meant by Michel Foucault — (self) disciplining practices, widespread surveillance, and the elimination of "antithetical" behavior (Dean 1999). In Burma, as elsewhere, it entailed interlinked processes of generating forest maps, resource inventories, and population censuses that provided a basis for draconian restrictions on popular access to forest products. These restrictions prompted popular resistance of many kinds.

Conflict was often most intense in teak forests. To reside in or near to them was to invite systematic state intervention because regulating teak involved promoting good conduct. The creation of reserved forests was particularly intrusive. Reserve borders transgressed existing settlements and land uses while disciplining people—*where* they could go, *when* they could go there, and *what* they could do while there. Yet such procedures were essential to the quality control measures that officials put in place in the forest to ensure that teak was extracted only under what were believed to be the best possible conditions, conditions in turn propitious for brand development.

One example that illustrates this process relates to the fire-prevention campaign mounted in reserves in the late nineteenth century in the (mistaken) belief that fire always harmed teak. This campaign sought to restrict such local practices as game hunting, cattle grazing, or honey gathering involving fire. Not surprisingly, it was bitterly resisted. Such resistance joined with growing scientific doubts over the utility of the campaign, leading ultimately to its demise in the early twentieth century (Slade 1896; Walker 1908).

Yet in its heyday fire prevention was one of *the* key ways in which foresters sought to safeguard the quality of teak. Still, it was a serious burden on villagers. On the one hand, fire-related restrictions were a source of considerable *individual* concern. Villagers needed to alter their practices or ensure that those practices were hidden from view. There was also the threat that they might be captured for violations, and ignorance of the law was no defense given publicity surrounding the campaign. Much depended on the response of forest officials and local magistrates. The latter were keen to moderate punishment mindful of potential unrest. Nonetheless, villagers were punished: a 500 rupee fine, six months in prison, or both.

On the other hand, fire prevention was also a serious *collective* burden. Villagers were required to help fight fires whatever their provenance. Not only did this involve entire villages in unpaid dangerous work, it also meant that they needed to drop at short notice their own livelihood activities. British complaints of peasant indolence and negligence were legion, suggesting just how much of a collective imposition it was.

The branding of teak linked violence to teak extraction as never before. Yet until very recently, teak's local reputation as a "blood" timber was kept from wider view thanks to artful marketing and consumer affection for the "king of woods."

A Little Bit of Empire: Consuming Teak

Teak consumption was profoundly shaped by how it was marketed, commencing with British rule. At first the focus was largely military. By the early nineteenth century, quality oak was all but gone even as the superiority of teak over oak in naval construction was clear (Lambert 1996). Burmese teak thus played a key role in producing warships built in British shipyards. By the 1920s, though, modern shipbuilding techniques and a small fleet meant that official requirements decreased sharply (Albion 1926/2000; Andrews 1930).

By then, however, teak was becoming a stylish consumer item. This process was related to the spread of imperialism: love of teak was linked to love of empire. Indeed, its utility and cachet was such that the British Empire cloaked itself in teak: its merchant ships and passenger liners, its railway ties, carriages, and bridges, its important public buildings, even its social spaces. Writing about Burma, where he served between 1922 and 1928, Orwell (1934/1987, 17–18) described the European Club as "a teak-walled place" that was "the real seat of the British power." There was also the symbolism of it all. By cloaking itself in one of the world's most expensive and durable woods, the message was clear: the British Empire too would endure.

Imperial symbolism and middle-class distinction (Bourdieu 1984) went hand in hand. Teak served as a marker of social distinction in the homes, gardens, and bureaus of the affluent. Four factors underlay this development. First, there was the demonstration effect, as elites became aware of the aesthetic beauty of teak from having seen it in colonial offices, railway carriages, and on board passenger vessels. That teak was embedded in symbols of imperial modernity (e.g., railway, passenger liner) added to its allure. Both of the world's largest liners in 1939, the *Queen Mary* and the *Queen Elizabeth*, for example, contained about 1,000 tons of teak used for decking, gangways, handrails, and window frames (Gardiner 1942). Second, there was the way in which the prospering British middle classes came to prize quality possessions as markers of moral respectability. As Cohen (2006, xv) notes, such consumerism gathered pace as the British became obsessed with shopping for goods (including luxury furniture and fittings) that formed part of a process of "middle-class self-fashioning." Third, there was the close association of Burma teak with a sense of adventure and the exotic, peddled in journalistic accounts (e.g., Geary 1886), guidebooks (e.g., Ferrars and Ferrars 1900; Kelly 1912), and personal memoirs (e.g., Williams 1950). To consume teak was to imbibe a glorious history of imperial adventure. Finally, there was sheer practicality of teak as a long-lasting timber that displayed little shrinkage in drying and very little subsequent movement in variable climates. These properties meant that it was ideal, for example, as a piece of garden furniture or even as decking and fittings for yachts (Scott 1945).

Teak first made its social mark in the tropics. Not only were supplies relatively close by, but also the wood was one of the few timbers immune to white ants. Teak was thus used in building the bungalows that were home to both colonial officials and indigenous elites: window frames, verandas, roof trusses and shingles, doors, paneling, flooring, and furniture. Similarly, the social world in which colonial and sometimes local elites mingled was often framed by this wood: in the case of the Royal Bombay Yacht Club, the ceiling, flooring, paneling, staircase, and tables were all made from teak (Gardiner 1942).

The imperial heartland also fell under the spell of teak. Teak paneled and floored offices became a hallmark of quality and success in the business world. Even in America, where supplies of local hardwoods were still available, the Ford Institute Museum in Detroit completed just prior to the Second World War was decked out in teak—covering an area of 340,000 square feet, this was then the largest single expanse of teak flooring in the world (Gardiner 1942). But the affluent household also acquired a taste for teak. Interiors were kitted out with teak flooring, paneling, stairways, window seats, doors, and fire surrounds (Morehead 1944). Yet teak really came into its own in the back garden. Here, teak furniture added an exotic touch of class to "landscapes of privilege" (Duncan and Duncan 2004) already molded by a British passion for exotic plants (Casid 2005; Wulf 2008). At a more modest level, teak ornaments added an exotic touch to the imperial living room: at the 1938 Glasgow world exhibition, "many thousands of hand-carved teak elephants, carved ash trays, flower bowls, labelled 'genuine Burmese teak,' [were] eagerly purchased by the public" (Myat Tun 1938, 808). There was thus "a little bit of empire" in many a home.

The Second World War interrupted the steady march of teak into the offices, homes, and gardens of the imperial heartland. Indeed, and as noted earlier, violence begat violence in the post-colonial era such that the supply of Burma teak to the international market was fitful and much reduced from prewar levels (Bryant 1997). By the early 1990s, though, conditions were propitious for the return of teak. The Burmese army had defeated most of the insurgents arrayed against it, while ever-closer links with China enabled large-scale timber exports despite a Northern boycott campaign against Burma's brutal military regime (Global Witness 2003). Meanwhile, middle-classes around the world rediscovered the aesthetic beauty and practicality of teak, above all in the garden and yachting sectors. Imports of (mostly Burmese) teak into affluent countries climbed rapidly and amounted to an estimated $37 million per year to the United States alone after the turn of the millennium (and worth $100 million per year in consumer purchases there) (Earthrights International 2005, 2).

Seemingly in only a few years, a comfortable outdoor lifestyle was predicated on the stylish teak dining set, garden bench, deck, and lounge chair. Teak was thus once more part of elegant modern living (fig. 16.3).

As one recent marketing pitch suggested, it was all in the attitude:

16.3. Teak and stainless steel sun lounge chair. Image courtesy Gloster Furniture.

Choose a space under the open sky and make it your own. In a hectic world, Gloster lets you create an environment that is not confined by walls, but defined by a sense of personal space, an oasis of peace, relaxation and freedom. View the outside as an extension of your home, an expression of your individual style, every bit as important as any other room in the house, and then furnish it with beautiful things. Choose from teak, metal, sling or woven furniture in a variety of styles, from the traditional to the contemporary. And remember, outside is a far tougher environment than inside, so accept nothing but the best. (Gloster, n.d., 1)

Meanwhile, Asia's hard-working middle classes were enjoined to pamper themselves with a bit of well-earned luxury. In Singapore, for instance, one marketing campaign encouraged them to buy teak furniture so as to be able to "enjoy resort living in your own home" (Teak and Mahogany 2005, 79).

For those with money and a maritime disposition, the yachting industry is awash with teak. As in colonial times, yachts are today kitted out with teak fittings, cockpits, doors, decks, and trimming supplied by specialist suppliers such as Jamestown Distributors and East Teak Trading Group (both in the United States) and Teak Decking Limited in the United Kingdom. And yet, the industry is much bigger than in earlier times as yachts (and power boats) become integral to the lives of the affluent. Indeed, the size of yachts gets ever bigger, with the rich and famous at the top end of the market outdoing one another in the race to have the biggest and best. Teak remains a favorite despite a wide array of nonwood substitutes. As Teak Decking Limited

(2007) asserts, it will uniquely "provide unmistakable lasting beauty, adding value and character to any boat." Here, the teak brand shines brightly among the marketing agents, ships chandlers, and yachting enthusiasts, as debate happily revolves around technical and aesthetic merits highly reminiscent of the imperial scientific foresters of old. Here, thoughts about the dark side of teak are banished in a modern romanticized narrative designed purely to enhance the consumption experience.

Conclusion

This chapter suggests that the fate of the branded forest is as multifaceted as the brand dynamics that swirl around it. The branding of nature is not new, and in its elemental form, long predates the modern industrial brand that riles the antiglobalization movement so (Klein 2000). Yet in certain cases, such as teak, the modern world of branding—with all of its culturally resonant connotations of consumer distinction, differentiation, and privilege—has also penetrated the world of nature. Diverse tools of branding—herein described as science, violence, and marketing—are brought to bear in the effort to build the brand. The social life of the forest is thereby irrevocably altered.

If this chapter mainly served to illustrate the dark side to branding the forest, it remains to be seen whether this multidimensional phenomenon that helps to shape many aspects of the modern world can be a force for good in an era of growing social and ecological disruption. Is modern branding antithetical to the promotion of social equity and sustainable ecological practices? If so, and there is certainly evidence to that effect as this chapter indicated, then antibranding must be part-and-parcel of antiglobalization struggles. Other examples of branded nature, ranging from mahogany wood in Brazil's Amazon to Blue Mountain coffee grown in the highlands of Jamaica, do not augur well (e.g., Booth 2008). And yet, other cases—for example, in Italy linked to the branding of fabled agricultural products such as Parmesan cheese and Parma ham—suggest the possibility of a more nuanced picture based on location-specific factors such as well-organized trade cooperatives and sympathetic laws like the European Union's Protected Description of Origin Law. Indeed, is there even room, for instance, to deploy in a systematic manner modern branding techniques to the promotion of fair trade and organic products, thereby giving a sector still plagued by amateurish marketing a much needed boost? (See, for example, chapter 18 by Menzies, this volume.) Clearly, much work remains in order to grasp the full implications of the brand, including the ways in which it marks, for good and bad, both people and nature in the modern era.

17 ✳ Gendered Knowledge and the African Shea-Nut Tree

JUDITH CARNEY AND MARLÈNE ELIAS

Introduction

Shea butter is a favored ingredient found in contemporary skin-care products marketed as "natural." Known for its exceptional healing and moisturizing properties, shea—or *karité* as it is known in French—is commercialized in moisturizers, sunscreens, lip pomades, and hair products. What Western consumers may not know is that the shea commodity chain begins in the open woodland forests of Africa south of the Sahara Desert. The tree that bears this valuable fruit thrives in the drought-prone Sahel, an environment that is commonly viewed in the West as overpopulated, deforested, and threatened by the Sahara's southward expansion. Besides the shea-nut tree (*Vitellaria paradoxa*), West Africa's Sudano-Sahelian savannas host other economically valuable species, notably gum Arabic (*Acacia senegal*) and the locust bean (*Parkia biglobosa*).[1] Since the 1990s shea agroforestry systems are benefiting from new markets for the nut butter, which is made exclusively by women. While this represents a small market share of the estimated 150,000 tons of shea nuts annually exported from West Africa—most is sold as a cocoa butter substitute in the chocolate industry—the ability of the savanna to supply the international market with ample quantities of nuts and butter is at variance with long-standing narratives of the Sahel as a degraded landscape with few trees. In providing rural women new income opportunities, the demand for shea butter actually contributes to the conservation of an indigenous Sahelian tree species. This chapter draws attention to the ways female butter makers influence the selection of trees and the management of the shea agroforestry system.

Amid growing scientific concern over the long-term effects of climate change on sub-Saharan Africa, recent scholarship is challenging negative perceptions of the role of Africans in the vegetation history of the Sudano-Sahelian woodlands (Fairhead and Leach 1996, 2003; Leach and Mearns 1996; Reij, chapter 23, this volume). Such findings demolish conventional views that hold traditional savanna agricultural practices to be detrimental and condemn Africans as destroyers of forest resources (Duvall

2003). In Guinea, Fairhead and Leach (1994a, 2003) show the ways farmers actively shape the environment to expand regional forest cover rather than to deplete it. In Niger (Reij, this volume) peasant households are actively contributing to greening the Sahel by protecting and managing trees that volunteer naturally. Research in Burkina Faso and Mali indicates that farmers select and safeguard trees that occur in cultivated fields if they have desirable characteristics (Maranz and Wiesman 2003).

It is now understood in African forestry systems that silvicultural knowledge within a rural population varies with ethnicity, socioeconomic standing, age, and gender (Rocheleau 1987; Schroeder and Suryanata 1996/2004; Schroeder 1999). In rural West Africa, gender roles are highly differentiated, and gender represents a key factor mediating access to agroforestry resources (Fortmann and Bruce 1988; Carney 1988; Rocheleau, Thomas-Slayter, and Wangari 1996; Freudenberger, Carney, and Lebbie 1997). Owing to differential access to, and use of, local vegetation, women and men develop knowledge about distinct resources, sometimes revealing dissimilar knowledge of the same environmental resource. This gender-specific knowledge informs local biodiversity management and is often critical to understanding the presence and diffusion of economically valuable trees (Norem, Yoder, and Martin 1989). The differential knowledge held by women is especially important for the shea-nut tree's management and selection in Burkina Faso, West Africa's largest shea exporter and one of the world's poorest nations.

The Shea Landscape in History

Africa's shea tree (*Vitellaria paradoxa* C.F. Gaertn.) grows in nineteen countries across 5,000 kilometers of Africa's Sudano-Sahelian savanna (fig. 17.1). The natural range of shea includes Benin, Burkina Faso, Cameroon, Central African Republic, Chad, Côte d'Ivoire, Ethiopia, Guinea, Mali, Niger, Nigeria, Senegal, Sudan, and Togo.[2] This region receives between 600 and 1400 mm of precipitation per annum, with rainfall increasing in a southward direction (Boffa 1999; Hyman 1991). Shea trees are resistant to bush fires and may live up to 300 years. Mature trees produce an edible, nutritious fruit that encases an oil-rich nut. Shea forms part of a managed savanna landscape of mixed vegetation and protected trees. Cultivated areas where one or a few well-spaced arboreal species are maintained due to their value to the local population are termed "parklands" in the Sahel.

Vitellaria paradoxa is conventionally described as a wild indigenous species, as its regeneration is chiefly natural, and its presence on the landscape is thought to result from minor modification of the surrounding environment (Chevalier 1948; Hall et al. 1996; Boffa 1999). Yet, this perception

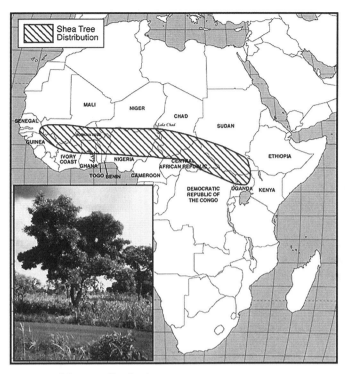

17.1. Map of shea tree distribution

conceals the significant and deliberate human influence on shea tree incidence and arboreal traits. In fact, the species' biogeography, density, and fruit characteristics are shaped by human influence as well as climatic and environmental factors (Boffa 1999). An overview of historical sources, archival materials, paleobotanical evidence, and oral accounts underscores the anthropogenic role in the tree's distributional range.

Palynological and archaeobotanical evidence from northern Burkina Faso indicates that *V. paradoxa* tree management was already occurring by 1000 AD (Neumann, Kahlheber, Uebel 1998; Kahlheber 1999). The shea-nut tree is today abundant on the Labé and Mali plateaus of Guinea's Fouta Djallon highlands (Maranz and Wiesman 2003), where it was reportedly rare during the colonial period (Ruyssen 1957). In Nigeria, the abundance of shea trees where wooded grasslands have replaced forests suggests *V. paradoxa* dispersal beyond its original biogeographical range (Hall et al. 1996). The southern spread of shea involved the suppression of other species, such as the doka tree (*Isoberlinia doka*), and their substitution by savanna species (including shea) esteemed by local populations for their nutritional, economic, ecological, and medicinal values.

Human influence on shea distribution and range accelerated during the 1930s, when British colonial officials launched an initiative to establish on a massive scale the shea-nut tree in the Gambia. They intended to satisfy the burgeoning nut demand by British firms manufacturing margarine and chocolate. Although Gambia shares a similar climate and environment with traditional shea-producing areas, the tree was not found at that time in the colony. At the end of the eighteenth century, the explorer Mungo Park identified the western boundary of shea as northeast of Gambia in Senegal (Park 2000, 84–85, 201). Hoping to meet the metropole's commercial demand for shea nuts, agricultural officials imported 14,000 shea seeds from Nigeria. They planted groves in each of Gambia's agricultural stations and distributed seeds to male household heads in fifty-three villages from one end of the colony to the other (NAG *Annual Reports* 1937–1938). But their ambitious scheme failed because they did not consider the number of years it takes a fruit-bearing tree to reach maturity.

Unfortunately for such commercial objectives, shea is a slow-growing tree. It may take as many as fifteen years to produce flowers, with peak fruit production beginning after forty-five or fifty years and continuing for another fifty. The time from seed to maturity is equivalent to the life expectancy of the average Sahelian. This in itself presents an obstacle to cultivation, but there are other factors inhibiting its dispersal to new areas. The shea-nut tree bears fruit from the end of the dry season into the rains, at a time when farmers are occupied with field preparation and farming. A mature shea tree produces an average of 20 kilograms of fresh fruit annually, but the quantity and quality of the fruit can vary unpredictably over short-term cycles (Chalfin 2000, 992). Where the tree forms a part of agroforestry systems, the fruit is consumed, while nuts are set aside for butter making. With traditional techniques of production, 20 kilograms of fruit typically yields about 4 kilograms of dried nuts and between 0.7 and 1.5 kilograms of butter (Terpend 1982).

Shea Butter and African Women

The shea species has long been central to the subsistence of Sahelian peoples. When processed into vegetal butter, the nuts provide a cooking oil, medicinal, illuminant, and a moisturizer to protect skin from the region's desiccating harmattan winds (Dalziel 1937; Boffa et al. 1996). The butter, which maintains a solid state up to 50 degrees centigrade, does not go rancid. These qualities have given shea an important role in the economic history of the region. Fourteenth-century Muslim scholars, for example, noted its significance to the trans-Saharan trade (Lewicki and Johnson 1974; Hopkins and Levtzion 1981).

Legends that commemorate Sundiata, the thirteenth-century founder of the Mali Empire, suggest a deep historical association between shea butter and African women. Reportedly born a cripple, Sundiata regained the use of his legs after his mother effected a miraculous cure by applying shea butter to them (Sommerfelt 1999, 107). Sundiata went on to create sub-Saharan Africa's greatest empire. Today, across the Sahel, it is women who collect and process shea nuts into butter. Their expertise in the selection, collection, and transformation of the nuts encourages the preservation of specific trees in the landscape.

But the importance of women to shea production was lost on colonial administrators of the francophone and anglophone Sahel. European officials did not consider the agency of Africans in *V. paradoxa* distribution and selection (Elias and Carney 2005). Instead, they viewed the tree's biogeographical range as a natural consequence of climate and seed dispersal by domestic animals, birds, primates, and bats (Burkill 1984; Hall 1996). Local land-use practices were, in fact, seen as inimical and destructive of specimens that would otherwise contribute to export production. More strikingly, given the historical centrality of rural women in shea nut collection and transformation, the influence of females on the shea agroforestry system was not directly recognized.

Traditional Management of Shea Parklands in Burkina Faso

V. paradoxa represents the country's most common parkland species. It is found on both cultivated and fallow lands (Breman and Kessler 1995). Traditional agroforestry techniques do not involve deliberate establishment of the shea-nut tree. Instead, farmers protect individual trees when clearing farmland, thereby increasing its relative abundance on agricultural land with respect to other arboreal species (Boffa 1995; Maranz and Wiesman 2003).

The shea-nut tree extends across Burkina Faso from the semi-arid northeast to the wetter southwest, which, respectively, receive an average of 500 mm and 1200 mm of rain per annum. Shea trees have been recently estimated to occupy 6.5 million hectares of savanna woodlands in Burkina Faso, or one-quarter of the country's total land area (Kessler and Geerling 1994). The species' density and distribution is a consequence of both rainfall and human agency. One study from the late colonial period in Burkina Faso showed average densities of fifty-five trees per hectare in the country's southwest, twenty-five trees per hectare in the densely settled central region, and thirty-five trees per hectare in the north (Terpend 1982).

Studies during the 1990s indicate that the number of standing shea trees had considerably declined as a result of the shift to orchard crops in the country's southwest, the intensification of agriculture in the central region, and cotton monocultures, ox-plowing, and fuelwood scarcity elsewhere (Saul, Ouadba, and Bognounou 2003, 159). These surveys report densities averaging between six to nineteen trees per hectare, a figure that rivals that of neighboring Mali, where the per hectare average is fifteen (Boffa 1991; Maranz and Wiesman 2003). Shea's prolonged growth and maturation period, unreliable production, recalcitrant seeds, and out-crossing breeding system favor the traditional method of sparing desirable trees during land clearance and burning for agriculture (Lovett and Haq 2000a).[3] However, in response to new markets for shea since the 1990s, pilot plantations are under development in Burkina Faso (Saul, Ouadba, and Bognounou 2003), and tree numbers appear to be increasing.

In southern Burkina Faso, the relative occurrence of shea trees in cultivated fields is five times greater than in uncultivated savanna. In northern Ghana and Burkina Faso, shea accounts for more than 80 percent of the woody specimens on farmed land (Boffa 1999; Lovett and Haq 2000a), but only 16 percent of those in uncultivated bush (Boffa 1995). Protected shea trees typically attain a diameter double those of the same age that grow in uncultivated areas (Boffa 1995). Selection has led to a larger proportion of mature productive trees in cultivated areas (Lovett and Haq 2000a). Because of the tree's slow growth rate and long life, landscapes dominated by aged specimens are indicative of long-standing protection by local populations (Boffa 1999).

Traditional management of shea trees also promotes the selection of preferred individuals within a population. Specimens with robust growth patterns and desirable fruit and nut traits are deliberately protected while undesirable individuals are culled for firewood or construction. This selection process has fostered over many centuries a substantial increase of indigenously valued traits in the shea gene pool (Maranz and Wiesman 2003).

Gendered Management of Shea Trees

In a recent study, Maranz and Wiesman (2003) describe genotypic changes in *V. paradoxa* due to tree selection in cultivated areas of Mali and Burkina Faso. They show how strong local selection for desired fruit and nut traits led to selective tree preservation and the culling of specimens with undesirable traits. As shea is exploited principally for pulp and fat, three economically valued characteristics were tracked: pulp sweetness (which is desirable because fruits are eaten), fat content of the seeds, and type of fat in the kernel. All three traits directly affect butter properties. The selection for arboreal specimens with fatty nuts is linked to the role of women in the shea agroforestry system. Women's expertise as nut collectors and butter makers informs the selection of trees for preservation on farmland, as females are the family members with the deepest knowledge about desirable nut qualities (Lovett and Haq 2000a).

Among studied trees, strong local selection has led to a convergence of preferred traits in certain shea populations, with a concentration of desirable characteristics appearing in the most intensively managed parklands. Shea populations in central Burkina Faso displayed the highest kernel fat and saturated fatty acid content. The percentage of stearic acid is a reliable measure of fat hardness, a quality that allows the butter to retain a solid state at temperatures that exceed 120 degrees Fahrenheit. Fat hardness is valued in the Sahel as it eases the molding of butter and the transport of shea products to market (Maranz and Wiesman 2003). Selection for high fat, harder vegetal butter in the Mossi Plateau region of central Burkina Faso has resulted in an increase in the stearic acid trait in managed specimens.[4] But a slightly higher percentage of oleic acid to stearic acid predominates in shea trees in other parts of the country (Maranz and Wiesman 2003; Maranz et al. 2004).[5] Such factors affect shea butter quality in foreign markets, since a lower ratio of stearic acid results in a runny, less viscous butter that does not hold its form as solid pats.

The impact of this type of centuries-long arboreal selection is so pronounced in central Burkina Faso that Maranz and Wiesman (2003) suggest that the Mossi plain may represent the center of domestication of *Vitellaria paradoxa*. This makes sense in light of the central role the species plays among the Mossi and the longstanding Mossi occupancy of the area. The centralized, hierarchical sociopolitical structure of Mossi society has long provided security and stability to the central plain and facilitated permanent residence, which permitted the evolution of highly anthropogenic and densely inhabited landscapes (Boffa 1999).

Lovett and Haq (2000b) describe a similar regional gradient in fat production among Ghanaian shea trees, due to anthropogenic selection for trees yielding fatty seeds. Trees bearing nuts with a high fat content are mainly located in the northern parts of their small and environmentally homogenous study area. While male farmers were unable to associate seed traits with oil content, women—who are experienced in nut gathering and shea butter production—understood that small seeds yielded more oil (Lovett and Haq 2000a).[6] This example exemplifies the gendered differential between the knowledge held by female shea butter processors and that of their husbands.

Brenda Chalfin's (2001, 2004) research on market women in northern Ghana also underscores female understanding of the regional and ethnic differences in nut and oil quality. There are recognizable shea butter traditions among localities, with some locales specifically renowned for the quality of their product. While in part related to different production methods, the shea products suggest that local knowledge systems promote the protection of trees with specific traits. The differences that affect the quality of shea butter have also been observed with another Sahelian food tree, the African locust bean (*néré*), whose pods women ferment and prepare as a condiment (Gutierrez and Juhé-Beaulaton 2002).

Nut and oil quality of shea butter is additionally affected by the locale where fruit collection takes place. Trees in cultivated fields have undergone considerable selection pressure to yield fruits with desired traits while those on uncultivated land often have not. Such trees are typically of smaller diameter (Boffa 1999) and produce fruits with nuts of lesser quality. Nonetheless, women also collect wild shea fruits. Wild shea trees act as genetic reservoirs while contributing to the species' biodiversity in Africa's Sudano-Sahelian region (Marantz and Wiesman 2003).

Conclusion

African management of arboreal resources embraces diverse and dynamic practices that are articulated through the differentiated knowledge of rural populations. This includes specific cultural repositories handed down through generations of women. The process of how these centuries-old interventions shape the ecology of African landscapes merits greater appreciation. Understanding the gender-specific knowledge and expertise involved in the selection, protection, and conservation of desirable species such as shea affects not only local environments, but also the social and economic well-being of women. Making African women visible in agroforestry systems is, moreover, crucial to efforts that aim to improve their abilities to benefit from new market opportunities.

18 ✳ Ancient Forest Tea: How Globalization Turned Backward Minorities into Green Marketing Innovators

NICHOLAS K. MENZIES

Until very recently, the region that includes Xishuangbanna prefecture in China's Yunnan Province, northern Laos, and northern Thailand, was perhaps best known as the Golden Triangle, notorious for production of and illicit trade in opium. The border region, referred to here as Montane Mainland Southeast Asia (MMSEA), is physically distant from political centers of power, a space over which states have struggled to extend their control over people and resources. It is and has long been a place where boundaries are contested and porous, where allegiances have shifted, and where local power brokers have acted as intermediaries to profit—often personally— from their marginality to translate the natural assets of the landscape into personal or local wealth accumulation (Sturgeon 2005).

Competing colonialisms during the nineteenth and twentieth centuries, the extension of the Cold War into the Indochinese wars of the late twentieth century, and state suppression of opium growing supported financially and materially by the global community have meant that remote and marginal as it is, MMSEA has been a part of the global political economy for a long time. Clandestine and informal networks of cross-border exchange, however, are giving way at the beginning of the millennium to the open, legitimate movement of goods, capital, people, and knowledge along the transport corridor being carved out by a new international highway linking the three countries.

This chapter reports on one component of an ongoing study of land-use changes in the MMSEA region funded through grants from the National Science Foundation (NSF) and the National Aeronautics and Space Administration (NASA) to the University of Hawaii's East West Center.[1] In separate contributions to this volume, Fox and Fujita use data from the project to trace land-use changes at a regional scale against a backdrop of economic reform, China's entry to the World Trade Organization (WTO), the recent free trade agreement between the three countries, and the construction of an all-weather road linking the region. This chapter shifts the scale from the regional perspective to the level of Xishuangbanna (XSB) prefecture. It adopts an ethnographic approach, constructing narratives of the different actors in one sector of the prefecture's rural economy—tea cultivation and processing—to examine how they are making decisions about securing and improving their livelihoods in response to the changing economic, political, and social dynamics of globalization. Focusing on the practice of growing tea under the forest canopy in Xishuangbanna, I will show how a land use that was formerly ignored or dismissed as backward has been recast as a sustainable, indigenous technology for the production of a marketable niche product. The growers of forest tea, most of whom are from ethnic minority communities, are challenging established categories of modern and advanced in the discourse of development in China, showing themselves to be far more attuned to the subtleties of global marketing than the official planners and development agencies that are, in principle, leading them into the modern world.

Land-Use Policies and the State in Xishuangbanna

Over the last fifty years or so, the landscape of Xishuangbanna prefecture has been transformed from a mosaic of tropical forest, paddy rice, and swidden cultivation, to extensive tracts of cash crop plantations, mostly rubber and tea—the two key cash crops in the prefecture. In the history of that transformation, the state has played a major role in the promotion, expansion, and marketing of both rubber and tea, advancing a vision of modernity in which large-scale plantations of cash crops effect the transformation of society from a peasant mode of production to a modern, industrialized, and ultimately socialist model.

The decision to introduce rubber to the area was made at the highest levels of the central government in Beijing for strategic reasons during the Korean War. Since the 1950s, strategic and ideological considerations have shaped who grew the crop, where, and how (Sturgeon and Menzies 2008). Physically, the outcome has been a dramatic transformation of the XSB ecosystem with monocropped plantations of rubber replacing what had formerly been composite swidden systems (Rambo 1996) on most of the sloping land between 600 m and 1,100 m elevation. Over the last twenty years, the social and institutional landscapes of rubber have evolved into a continued but increasingly sclerotic State Farm sector, rebranded now as corporations, and a rapid expansion of entrepreneurial smallholder rubber responding rapidly to changing markets, new opportunities for cross-border trade and production, and a growing repertoire of innovative labor and tenure relations (Xu 2006).

The state orchestrated the introduction of agronomic research into rubber, a new and inherently industrial crop, to Xishuangbanna. Tea, a product with a long and distinguished history in Xishuangbanna, has also been the target of official efforts to apply science and technology to increase production in order to promote development and to transform society. From the founding of the People's Republic of China (1949) until the 1990s, the state-supported Tea Research Institute at Menghai conducted research and provided technical expertise for tea plantations on State Farms and for the State-owned Menghai tea company, a subsidiary of the huge, national State-owned trading company responsible for the export of Chinese agricultural products. For the last decade or more, as political and economic reforms have dismantled the planned economy and decollectivized production, the prefectural arm of the central government's Poverty Alleviation Bureau has been the primary source of investment, channeling funds to local governments and households in concert with another national program to reforest sloping land. To the planners and technical agencies responsible for this task, tea should be grown in densely planted monocrops on terraces, requiring official intervention to mobilize villagers to build terraces and to teach them scientific planting techniques, such as the selection of high-yielding varieties, correct pruning, and the use of fertilizers and pesticides. These agencies have consistently—even doggedly—pursued a strategy for development of moving farmers from what are deemed to be backward, unscientific practices, such as swidden or mixed cropping smallholdings, to officially sanctioned modern, scientific, large-scale production.

While official landscapes are composed of extensive tracts of intensively cultivated tea, there are still significant areas at higher elevations with a diverse patchwork of crops including gardens of old, scattered tea trees that have been grown for generations under the forest canopy. Until very recently, the existence and history of this forest tea, also known as "ancient forest tea (gu shu cha 古树茶)," have been invisible to—or deliberately erased by—the government agencies responsible for land-use planning and poverty alleviation. Over the last five years or so, however, urban and international demand for rare and high-quality teas has soared, creating markets for forest tea, a product that has the added cachet of being organic and carries an aura of history and cultural or ethnic authenticity. The website of one gourmet tea and coffee company in the United States lists "Ancient Trees Organic Pu-erh" tea among its rare teas:

> In the mountains of southwest Yunnan is a forest containing semi-wild tea plants, many of them centuries-old and as tall as trees. The local people

have been making organic Pu-erh tea from these old growth tea trees for many generations.[2]

The rediscovery of forest tea is confounding state agencies' conventional constructions of scientific development and introducing a new vocabulary of modernity that accommodates market concepts such as value added, differentiation, niche products, and organic produce.

ENVIRONMENTAL AND SOCIAL HISTORY OF TEA IN XSB

Most of the tea produced in China is small leaf tea (*Camellia sinensis*, or short leaf tea). Xishuangbanna and other regions of southwestern Yunnan, however, cultivate and process big leaf tea (*Camellia assamica*, also referred to as long leaf tea) with a flavor closer to the tea produced in India. Pu'er tea is the main variety planted and produced in southwestern Yunnan. Historians of tea note that despite the name, Pu'er tea was not grown in Pu'er county (now in Simao prefecture), which was historically the center for processing and packing, but from a region of southeastern Xishuangbanna referred to as the "six ancient tea mountains" (*gu liu da shan* 古六大茶山): Yibang shan 倚邦山; Youle shan 攸乐山; Mangzhi shan 莽枝山 ; Gedeng shan 各登山; Manzhuan shan 蛮砖山; Yiwu shan 易武山 (Wu 1990; Zhang 2006).

There is a long history of production and export of tea from Yunnan to other parts of China, Tibet, Burma, and India along what is variously referred to as the Southern Silk Road, or the "ancient horse caravan route" (*cha ma gu dao* 茶马古道) (Yang 2004). In the documented historical record, the first specific reference to tea trade in Xishuangbanna dates from 1398 (Hill 1989, 324–25; Li 1981, 73). The high point in production was during the middle late eighteenth to mid-nineteenth centuries (Jiaqing to Daoguang periods of the Qing dynasty[3]). At this time, there were 170,000 residents and a labor force of 60,000 in the area. After some 130 years of prosperity, a period of warfare and revolution disrupted production and trade and people gradually lost the art of making tea.[4]

Central government interest and involvement in tea production predate the founding of the People's Republic in 1949. In 1939 the Guomindang government sent a researcher from the state-owned China Tea Company to investigate the potential of the area for tea production. By the end of 1940, he had brought twenty-one people from Kunming (where the parent company was established) to set up the factory at Nannuo Shan.[5] The shells of the KMT research station buildings still stand at Nannuo Shan today, and the institution itself became what is today the Menghai Tea Research Institute.

After the founding of the People's Republic in 1949, Menghai became the center of an experimental area for tea cultivation. The area included what became the Li Ming State Farm.[6] The State Farm and, after the establishment of communes in the 1960s, production brigades had to meet production targets and quotas set by the state plan.[7] The Menghai Tea Factory, a part of the state-owned China Agricultural and Livestock Produce Corporation, was assigned responsibility for all processing, purchasing and sales of tea. "The model at the time was to establish 10,000 mu tea plantations . . . with 800 to 1,000 plants per mu."[8]

The large-scale plantation model of tea production dominated through the collective period and beyond. As late as the mid-1980s, some 10,000 mu (625 ha) of "wasteland" (huang shan 荒山) in northern Xishuangbanna were turned into the Dadugang tea plantation. In 1987, the Menghai tea company established "tea production bases," each covering several hundred hectares (chaye jidi 茶叶基地) in high elevation areas in the southwestern part of the prefecture. Throughout the 1990s and into the present, the Jinghong Poverty Alleviation Bureau has allocated government funds to establish tea plantations in villages with incomes below the poverty line, nearly all of which are ethnic minority communities. In Xishuangbanna, tea production came to mean terraced monocultures, with the bushes densely planted and pruned to waist height for easy picking. Success was measured in the area of land planted and in volume and weight of output. The historical legacy of tea received cursory acknowledgement, but any recognition of any other way of planting or producing tea had effectively been erased from view, if not from memories.

REDISCOVERING FOREST TEA AND TRADITION

Zhang Yi, a local historian of tea based in the XSB prefectural capital of Jinghong, describes the "rediscovery" of forest tea in these words:

In August 1994, the Head of the China Tea Research Institute held a conference on China's tea culture. Some researchers from Taiwan wanted to visit the original tea mountains. People told them not to bother because the only real Pu'er tea comes from Pu'er (in Simao prefecture). The Taiwanese said that they had read in old texts that the best tea came from southeast of Pu'er. Officials in Jinghong also told them not to bother going there (to the tea mountains) because there is nothing there. But they insisted and said they wanted to go anyway. . . . But once they got to Yiwu, they saw what we were still doing there. They talked to me and I told them what I knew about the old methods of planting and processing. They were very excited and wrote reports about what they had seen.[9]

The people of Yiwu were still growing and producing tea as large trees under the forest canopy, in "ancient forest tea" gardens.

Forest tea grows on individually managed plots under the forest canopy, generally at elevations above 1,000 m, in areas where most communities are ethnic minorities, such as the Yao, Jinuo, Hani (Akha), and Yi. Some of the tea trees may be naturally occurring wild trees, and others appear to be enrichment plantings using seedlings from natural regeneration. The trees are allowed to grow to a height of several meters (at least 2–5 m), and then they are pruned annually to keep their height and to promote an open structure that allows light and air in. Management also involves some thinning of the forest canopy to maintain an ideal of 20 to 30 percent cover. Tea gardens are not formally marked out in the forest, but families recognize their own plots, which can be inherited or alienated. Contiguous plots in one forest area can be quite extensive. Man Zhuang (Xiangming County) has more than 1,000 mu (62.5 ha) in one extended, connected piece of land in the forest, and some places have a great deal more.[10] Rights to these plots are not legally recognized, but they are recognized and respected in practice. In Yiwu county, informants said that even during the years of collective production under the People's Communes, when all land officially belonged to and was managed collectively by the commune, in practice, individual households were assigned to manage the plots that had been theirs in the past. When land was reallocated to households in the 1980s, their original plots were returned to them.

At the time the Taiwanese specialists visited Yiwu, forest tea, when it was sold at all, fetched no more than seven yuan for one pressed tea cake (*cha bing* 茶饼; 357 g). In early 2007, the average price of 1 kilo of forest tea reached a peak of ¥1,200, while just one cake of a particularly fine, well-aged tea could fetch as much as ¥3,000. Business boomed for small, artisanal family enterprises producing forest tea for sale to buyers who came in person from Taiwan, Hong Kong, and elsewhere to select and purchase a product that had literally been invisible for decades beneath the forest canopy and erased in favor of terraced tea from the repertoire of recognized land-use practices.

While neat, "legible," highly productive terraces of tea remain officials' preferred landscape, demand in domestic and international markets is moving rapidly away from bulk or blended red (known as black tea in the west) or green teas toward a preference for varietal teas, organic teas, and specialty teas. In 2006, officials responsible for agricultural planning, rural development, and poverty alleviation in Xishuangbanna, from the Prefec-

Table 18.1. Tea prices in Yunnan Province, 2005–2008

	Prices quoted by Yunnan Provincial Government (RMB/kg)		Prices quoted by the Menghai Pengcheng Tea Company (RMB/kg)	
	Forest tea	Terraced tea	Forest tea	Terraced tea
2005	60~80	17	40~70	8~34
2006	250	25	80~120	10~38
2007	480	40	800~1200	40~110
2008	Not yet available	Not yet available	300~600	9~43

Source: Data provided by Xu Jianchu, ICRAF, Beijing (March, 2008).

ture government to county governments, all assumed without question that market demand would continue to rise to absorb all of the Pu'er tea from the terraces they continued to plan for and to push farmers to build. In 2007, however, overproduction finally caught up when the prices for Pu'er tea collapsed (table 18.1).

While the price for both varieties of Pu'er (forest and terraced) tea fell, the marketing instincts of the "backward minorities" proved to be sharper than the planners' misplaced optimism. The price of forest tea has dropped by some 50 percent, while the price of terraced tea has dropped to about 35 percent of its value in 2007.

OLD-STYLE MODERNIZERS AND BACKWARD MARKET INNOVATORS

True to the deeply rooted socialist paradigm of development requiring the transformation of the mode of production along a trajectory from a backward and even primitive state to rational and scientific, planners continue to describe upland minorities as obstacles to development:

> Jinuo Shan now has roads, but the Jinuo people are not making progress. They need to improve their quality. They prefer to sleep, play, and enjoy themselves. People like this need to improve their approach and to take advantage of their forces of production.[11]

The trope of ethnic minorities as people of inferior quality persists even as those same minority communities are at the vanguard of producing and marketing the traditional forest tea in the international markets that the planners are actively courting. To bring development and modernization to the prefecture, officials articulate their mission in terms of the current market-oriented development discourse, emphasizing the need to implement policies that will attract investment. Moving from the mission to its implementation, however, the same official explained in the same interview that the Commission planned to direct government funds to increase the area of terraced tea, slipping back into the older, familiar pattern of increasing production through scientific management—of a product for which market demand is falling.

Meanwhile, in the villages of the six ancient tea mountains, the opening of new markets has subverted the old paradigms of modes of production, ethnicity, and development. A portrait of one forest tea producer—one among many—demonstrates just how readily the "backward minorities" have spotted and taken advantage of new market openings.

Meng Zhongcun used to be the Party Secretary of a village on Manzhuan mountain (one of the six tea mountains). He founded his tea company several years ago, after retirement, giving it a name reminiscent of the companies that had produced tea in the region in the late nineteenth century at the end of the Qing dynasty. He now produces more than ten varieties of processed tea. He has no formal contracts to supply any single company but sells directly to buyers who come mainly from Hong Kong and Taiwan, tasting his wares in a purpose built, traditional-style tea garden.

Meng Zhongcun grows forest tea in an area of forest where his father and grandfather produced tea. He produces all his forest tea to match incoming orders, buying loose tea from other growers in the region if needed. Some of the old forest tea is of such high quality that in the past it used to be made into cakes as "tribute tea" (*gong cha* 贡茶) for the imperial court in Beijing. Taking advantage of this history, Meng Zhongcun makes one brand of tea cakes that he calls Man Song Tribute Tea. He is also experimenting with new products to diversify his range. His most recent creation is called *de shui huo cha* 得水活茶, "Tea that comes alive in the water." The tea comes in little bundles with a tea flower inside. When water is poured over the bundle in a glass, the leaves slowly unfurl and the flower emerges, floating in the middle.

Meng Zhongcun also processes "new" tea picked from trees planted in openings in the forest canopy and from an area of land converted to terraces as part of the sloping land conversion program planted to tea with seedlings provided by the government. He carefully keeps new tea and forest tea separate in processing, selling them as different products, with the forest tea fetching much higher prices. He is also experimenting with the management of his terraced tea to improve its quality, by which he means making it closer to forest tea. He is thinning the "scientific" densely packed hedgerows of tea bushes, allowing individual bushes to grow into trees as tall as 2 m and pruning the branches to open the structure to light and air. He is clear that this modified new tea will never be the same as forest tea, but he believes that it will, nevertheless, fetch a higher price, since it will be closer in quality and flavor than normal terrace tea.

In an interesting development, Meng Zhongcun, other tea growers, and even some academic researchers studying the history of tea are recovering (or perhaps constructing) the memory of an ethnic group called the Ben (本) that is said to have preceded the arrival of the Han and even of some of the groups currently recognized as minority nationalities by the Chinese authorities. Although the character *ben* (本) can mean "original" and the term "Ben people" could simply mean "the original people," in its current use in this area, informants were insistent that they were not talking in general terms about the original inhabitants of the area, but about a particular ethnic group called "Ben." The Ben are said to have had a distinct set of customs and beliefs, and they spoke a language called Ben that some people say their parents or grandparents still use to speak.[12] The Ben are becoming associated with a notion of authenticity, as the original producers of forest tea: "Ben people all grow tea so we sometimes use the term 'the Ben people of the tea mountains' (*cha shan ben ren* 茶山本人)."[13] Farmers have been quick to recognize that the cultural history of forest tea is, in itself, a marketable quality.

Meng Zhongcun's story twists the official narrative of state-driven progress and modernity in Xishuangbanna. State farms and expanses of terraced tea were emblematic of progress and socialism. State farms are now struggling to survive as corporations, and it is ancient forest tea that is opening new markets and opportunities. Farmers producing forest tea, village officials, and managers of the commercial enterprises now marketing tea are all keenly aware of the value of tradition, ethnicity, and history in making forest tea a recognized brand with a competitive edge in the market. Their concern is not with backwardness and primitive societies, but with guarding against fraudulent claims of authenticity, quality control, and guaranteeing the credibility of their product. Innovation in production, processing, and marketing is not coming from the planners of development, but from the farmers themselves, and from new entrepreneurs making connections between smallholder artisanal producers and the companies that sell forest tea in the cities and the global marketplace.

Conclusions

Unlike rubber, tea in Xishuangbanna has been a part of local economies and cultures for a long time. The modernizers have to make a case for why the plantation model is superior to the smallholder model. In the past, the case was ideological—the innate superiority of collective production as part of the transformation of the mode of production that leads to socialism. Now, the measure of progress is economic, but the rise of green markets and niche markets gives lie to the proposition that large-scale, intensive, high-volume production is the most efficient and cost effective. Household production is booming.

For all tea growers in Xishuangbanna, marketable now means competitive in a global market. Competitive means that which distinguishes one product from others. In an ironic reversal, what distinguishes Xishuangbanna tea from others are its historical and cultural attributes. The same attributes that had made forest tea production "backward" now make it "authentic" and "far above the usual quality" (in the words of the Peet's Coffee website); a product for which there is high demand in international markets. Natural, traditional, and indigenous used to be qualities officials associated with "backward" and "low quality." They are now the very qualities for which buyers from Taiwan, Hong Kong, and elsewhere will pay a premium. Entry into the world market has transformed the "backward minorities" of the six ancient tea mountains into entrepreneurs defining the cutting edge of green marketing.

19 ✳ The Production of Forests: Tree Cover Transitions in Northern Thailand, Northern Laos, and Southern China

JEFFERSON FOX

In the former opium-growing region of the Golden Triangle a new road was recently opened that is bound to change the social, economic, and environmental fabric of the region forever. The road, inaugurated with great fanfare in early April 2008 by the prime ministers of Cambodia, China, Laos, Myanmar, Thailand, and Vietnam, links the city of Kunming in southern China through Northern Laos to Chiang Mai in Northern Thailand and then on to Bangkok, nearly 700 kilometers away. Route 3, the first modern ground link between China and Thailand, cuts directly through the formerly isolated high mountain areas of the region—the famed Golden Triangle.

In 2004, with funding from US National Aeronautics and Space Administration (NASA)[1] and the US National Science Foundation (NSF)[2] we began two projects to look at ecological and social implications of this road corridor. The NASA-funded project simulated land-cover and land-use changes in the region and modeled how these changes would affect local and regional energy and moisture fluxes, and the consequences of those changes for continental-scale atmospheric circulation and climate. The NSF-funded project examined the political, economic, and social drivers of these changes and their implications for people's livelihoods.

At the close of World War II, the landscape along this corridor and the daily lives of its inhabitants were relatively similar. Shifting cultivation (also called "swiddening") had been practiced for at least a millennium and had greatly influenced land cover and land use across this transect. Most people tended to be defined as and to define themselves as ethnic "minorities"; at national scales their swidden cultivation practices and histories identified them as minorities. Over the past five decades, however, these countries have been under vastly different economic and political regimes, and these differences have influenced land use and land cover in the region today. Thailand has had an open market and democratic government. In contrast to its neighbors, Thailand never experienced any period of land collectivization and has been the most reluctant to provide any form of legal recognition of land title on sloping lands (Thomas et al. 2008). China and Laos are socialist states that have differed in the timing and ways in which they opened their economies to world markets and provided private usufruct rights to natural resources (Xu et al. 2005; Thongmanivong, Fujita, and Fox 2005). Cultural differences in the ways in which different ethnic groups use lands, their customary trading practices, and their relationships with other groups have also affected how they use land and their responsiveness to different government polices and market pressures.

Land-use and land-cover changes are often viewed through two broadly defined lenses, those of structure and agency. In this chapter I review the role played by structural variables — state polices and markets. State policies, of course, do not exist in isolation; they are often overlain, undermined, or reinforced by other factors, such as historic power relations between ethnic groups, local economies, or informal patterns of exchange and communications across the region. Human agency, how different actors at different locations understand, interpret, ignore, or enforce policies at different times, plays an important role in how land is used. In this chapter, however, I examine narratives from Xishuangbanna Prefecture, the most southern prefecture in Yunnan Province, Northern Laos, and Northern Thailand on how an assortment of government policies on issues ranging from forest classification, opium eradication, stabilizing shifting cultivators, to promoting trade and developing infrastructure and markets, affected land use and land cover in different ways in each of the three countries. I also examine what the different types of tree cover we find developing mean for how forests are defined and how people live their lives.

Narratives of Change: Xishuangbanna, Yunnan, China

In response to military needs highlighted by the Korean War, the Chinese Central Committee introduced rubber to Xishuangbanna in the early 1950s as a strategic, industrial product to be produced on large-scale state collective farms.[3] In 1955 the Bureau of Reclamation began to organize extensive land clearance using demobilized soldiers, often veterans of the Korean War, almost all of whom were Han Chinese. In 1960 as the pace of rubber planting quickened to fulfill the targets of the Second Five Year plan, State Farms began to experience labor shortages. In response, the Central Committee decided to bring migrants from rural areas of Hunan. The next wave of State Farm workers arrived during the Cultural Revolution when educated youth from the cities were sent to rural areas to learn from peasants and to bring the revolution to backward and rural border areas. Between 1968 and 1978 more than 100,000 young men and women from cities across

China were sent to work on the State Farms in Xishuangbanna (Yunnan Province Bureau of Reclamation 2003; Deng 1993).

Throughout the period when the State Farms were being established, local minority farmers labored on agricultural communes. In 1982 the Chinese government dismantled the farming communes and introduced a new ideology of land use, turning farm households into entrepreneurs responsible for caring for their own needs. Farmers received commune land, but needed to pay for the educational, health care, and local services once provided by the communes. The farmers' new needs for cash caused them to convert their available land to cash crops. Initially, lowlands became wet rice fields and uplands were used for livestock. But a major state campaign encouraged upland farmers to plant rubber at elevations below 700 m in fields used for swiddening. State Farm personnel provided seedlings and technical training. Later, a subsidized state antipoverty campaign encouraged farmers to plant rubber on sloping lands.

After 2002, the incentive for planting rubber became even stronger. A new "Grain for Green" campaign, intended to promote the development of China's western provinces and protect the environment, provided farmers grain for eight years if they planted forest cover on degraded slopes. In Xishuangbanna, the authorities decided to count rubber trees as forest cover. About the same time, a rapid rise in rubber prices occurred. Eager for wealth, households began planting rubber in their traditional woodlots, in village forests, and on the remaining and steeper slopes. Below 700 m, but even higher, rubber became ubiquitous.

Today, rubber farmers in Xishuangbanna have achieved unprecedented wealth. Janet Sturgeon (2010, 325) quotes an Akha rubber farmer as noting that "Money is the most important thing; money makes everything possible." That "everything" includes sending their children to high school (and for some, on to university), buying insurance for retirement and health care, and even a holiday in the city for an entire village. Indeed Sturgeon argues some ethnic-minority rubber farmers in Xishuangbanna have achieved a standard of living today that has more in common with middle-class urban residents than with most fellow farmers.

The exact reason for the success of these farmers, other than the price of rubber, is not totally clear. I speculate that it has to do with clear tenure rights (long-term leases), previous knowledge about how to grow rubber gained from working on State Farms or from friends and relatives who have worked there, the existence of the infrastructure necessary for collecting and processing rubber, and perhaps a relatively sophisticated population in comparison with ethnic minority populations elsewhere in mainland Southeast Asia. Regardless of the reason, many small ethnic minority farmers in Xishuangbanna have become financially successful rubber cultivators over the last two decades.

The environmental implications of the rubber story, however, are less positive. Xishuangbanna has the highest biodiversity in China; it is included in the Indo-Burma biodiversity hotspot identified by international conservation organizations (Myers et al. 2000). The Chinese Central Government set aside approximately 240,000 ha or 12 percent of the total land area of the prefecture as the Xishuangbanna Biosphere Reserve (Guo et al. 2002). Yet almost all land not protected in the reserve (and perhaps some within) that lies between 400 m and 1,100 m has been converted to ordered rows of rubber. Sturgeon and Menzies (2008) note that state rubber plantations count as "forest cover," even though State Farm personnel acknowledge that tropical rainforest were removed to clear land for rubber.

Hydrological implications of rubber are also not benign. Based on long-term observations, Liu (1990, 55–111, reported in Wu, Liu, and Liu 2001) showed a negative relationship between the presence of fog and the increase of rubber plantations. Wu, Liu, and Liu (2001) showed that surface runoff increased by a factor of three, and soil erosion increased by a factor of 45 as a result of conversion from tropical forest to monoculture rubber. Our NASA-funded research (Guardiola-Claramonte et al. 2008) suggests a striking difference in the timing and rate of dry season root-water extraction under rubber as compared with native vegetation. We have documented that rubber's water demand is concentrated around the equinox (the peak of the dry season), when soil water availability is at its lowest and atmospheric demand is at its greatest. In similar settings (where precipitation and atmospheric demand are "out of phase"), changes in native vegetation have resulted in dramatic decreases in streamflow and/or groundwater levels (Wilcox et al. 2006).

Narratives of Change: Northern Laos

Across the border in Northern Laos a different story has unfolded.[4] Postwar political insecurity until the mid-1980s prevented active development efforts in the uplands; government agents, however, encouraged upland people to move to lower elevations and to consolidate their villages into larger centers that could be assisted by government service providers. In the early 1990s, the Lao government introduced measures to demarcate forests and protected areas (National Biodiversity Conservation Areas Law of 1993 and Forest Law of 1996). The National Land and Forest Allocation Policy (LFA), proclaimed national policy in 1996, supported the delineation of village boundaries, and recognized village's rights to manage and use forest resources as well as farmers' rights to use agricultural land (Fujita and

Phanvilay 2008). The LFA categorized forest areas and agricultural lands at the village level, and in the process, sought to stop shifting cultivation in upland areas.

In the mid-1980s, the Lao government began to liberalize the market and to promote private sector activity. This was followed by the removal of agricultural price regulations, production quotas, and agricultural taxes, actions that allowed farmers to sell their products freely without government intervention. In the mid-1990s the goal of the government's agricultural policy switched from one of improving food production to an emphasis on integrating rural farmers into the market economy.

In the early 1990s the government opened international borders with neighboring countries and began constructing improved roads. These changes launched new economic opportunities, particularly for farmers living near roads. These changes also released an influx of people and goods. Chinese investors initiated new projects including investments with relatives who live across the border as well as small-scale private investment that encouraged farmers to transition from subsistence to cash-crop production. A Chinese national government program provided Chinese investors with government funds for projects that sought to eradicate opium in Laos and Burma. Under the guise of opium eradication, these investors supplied Lao farmers with planting materials for cash crops ranging from rice, watermelons, chilies, and pumpkins in the lowlands, to maize, sugarcane, and rubber in the uplands. The combination of investment capital and market opportunities encouraged Lao farmers to plant these crops in their dry season paddies and upland agricultural fields (including both active swiddens and forest fallows). These incentives also created agricultural wage labor opportunities in areas where cash crop production became prominent. These new economic opportunities motivated upland households to move to lower elevations near roads to access both agricultural land and wage labor opportunities.

In 2002 the national government initiated a national campaign to eradicate opium cultivation in upland areas by 2005. Villages found cultivating opium were fined and their leaders detained in district centers for "reeducation." These activities triggered another exodus of upland people to lower elevations, where they often settled near relatives or earlier settlers. Fujita et al. (2007) documented a 20 percent decline in the number of Akha villages in Sing District between 1995 and 2005, as villages were relocated at lower elevations and consolidated. They also showed that the total population in mountainous subdistricts declined, sometimes by as much as 30 to 50 percent. They documented the expansion of lowland agriculture in Sing District at annual rates ranging from 4 to 11 percent. They showed permanent agriculture (rubber) increasing in the uplands at rates of 3 to 7 percent between 2000 and 2004.

Phanvilay (2010) documented the expansion of rubber in an Akha village on the Laos side of the Laos–Chinese border. He found pioneering settlers first planted rubber in 1994 and 1995. These early adopters planted the crop because they had gained experience growing and tapping rubber working with relatives in China, where small holders started planting rubber in the mid-1980s, shortly after China dissolved the communes and allocated agricultural and forest lands to farmers. The privately owned rubber trees in China began to produce latex in the early 1990s, making the farmers who owned them comparatively wealthy. Soon after, Laos farmers decided to start planting rubber too.

While farmers in the village observed by Phanvilay have an LFA plan, they do not have formal land titling; an informal land market, however, does exist. A few of the early adopters capitalized on the vulnerable economic status of late adopters by buying their land. Thus, those who missed the first waves of conversion to rubber and then sold land are relegated to receiving only secondary benefits (wage labor opportunities). The village committee that is responsible for implementing the LFA plan is under significant pressure to help poorer households find additional land for agricultural production. More importantly, in 2007 the Laos military and a private company both planted in excess of 200 ha of rubber within the village's boundaries.

Land use is changing rapidly in Northern Laos, reflecting broader patterns operating in other parts of the uplands of Southeast Asia (Li 2002; Fox, McMahon et al. 2008). The commercialization of farming systems has created a new source of income for many families, at the same time stimulating land markets and accelerating land alienation. Unlike in neighboring Xishuangbanna, where land-use and tenure policies are clear and communities have maintained legal rights to their land, some communities in Northern Laos are selling their land and beginning to experience a chaotic pattern of land-use and tenure change. The future may depend on the extent to which these communities receive support from outside agencies, including both nongovernmental organizations (NGOs) and government programs, to protect their lands from illegal land speculators. Another key input in rubber cultivation is labor. With plantations expanding beyond local labor capacity, labor shortages, and migration, both internally from other parts of Northern Laos, and externally from China, is already under way and will continue to rise in the coming years.

The environmental implications of land-cover change in Northern Laos are not yet clear. Fujita et al. (2007) documented changes in forest cover between 1973 and 2004 in Sing District. They observed that dense forest cover declined from 68 to 42 percent between 1988 and 2004, and secondary forest increased from 16 to 35 percent. Overall they documented a loss of 7 percent of total forest cover during this sixteen-year period and an increase in forest fragmentation. Concurrent with the loss of forest cover and forest degradation has been the rapid expansion of rubber plantations near towns and roads.

Narratives of Change: Northern Thailand

Our research in Northern Thailand focused on the Ping River basin, a major source of water for Central Thailand and Bangkok.[5] The Ping basin includes the provinces of Chiang Mai, Lamphun, Tak, Kamphaengphen, and Nakhon Sawan. Thai national policymakers have been concerned for years about land-use practices and their implications for watershed management in the Ping basin. In the 1960s and 1970s, large areas of the basin were set aside as forest reserves in order to protect forests and watersheds; many of these areas were subsequently declared national parks and wildlife sanctuaries. In the mid-1980s, a national program was launched to classify all lands in the country according to their watershed characteristics, which, under the guise of science, placed new restrictions on the ability of people to use land (Thomas 2005).

In addition to constraints on land-use practices, people living on these protected lands are not eligible to apply for official land-tenure documents. Most people in midland and highland zones of Northern Thailand have no form of official recognition of their rights to use land for any purpose whatsoever because their land is claimed as state forest land, and/or because use is restricted by Class 1 or 2 protected watershed status, or at least limited by Class 3 or 4 status. Thus, land cannot be used as collateral for access to normal institutional sources of credit, it is often difficult to defend against encroachment by outsiders, and elected local governments have no legal basis for imposing taxes or otherwise regulating land use. While local land-use practices in many areas still include various forms of agriculture, state agencies have continued to expand protected forest areas (national parks and wildlife sanctuaries) that legally exclude all other forms of land use. Resulting tenurial insecurity is an important disincentive for livelihood strategies that include longer-term investments at either community or household levels.

Although land use is officially forbidden or heavily restricted in most upland areas, many forms of land use are present and are continuing to evolve. These forms of land use are sanctioned and governed by local institutions.

Thus, there are substantial and growing discrepancies between what land use is recognized by institutions at these different levels.

A study by Benchaphun et al. (2005) found that farming systems at lower elevations in the Ping basin are diversified. While farmers still grow rice in the wet season, they have adopted diversified cash crops in the dry season, including soybeans, garlic, shallots, tobacco, baby corn, potatoes, tomatoes, sweet corn, onions, cabbage, and other vegetables. Contract farming and processing of farm products is common. High-income markets are tapped with good packaging technology and marketing skills. Speedy transportation and communication via good roads, railways, airline, postal, telephone, and Internet services have expanded market outlets for crops and products.

Since 1969 the Royal Project Foundation (RPF) has promoted new crops to replace opium in the highlands of Northern Thailand (Thomas et al. 2008). In the early days the RPF solicited wide cooperation from government and nongovernmental agencies, including foreign donors and technical assistance agencies. Through these partnerships RPF conducted on-farm research in new, alternative crops suitable for the highlands. Most of the crops introduced were temperate, high-value, capital-intensive crops, such as vegetables (lettuce, strawberry, Japanese pumpkin, zucchini, bell pepper, carrot); flowers (statis, gypsophylla, carnation); and fruits (avocado, peach, pear, apricot, apple). Later, the RPF launched marketing strategies, including the creation of Royal Project outlets, shops, and a brand name (Doi Kham). Technical assistance was contributed by a range of professionals with experience in the private sector. Resulting lines of products have successfully tapped well-off urban segments of the market. One of the least discussed, but perhaps most important, aspects of what the RPF has accomplished has been the establishment of systems and operations that are generic enough that the same facilities and systems can be used for product lines that include a relatively diverse and evolving set of products, rather than the single commodity or narrowly focused chains that are more common in the region (i.e., rubber).

A survey of farm households in Chiang Mai, Lamphun, and Chiang Rai conducted by Benchaphun et al. (2005) found that nonfarm income constituted approximately one-third of total household income. This income comes from a very diversified range of activities, including trading, handicrafts, food processing, wood carving, construction, wage labor in neighbors' farms, factories, remittances, and so on. As the nonfarm sector has grown, employment opportunities for farmers and their household members have also grown. Moreover, the growth of the tourist sector in Northern Thailand has provided many villagers with income earning opportunities, including those related to provision of guest houses, restaurants, rafting, trekking, and souvenirs. In Mae Wang watershed, for example, the research team found that ecotourism services as well as handicraft production for tourists have flourished, and now provides very substantial supplementary income for villagers.

Among the government policies that have, perhaps, had the greatest impact on the livelihoods of people in Northern Thailand are those that have denied recognition and security of local claims to land resources and those that have expanded protected forest areas where all other forms of land use are legally excluded. Thailand finally passed a community forest bill in November 2007, after eighteen years of drafts and revisions; the bill allows people living in forests to participate in forest management and to support government forestry efforts to protect forests. Two provisions in the law (articles 25 and 35), however, are viewed by many as depriving communities of their rights to manage forests and to use resources that they have been protecting for generations. These articles exclude from eligibility to community forest rights 20,000 forest-dependent communities living outside protected areas and prohibit even eligible communities from cutting any trees in these forests. Some say the community forest bill thus deprives rural populations of traditional rights on which their livelihoods depend. Forest policies, however, have often been implemented in a manner where some illegal land-use practices have been tolerated and in an environment where off-farm employment opportunities have been available. Thus, local people's livelihoods have not deteriorated and in many cases have improved significantly.

The environmental impacts of these policies are not known. Thomas (2005) has documented the expansion of tree cover in many villages where swidden agriculture used to be practiced or was practiced more widely. There has not been a comprehensive study, however, that suggests that tree cover has increased significantly in the region. Nor, to my knowledge, has there been a study of the ecological characteristics of this newly regenerated tree cover.

Policy and the Production of Forests

In this chapter I have reviewed narratives of how public policy has affected the production of forests over the last half century in the three nation-states that lie along the new highway corridor linking Chiang Mai, Thailand, with Kunming, China. These narratives suggest that in the early 1950s tree cover across this transect was fairly homogenous. Over the past six decades, however, natural resource management policies in the three countries have differed significantly. This is apparent in terms of policies related to conservation, watershed management, and the promotion of agroforestry crops. Policy objectives in Thailand and Laos have been similar with respect to conservation and the expansion of protected areas; both countries have attempted to segregate agriculture from forested landscapes. But Laos is seeking to implement localized zoning processes aimed at providing security for land use by local households and villages, including recognition of village forest areas and local commercial production of both nontimber forest and agroforest products. Thailand, in contrast, seeks to limit any local use of forest products to subsistence purposes. Policies in China have been considerably less aggressive in expanding exclusionary protected areas or restricting production of nontimber forest products; China has also been considerably more flexible and willing to accept multifunctional agroforestry landscapes.

In Xishuangbanna we have seen the conversion of tropical rainforests into a nearly homogenous carpet of rubber between 400 m and 1100 m. In Northern Thailand, natural forests (as defined by the Royal Forest Department) still cover approximately 55 percent of the landscape. In Northern Laos, the story is mixed with a large portion of the landscape still covered by native trees, but the conversion to rubber occurs at a rapid pace. In all three countries, however, the tree cover is anthropogenic. In Xishuangbanna this is obvious, with man having forcibly removed the forest to replace it with rubber. In Northern Thailand, natural forests have not been removed, but they been affected both by the traditional swidden cultivation practices that shaped them for a millennium or longer, as well as by modern-day conservation practices. In Northern Laos, tree cover has been affected by all these factors.

Accessing the success of these policies depends on who is telling the story. Clearly from the perspective of many small ethnic minority farmers in Xishuangbanna the conversion of forests to rubber agroforestry has been an overwhelming success. Their household income has multiplied, and they count themselves among the middle class in a rapidly changing market economy. From the perspective of a conservationist, however, the conversion has been an unmitigated disaster, destroying a biodiversity hotspot, one of the most biologically rich landscapes on the face of the earth. The farmers have also made themselves vulnerable to market and environmental forces. A long-term fall in rubber prices or an outbreak of a new virus affecting rubber trees could wipe out the good fortune of many of them quickly. Likewise, in Thailand, the protection of more than half the landscape under some form of conservation or watershed management has probably been a great success from the perspective of a conservationist (although we still know little about the ecology of these landscapes). But it has come at the cost of people's livelihoods being upset and many people being denied legal right to their ancestral land. The impact on people's lives could have been much worse if Thailand had not been doing so well economically during this period, making alternative futures possible. Thailand has also been unusually successful in developing a diverse and evolving set of agricultural products, rather than relying on a single commodity. In Laos, the future is not yet clear. Will it follow the example of Thailand, a protected landscape where people have been excluded and land use curtailed, or the example of Xishuangbanna, an agroforestry landscape where the role of nature has been minimized? Or worse yet, will it convert its landscape to a rubber monoculture and fail to protect the interests of small farmers? Laos has the opportunity to learn from the lessons, both inspirational and cautionary, from Thailand and China. Ideally, the Lao people and government will manage to locate themselves in a diverse multifunctioning landscape that includes areas for growing food, commercial agroforestry products, and sustainable forest management, and where small farmers can continue to find a home. For better or worse, government policies will play a major role in how this future plays out.

20 * From Swidden to Rubber: Transforming Landscape and Livelihoods in Mountainous Northern Laos

YAYOI FUJITA LAGERQVIST

Introduction

Rural production system in Lao People's Democratic Republic (hereafter Laos) is rapidly moving away from subsistence to market-based agricultural production system. Development of commercial agriculture and trade are particularly prominent in border areas as regional trade prospers. This includes both formal and informal trade, which involves activities of both registered investors and private individuals transporting goods across the border.

Northern Laos is predominantly mountainous. It is a region that is one of the poorest, where rural farmers depended on shifting cultivation or swiddening in the upland fields. However, upland farmers are beginning to convert swidden and fallow lands to cultivate cash crops such as sugarcane, maize, cassava, and rubber (Lao Extensions for Agriculture Project [LEAP] 2007; Thongmanivong and Fujita 2006). Natural rubber is especially spreading rapidly in the upland areas of northern Laos since 2000.

This chapter summarizes the main results of a study conducted in Sing District in Luang Namtha. It is a part of a regional multidisciplinary research project that received funding from the National Science Foundation (NSF; see also Fox, chapter 19, this volume). The regional project involves a range of researchers from the Chiang Mai University (Thailand), the Kunming Institute of Botany (Yunnan province, China), the National University of Laos, the World Agroforestry Centre (Thailand and China), and the East West Center (United States). A joint team of researchers, including the current author, as well as teachers from the Faculty of Forestry of the National University of Laos, carried out the research in northern Laos.

The current chapter focuses on the changes taking place in Sing District, one of the research sites in Laos. The main aim of the research is to understand patterns of land-use change in the upland areas where commercialization of agriculture is rapidly taking place. It also aims to understand the driving forces of change and the implications for local people's livelihoods and their relationship with land. The study shows expansion of agricultural lands in the upland areas replacing old swidden and fallow lands, as well as encroachment of commercial agriculture into forest. The study also indicates that there is complex interaction between local stakeholders and evolution of policies that influence households' decisions regarding land use and their livelihoods.

Research Site

Sing District is a northern district of Luang Namtha province bordering Burma and China (figure 20.1). Prior to French colonial intervention, Tai Lu people ruled this region and called it *Xieng Kheng* (later *Muang Sing*). The small principality thrived on trade. Yunnanese Chinese caravans traveled to and from China on their way to Burma and Siam through *Muang Sing* (Walker 1999). It was also a critical region that determined the British and French colonial boundary along the Mekong in the late nineteenth century. Records of Europeans who reached this region during the late nineteenth century and early twentieth century describe its lively markets, where not only the Lu people in the lowlands but also the upland Akha traded forest products (Izikowitch 1944/2004). Lives of people in this region were devastated during the period of the Indochina War until the establishment of the socialist regime in 1975. However, as the regional borders reopened in the early 1990s, it once again became a thriving trading point.

FOREST DEGRADATION AND EXPANSION OF AGRICULTURAL LAND

A team of Geographic Information Systems (GIS) experts based at the National University of Laos analyzed satellite images to assess changes in agricultural and forestlands in Sing District. Figure 20.2 shows that while percentage of total forest area remained close to 80 percent in 2004, the proportion of dense forest dropped from more than 60 to 40 percent between 1973 and 2004. Table 20.1 also indicates that between 1973 and 2000, numbers of forest patches increased by 38 times from 164 patches to 6,306. The mean size of forest patch has also declined drastically from 446 ha to 7 ha between 1973 and 2000. These results suggest loss of forest areas and fragmentation into smaller patches.

However, figure 20.2 also indicates expansion of agricultural land. The lowlands experienced cycles of decline and growth. In the lowlands, development of irrigation contributed in expansion of paddy fields. In the upland, however, agricultural production shifted from being rice-based swiddening to commercial cash crop production. Table 20.2 indicates decline

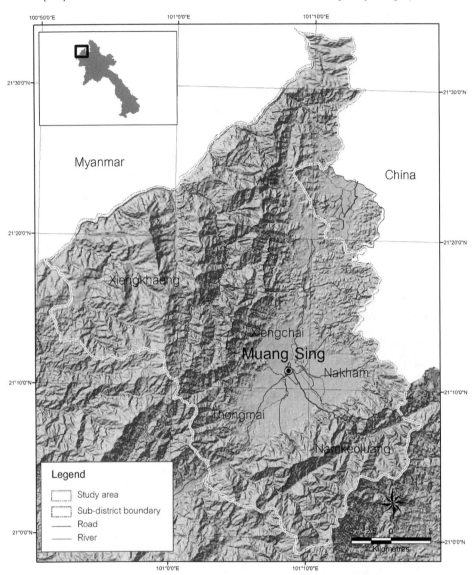

20.1. Map of research site

of upland rice cultivation from 1,515 ha to 500 ha during the period from 1999 to 2006. The sharp decline of rice cultivation area occurred between 2002 and 2003. At the same time, figure 20.3 indicates increasing areas of upland cash crop production during 2003 and 2006. Sugarcane production steadily increased during this period, while rubber skyrocketed between 2005 and 2006. The District Agriculture and Forestry Office (DAFO) re-

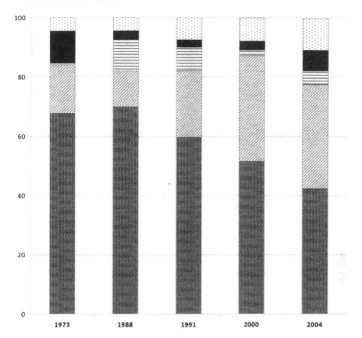

20.2. Land-cover change in Sing District, Laos. Source: Spatial analysis by Thongmani-vong (2007).

Table 20.1. Forest fragmentation

Years	1973	2000
Numbers of patches	164	6,306
Average size of patch (ha)	446	7

Source: Spatial analysis by Thongmanivong (2007).

corded fifty-three villages planting rubber in 2004; this jumped up to seventy-three villages in 2005.

COMMERCIALIZATION OF AGRICULTURE AND HOUSEHOLD LIVELIHOODS

Upland farming systems based on traditional swidden cultivation are highly dependent on forest resources and allow farmers to diversify their livelihood activities (Raintree 2001). However, Bouahom, Douangsavanh, and Rigg (2004) claim that increased market opportunities and farmers' need for cash income are pushing them to become involved in commercial agricultural production.

Table 20.2. Rice production in Sing District

Year	Lowland rice			Upland rice		
	ha	ton	ton/ha	ha	Ton	ton/ha
1999	3,652	13,747	3.8	1,515	2,575	1.7
2000	3,829	14,219	3.7	1,379	2,506	1.8
2001	4,302	15,082	3.5	1,326	2,254	1.7
2002	4,511	16,290	3.6	1,326	2,259	1.7
2003	3,007	7,530	2.5	651	520	0.8
2004	5,444	21,626	4.0	530	901	1.7
2006	5,115	20,460	4.0	500	1,000	2.0

Source: Lyttleton et al. (2004); District Agriculture and Forestry Extension Office of Sing District (2005, 2007).

In Sing District, we interviewed households from seven selected villages in three subdistricts to understand farmers' production activities. In the mountainous Xiengkheng subdistrict bordering Burma, road access is limited. However, people traveled to and from Thailand and China using the Mekong River. Mom subdistrict, adjacent to Xiengkheng, is a hilly region bordering China. A new road connects Mom subdistrict directly to Meng Peng County in Xishuangbanna prefecture, which is located to the west. In the last region, Nakham subdistrict, gently sloping hills surround extensive areas of lowland paddy fields; this region borders Meng Mang county of Xishuangbanna and is where the regional border between China and Laos is located.

Figure 20.4 indicates household production in three subdistricts. In all villages, main household production includes paddy, upland rice, cash crop (maize, sugarcane), livestock, and other activities, including collection of nontimber forest products. Figure 20.4 indicates the relative importance of livestock in two upland subdistricts (Xiengkheng and Mom). Out of all three subdistricts, household production in Akha village of Mom sub-district are highest, while production is the lowest in Tai Lu village of Xiengkheng.

Although Nakham subdistrict along the road to China was initially expected to have the highest household production, given its accessibility to market and road, as well as household access to lowland paddy fields, their average production was lower than those in Mom subdistrict. This is due to low livestock production in Nakham. Farmers from this region claimed that numbers of large livestock declined in the last decade due to theft of cattle and loss of grazing land. Farmers also explained that population increase in the region and expansion of commercial agriculture had caused these problems.

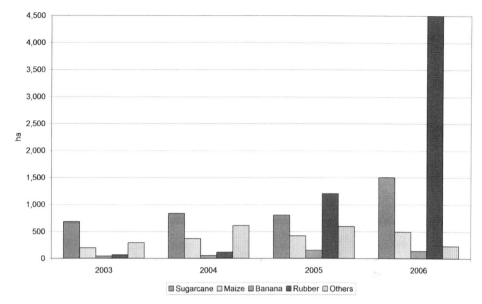

20.3. Areas of upland cash crop production (unit: ha). Source: District Agriculture and Forestry Extension Office of Sing District (2005, 2007).

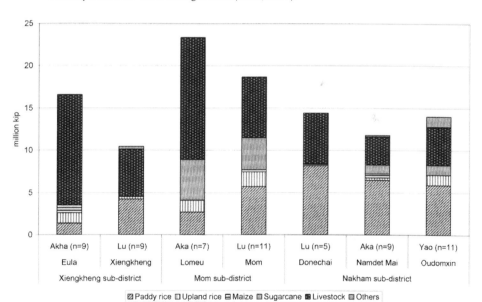

20.4. Household agricultural production (unit: million kip). Source: Fieldwork (2005).

Table 20.3 indicates shares of household inputs in agriculture. The total household input for agriculture is exceptionally high in Mom subdistrict, ranging from 26 to 31 million kip (approximately 2,750 to 3,300 US dollars), where more than 60 percent of the inputs are concentrated in sugarcane and rubber. In Nakham, it ranges from 9.8 to 16 million kip (approximately 1,000 to 1,600 US dollars). In Donechai, a Lu village concentrates on paddy production, while in Namdet Mai and Oudomxin, farmers invest in lowland paddy rice production, as well as upland crop production including sugarcane and rubber. In Xiengkheng, household agricultural inputs range from 8.8 to 13.9 million kip (approximately 930 to 1,500 US dollars) but are mainly concentrated in paddy and upland rice production for household consumption.

Table 20.3 also shows that farmers across Sing District are investing in a new crop: rubber. Input in rubber is the highest in Lomeu, an Akha village in Mom subdistrict. A village leader who was one of the first to plant rubber in his village claimed that the main motivation was the profitability of rubber and material wealth that it brought. He claimed that his relatives in China were poor like Akha people living in Laos before. However, after successfully planting rubber they began to accumulate wealth. A notable indicator from the village leader's point of view was possession of new homes and vehicles. Lomue villagers crossed the border on a daily basis to sell and exchange products in China. It was closer for them to cross the border than to head down to markets in the district center. Many women from the village also went for wage labor during agricultural seasons.

In a neighboring Mom village, farmers' inputs in rubber are also high. Here, the Lu villagers also cross the border on a daily basis. They began to cultivate rice and sugarcane for markets in China after the road leading to China from their village was improved in the 1990s. In the past, people used to raise livestock and take it to the district center in *Muang Sing* for trade.[1] The village lost their grazing lands as more Akha farmers resettled in surrounding areas, both spontaneously and because of government policies that relocated the upland population. The village also lost an area of forest along the Chinese border to the Lao military. This caused anxiety among some villagers who were requesting additional land for rubber plantations and negotiating land-use plans with the district authority.

Even in the seemingly remote parts of Sing District, where farmers were principally engaged in rice production, farmers are now planting rubber. There is a Chinese Akha trader living in Xiengkheng village who collects forest products from villagers and ships them to Jinghong by boat. She also supplies local villagers in the region with agricultural inputs and other goods. Although the German bilateral agency Deutsche Gesellschaft für

Table 20.3. Shares of household inputs in agriculture

Subdistrict	Village	Average household input in agriculture (million kip)	Paddy (%)	Upland (%)	Maize (%)	Sugarcane (%)	Rubber (%)	Others (%)
Xiengkheng	Eula	8.8	23	57	17	0	1	1
	Xiengkheng	13.9	75	8	0	0	8	9
Mom	Lomeu	26.1	10	26	0	21	41	3
	Mom	31.3	16	4	3	39	37	0
Nakham	Donechai	9.8	60	0	6	0	9	24
	Namdet Mai	16.0	34	2	7	47	8	2
	Oudomxin	15.0	40	15	0	22	22	0

Source: Fieldwork (2005).

Technische Zusammenarbeit (GTZ) and DAFO have been working in this region since the 1990s, introducing different cash crops as an alternative to opium, production of alternative crops did not take off. Sugarcane factories in Meng Peng also did not reach this region because of the difficulties of the road access. However, Lu and Akha villagers in Xiengkheng are planting rubber on their own, purchasing seedlings from the Chinese Akha trader. Seeing their relatives' success in China many of the Lao farmers felt more confident that they could also grow rubber in Laos.

Rubber is also a prominent agricultural investment in Nakham region, inducing competitive land use. Yao farmers of Oudomxin were repatriated from Thailand in the early 1990s. The district government designated the residential area for the people while the United Nations High Commissioner for Refugees (UNHCR) provided them with housing materials. Some farmers purchased paddy fields from nearby villages. Others practiced upland swidden cultivation in the gently sloping land surrounding the village, which the district government recognized as part of the village under the land and forest allocation process (LFA). However, the rotational cycle was short because of limited access to swidden fields. This forced many farmers to switch from upland rice production to cash crop production. Farmers began contract farming with Chinese investors, first producing sugarcane in their upland field. Some farmers rented lowland paddy fields from neighboring villages to cultivate watermelon and other vegetables during the dry season. One of the farmers who started planting rubber says he observed his relatives in both China and Thailand. "My relatives in China were successful with rubber. I compared it with other crops. Relatives in Thailand had success with fruit trees. I visited both places and observed the market. They were both successful but rubber seems to make more economic sense here."[2] The farmer taught himself everything about rubber by working in rubber plantations in China and in Thailand. "Unlike in China or Thailand, there's no support for rubber planting in Laos. Banks do not give you a loan. There are no places where you can purchase good quality inputs."[3] He and his family acquired upland fields outside of the village and planted rubber. He is a local expert equipped with knowledge based on his own experiences and was expanding his business. For instance, he was selling his seedlings to relatives living in northern Lao province of Luangpabang. He not only sold the seedlings, but also provided technical advice for his relatives on planting rubber.

In the same Yao village, there was a growing tension with the neighboring Akha village over agricultural land use. For the second time, the neighboring Akha village encroached into the watershed protection forest of Oudomxin. The first time, they were clearing forest for sugarcane. This time they were clearing the forest for rubber. The district authorities mediated village meetings, inviting leaders of both villages. Throughout the meeting, they discussed the boundaries recognized by LFA. After the meeting, Akha village continued with their plan to plant rubber on the basis that the land was part of their customary land. "Nobody was here when we first began to move down here. We have households without any land for rubber. We are going to share the new area for each household."[4] The incidents left leaders of Yao village dissatisfied and they questioned the meaning of carrying out LFA, which produced official documents that recognized rights.

In the same region, a lowland Lu village lost legal access to upland agricultural fields that surrounded the village as more people moved in. "Our new village boundary based on LFA doesn't include the hills. They are all occupied by other people."[5] Farmers in this Lu village seemed too busy to be preoccupied with loss of upland fields where they once grazed their cattle. Households were busy with paddy field cultivation and dry season vegetable cropping. They no longer had large livestock, as they lost their grazing land, and they began to cultivate dry season crops. Those who were not cultivating leased the land during the dry season for others to cultivate. Neighboring Yao villagers as well as Chinese people came to rent their land. "Our lives in the village changed dramatically after the trade with China resumed. People in our village are always busy nowadays. We also have to hire Akha laborers for our agricultural activities because people no longer have time to share labor in our village."[6] They have also tried sugarcane, but quickly realized that they did not have sufficient labor, and it was not as profitable since they had to purchase all inputs from a Chinese company. Farmers quickly learned about rubber from their relatives in China. Some of the villagers also became local experts as they worked on state rubber farms in China.

Although Donechai villagers lost access to customary lands surrounding their village, finding new access to upland fields was not a significant problem for them. They found ways to work with Akha villagers who settled near their village. They planted rubber for Akha farmers on their land, and shared the ownership of trees. However, even before they could tap the latex, most Akha farmers borrowed money or asked for rice on credit. This made them indebted, which eventually led them to relinquish their shares of trees to Lu farmers. The ownership of the tree gives Lu farmers de facto rights to land outside of their village. The village leader claimed that when the tree is ready for tapping latex, they will train and hire Akha people as wage laborers on their own land.

PRIVATE INVESTMENT AND DIFFERENT PERSPECTIVES
ON RUBBER

While farmers experimented with rubber, investors also played a role in
promoting rubber in Sing District. An owner of a small agriculture trading
company established his business in Sing District during the late 1980s. He
initially sold furniture in Sing District. With the profit he made, he then built
a guesthouse in Sing District serving Chinese traders who began to frequent
the region for business. He made connections with the State Seed Company
from China and then began to trade agricultural products. He brought new
high-yield rice varieties from China to the local farmers. Farmers produced
the rice, and he purchased the products from the farmers, sharing profits
with them. Gradually, he expanded into other crops but kept his trade on
a small scale to avoid paying excessive tariffs at the border. During the late
1990s, he began to expand his business into rubber. He rented land from local farmers and began to develop a nursery. He also contracted local farmers
who wanted to plant rubber. He offered to provide inputs and then planted
rubber for them on their agricultural land. Instead of sharing profits from
the sales of latex, he made agreements with villagers to share trees on the
field. "Villagers keep their land and trees while I maintain my share of trees
on their land. This allows my company to take care of rubber with our
people."[7] He was well aware of the complex documentation process of seeking approval for business in Laos and sought direct business with farmers.
This direct negotiation with farmers allowed him to set up a rubber plantation although rubber was not a part of the registered investment activities
of his company. Officially, the rubber plantation set up by the Chinese trader
was registered as the farmers' own.

In contrast, another Chinese investor in Sing District is struggling to
set up a rubber plantation. The company plans to plant rubber in northern
areas of Sing District (Mom and Xiengkheng subdistricts) covering up to
500 ha. They have surveyed the region and registered their business through
the Provincial Investment and Planning Office. Their investment activity
also includes a latex processing factory. The company is waiting for DAFO
to designate the land for his company. However, DAFO responded that they
needed to complete LFA for local villages before they could decide where the
company could operate. The Chinese Lu representative of the company was
frustrated. "Our company has many years of experience, so the government
should not be too concerned with zoning forests. We are bringing development for rural people in Laos."[8]

While different investors approached farmers, most farmers felt that contract farming was not preferable. If they could afford to invest, they would

rather plant the rubber trees on their own. For those farmers that ventured to work with investors, they also hoped to pay back their debts early than being bound to a long-term profit-sharing arrangement. The rapid expansion of rubber was also stirring a sense of anxiety among many farmers in Sing District. The majority of farmers in Sing District stated that planting rubber was a long-term investment for their children and that they felt it was necessary for them to make claims to land by planting rubber before others made their claim.

Staff at DAFO expressed their powerlessness to control and manage the spread of rubber. DAFO's primary role was encouraging farmers to stop shifting cultivation and opium cultivation. They have led government policies and implemented land-use planning with villagers since the mid-1990s. The head of DAFO sees the current expansion of rubber plantation as forest clearance and conversion of forest into nonforest land. "There are [sic] more clearing and burning because of rubber."[9] He further claimed that a rubber plantation is "not a forest because, you don't see many animals or other things in them."[10] However, DAFO continued to register rubber as tree and not as agricultural product. As an organization, DAFO did not promote rubber, but farmers themselves were making decisions to plant rubber because "They listen to their relatives in China."[11] The GTZ, which has been working with DAFO to develop land-use planning process in upland villages, also did not promote rubber, although it supported alternatives to opium production. Rubber mono-cropping was not an alternative that they envisioned as being sustainable. Expansion of rubber also caused farmers to encroach into forest areas that farmers agreed on during the land-use planning.

At the provincial level, organizations such as the Provincial Agriculture and Forestry Office (PAFO) support rubber and see its potential for household economy. The PAFO was also a leading agency that implemented policies regulating expansion of shifting cultivation in the upland area and supported the development of conservation forests with assistance from the international agencies. In spite of the organizational support for rubber, the technical staff of the PAFO, which had been enforcing land and resource management plans, had doubts about the rapid expansion of rubber. "In the past, land was abundant and cheap, people traded and transferred without payment. Now with increased interest in rubber, agricultural land value has increased. Poor households are often disadvantaged in making claims to land. Communal land areas and forests are being lost."[12]

Another government agency at the provincial level supporting rubber is the Provincial Planning and Investment Office (PPIO), which screens and approves investment projects. Their main mandate is to promote foreign investment in the province and to boost its economy. The office registers investment proposals for rubber as agricultural activities. Under the current process, identification of the actual investment sites takes place after the approval by the investment project. "It is PAFO and DAFO's role to identify land for the investors."[13] Agencies such as PAFO and DAFO both express the need to review and revise the land-use plan. The technical staff at the PAFO believe that they need to carry out another zoning taking a "scientific approach" and considering aspects such as climate, temperature, slope, elevation, and soil type to identify suitable areas for rubber. They also claimed the need to revisit the status of protected forest areas, and make sure that they were mapped properly.

Conclusion

The current study shows decline of dense forest areas and overall increase of degraded areas in Sing District during the last three decades. Expansion of agricultural land is the main cause of forest loss and degradation. In the last decade, there has been a significant increase of upland agricultural land, especially as commercial crops such as rubber and sugarcane expand and replace old swidden and fallow lands. The study also indicates encroachment of agricultural land into forests. Intensification of land use for commercial agriculture is especially prominent in areas along the road where population is concentrated due to government policies that encouraged relocation of upland people. However, the study also indicates that farmers across the district are making investments in rubber.

Narratives of farmers indicate different factors influencing farmers' individual and collective decisions to plant commercial crops, including concentration of population and loss of access to extensive agricultural lands, increased need for cash income to support families' needs and available access to capital and inputs through relatives and investors. It also highlights that regional political and economic environments that allowed for increased trade and exchange of goods and people across the border influenced farmers' decision to become engaged in commercial agriculture.

Expansion of commercial agriculture in Sing District is not only transforming the landscape but also bringing new relationships between people and land. Villagers are increasingly privatizing areas of communal land by planting rubber and making claims to land, as there is no formal land title issued for households. Competition for land is also increasing as investors and others make claims to land, which creates a sense of anxiety among farmers. While some farmers are succeeding in establishing smallholder rubber plantations, others are losing their rights to land by becoming further indebted to investors who offer capital and inputs to plant rubber. Loss of access to land and other basic means of livelihood are creating a new kind of poverty in rural Laos (see also Rigg 2005). There is also a growing tension between villages, as farmers encroach and clear forest areas.

The narratives also highlight the struggle of local agencies at provincial and district levels in adapting to the rapid transformation. The expansion of shifting cultivation receded in the late 1990s through government policies that regulated the expansion of shifting cultivations and opium production, as well as through relocation of upland populations. However, increased trade and investment with China created an environment in which farmers are actively engaged in a market economy without direct assistance of the government. Information on rubber gained through their relatives in China, capital accumulation through participation in market activities, and support from relatives and investors have empowered many rural farmers to make a decision regarding land use without seeking government support.

Local government agencies must also face and negotiate with private investors who seek access to land. While it is government agencies' organizational goal to promote commercialization of agriculture, this creates a conflict over access to land. Although agencies such as the PAFO and the District Agriculture and Forestry Office (DAFO) seek to introduce zoning and revise existing institutions on land and resource management, enforcing new resource boundaries alone does not seem to resolve the increasing conflicts over land and resources. It is imperative for local agencies to adapt themselves as an organization to the changing economic environment of the region and find new ways to work with farmers who are already participating in the market economy.

PART IV
Institutions

Institutions: The Secret Lives of Forests

SUSANNA B. HECHT

Institutional politics underpin many dimensions of the social lives of forests, and many of the chapters in this book deal with this question thematically, rather than explicitly. This set of chapters specifically focuses on institutions, their changes, and implications. These landscapes are imbued with symbolic meaning, traditional rights and powers, and diverging political economies that reflect past and emerging institutional frameworks that mediate bundles of rights and powers over forest lands. These, not surprisingly, are contested at different levels from household and clans, from localities and regions to the nation-state where different actors compete over resources through a kind of "lawfare" that is manifested in shifts in institutional frameworks. This complex terrain is at the nexus of living systems, ethnography, political economy, and histories, and much of it is quite invisible, at least to outsiders.

The changing politics of climate change, globalization, the new rurality, and the expanding politics of governance over forests make the questions of institutions especially significant. While agricultural institutions are complex, those pertaining to forests are far more complicated, as most of this book suggests. What is also quite striking is this: if agrarian reform galvanized rural politics in the twentieth century, forest tenurial reform may well be the axis of rural politics in the twenty-first.

This section begins with historical ethnographer Sara Berry's discussion of the institutions and politics in Ghana and highlights how ideas of forests, forest traditions and practices, and the array of political institutions that shape access use and power changed over time. She points to various uses of the "inventions of tradition" to consolidate control over land and how the shifting institutionalities of more centralized and then decentralized models have played off against each other historically. She describes how trees as symbols of the sacred, symbols of ownership and identity—and as an extremely valuable resource—have come to work in the new deployments of power over "customary common property" in current international development policy. This analysis of the malleability of institutional politics and frameworks speaks to the ethnographic and political complexity of forest economies.

Over the last twenty years, more than 250 million hectares of forest have been transferred to communities. While much of the twentieth century was galvanized by agrarian reform, "forest tenure reform" has increasingly taken the stage in rural tenurial debates. This transformation is the subject of the chapter by Barry and Meinzen-Dick. This has been a rather stealth process, not noted by many mainstream development analysts. How power, access, and use is structured is highly contested—and potentially disenfranchising—but emerging collective politics, as well as techniques of participatory mapping that map *rights* and not just land *perimeters* can make the invisible claims and practices more legible and valorize otherwise "invisible" practices, histories, and uses. The central idea is that a range of resources and resources histories are incorporated into the tenurial/institutional structures. This is an evolving arena, but a future with forests will require forest institutions, not just agrarian "legibilities" slapped on top of forested landscapes, the norm for much of the twentieth century.

Chris Reij analyzes the "basket case" of Africa, the Sahel, and points to the remarkable rewooding that has occurred there on more than 5 million hectares in some of the poorest regions like Niger and Mali. This region has been seen as a place of deforestation and desertification, processes generally laid at the feet of its human inhabitants, and as a cautionary lesson about Malthusian problems exacerbated by climate change. Reij points to an array of processes that brought about this startling transformation, including transferring property rights in trees from the state to producers, the catastrophic failure of a pure stand eucalyptus reforestation "expert system," where local people really only provided labor, to a more complex system based in indigenous soils management, water harvesting, and farmer systems of forest succession management. The expansion of wooded dryland systems will become crucial under current projections of climate change. Institutional change and the application and extension of indigenous knowledge systems in one of the trickiest planetary environments is a remarkable lesson about rethinking the agricultural matrix.

21 * A Forest for My Kingdom? "Forest Rent" and the Politics of History in Asante (Ghana)

SARA BERRY

In 2002, residents of A__—a suburb of Kumasi, capital of the Ashanti Region of Ghana—visited their chief to ask that he rescind a recently announced moratorium on building cement tombs in the town cemetery. A__ had grown rapidly in the preceding decade and the chief warned that the town was running out of space to accommodate the remains of its departed citizens in the local cemetery. In the past, he reminded his constituents, Asantes did not entomb their dead in cement, but laid them to rest in earthen graves, covered only with low mounds of soil. Graves were dug in the "sacred grove"—a patch of uncut forest left standing on the outskirts of town—allowing departed souls to remain in their ancestral villages, while their bodies were absorbed into the earth (Rattray 1927, 161; De Witte 2001, 76–77). By returning to the ways of their ancestors, the chief suggested, citizens of A__ could respect the dead, provide for their descendants, and protect the environment, all at the same time.

After talking it over, a group of local residents gathered at the chief's palace to ask him to change his mind. While acknowledging the historical as well as the environmental merits of traditional burial practices, they pointed out that times had changed. For the majority of the town's inhabitants, Christianity had supplanted older spiritual beliefs, and people wanted to provide Christian burials for their relatives. Moreover, Asantes today are highly mobile. Many citizens of A__ live and work in Kumasi, Accra, or outside the country, raising children who have never seen their ancestral communities.[1] By placing the deceased in cement tombs marked with their names and accomplishments, families seek to create permanent memorials "so that our children will know their grandparent(s)." In the fast world of today, they argued, permanent tombs would do more to preserve ancestral traditions and pass them on to future generations than would the literal reenactment of a traditional practice that rendered individual ancestors indistinguishable from the natural world around them. In keeping with his reputation as someone "who listens to the people," the chief graciously acceded to his "subjects'" wishes and agreed to their request.

This incident points to the complex interplay between past and present that runs through the contemporary social life of Ghanaians and Ghanaian forests. Funerals occupy a prominent place in Asante social life, marked by far more elaborate and expensive observances than births, marriages, or other life cycle rituals. Families spend weeks planning funeral celebrations, allocating tasks, negotiating protocol, assembling materials and equipment, and mobilizing contributions to cover funeral costs. Seeking to honor the living as well as the dead, families spare no expense, often incurring large debts in the process. Lavish funerals enhance the reputation of the bereaved family by demonstrating solidarity among its members and their collective respect for the dead, and earn special regard for individuals who sacrifice personal wealth for the collective honor of their kin (De Witte 2001, 76; Gilbert 1988; van der Geest 2000; Arhin 1994; Mazzucato, Kabki, and Smith 2006, 1057–58).

Since colonial times, chiefs, politicians, and government officials have made repeated attempts to curb funeral expenses, arguing that elaborate ceremonies waste resources. Their criticisms have been echoed by economists who complain that the bulk of emigrants' remittances are "wasted" on consumption, rather than invested in productive enterprises. (Azam and Gubert, 2005) Neither public nor scholarly criticism has done much to dampen the irrepressible economy of Asante funeral celebrations, however, and some argue that funeral expenditures actually stimulate the economy (Arhin 1994, 318; De Witte 2001; Mazzucato, Kabki, and Smith 2006, 1057–58).

As towns and cities have grown in recent decades, many forest cemeteries have become sites of contestation. If the cemetery in A___ filled up with cement tombs, the chief might have to set aside additional land to accommodate the dead—land that could otherwise be leased out or sold.[2] With building plots going quickly around the perimeter of Kumasi, and land prices on the rise, extending the cemetery would mean a significant sacrifice of prospective financial gain for whomever might have sold the land instead. Behind the chief's polite deferral to town residents over the relative merits of modern and traditional burial practices lay an unspoken subtext of tension over land and the place of traditional authority in contemporary social life.

Beyond cemeteries and sacred groves, debates over the prerogatives and responsibilities of traditional office permeate many domains of contemporary social life—land markets and local governance in particular. Challenged, if not eclipsed, by spreading commercialization in the nineteenth century (Austin 1996; McCaskie 2000), chiefly authority was reinforced during the colonial era by British practices of "indirect rule." Seeking to undo the colonial legacy, Ghana's first President Kwame Nkrumah stripped chiefs of their administrative and judicial roles under colonial rule, but stopped short of abrogating stools' authority over land altogether.[3] Since Nkrumah's fall from power in 1966, chiefs have regained a considerable amount of influence de facto, especially in matters relating to land allocation and local governance (see, e.g., Arhin 2001; Odotei and Awedoba 2006). Largely informal, rather than explicitly sanctioned in law, chiefly influence has been further reinforced by recent policy initiatives aimed at registering land, privatizing government assets and services, and decentralizing governance and power.

Questions of Property

Although "customary law" was formally abrogated after independence, chiefs' authority over stool land was reconfirmed in subsequent constitutions, and customary land arrangements are routinely recognized in the courts (Woodman 1996; Berry 2001; Kasanga and Woodman 2004). Under present arrangements, timber "trees belong to the chief and are vested in the state to manage . . . in the 'national interest'" (Amanor 1999, 68). Revenues from stool lands, including timber royalties, are divided between the state, the stools, and local governments, but

> [n]one . . . goes to the farmer on whose land the trees have been preserved. These regulations are established by appealing to tradition. But this is an invented tradition which has been changed constantly to fit evolving socioeconomic conditions. (Amanor 1999, 68)

Chiefs are not freehold owners, however. The constitutional stipulation that "all stool lands shall vest in the appropriate stool on behalf of, and in trust for the subjects of the stool in accordance with customary law and usage" (Ghana 1992, Art. 267(1)) is ambiguous, to say the least. In what sense are citizens of contemporary Ghana "subjects" of other citizens who also happen to be chiefs? If chiefs are legally empowered to act as trustees in charge of stool lands, what are their obligations with respect to the use and disposal of those lands? Who decides where to draw the boundaries between lands and/or subjects of one stool and another?

The Constitution's answer to these questions leaves them open to debate (Ghana, 1992, Art. 267(1)). Chiefs are required to turn over about 60 percent of the revenue they receive from land transactions to local governments and state officials, but the Constitution is silent on procedures for reporting the *amounts* they receive. In the courts, chiefs are said to hold allodial (or ultimate) title to stool lands which, in principle, are inalienable. In practice, chiefs regularly sell portions of stool land at going market rates while retaining ultimate title, and citizens are left to try to beg, cajole, or threaten them into spending some of the money to benefit the community. To paraphrase Jean-Pierre Dozon (1985, 259), the more a chief sells stool land, the more the stool owns it.

In the closing years of the twentieth century, debates over the status and future of Asante's shrinking forests intersected with pressures brought to bear by international lenders for African governments to "liberalize" their economies. Focused initially on economic objectives (monetary and fiscal stability, free trade, private property), the neoliberal agenda expanded in the 1990s to include political and institutional reforms, such as multiparty elections and "good governance," and environmentally sustainable development. In 1999, after years of pressure from the World Bank and other donors, the Ghanaian government published a National Land Policy for "the first time in the history of the country." Rather than proposing specific legislation, the policy offers guidelines for future policy actions. Invoking "the principle of participatory democracy," it calls for "involvement of the local community, opinion leaders, traditional authorities as well as government agencies in the land development process," and urges that land improperly acquired by the state should be returned to "the owners," adding that "the primacy of a land title derived from customary or common law sources takes precedence over any other interests" (quotations from Ghana 1999).

With start-up funding from the World Bank, in 2003 the government launched a fifteen-year Land Administration Reform Program to overhaul relevant state agencies, and modernize traditional land administration by establishing "customary land secretariats" throughout the country and training chiefs to use them effectively. The long-term impact of this initiative remains to be seen. It is not clear how it might affect the staggering backlog of land litigation pending before the courts — estimated by a source for the *Ghanaian Times* (October 16, 2003 and October 31, 2003) to include 60,000 cases pending on appeal before the Supreme Court in 2003 (compare Berry 2001).

What Is Being Sustained?

Efforts to manage resources in the present so as to preserve them for the future often draw on understandings of the past. Plans for sustainable development, including forest recovery, necessarily make assumptions about what is to be sustained. Ostensibly, objective debates over scientific evidence, economic priorities, and technological options often rest on imaginaries of bygone forests and implicit ideas about the relevance of knowledge about the past for present and future practice. In recent years, policymakers and pundits have joined cultural advocates in stressing the value of indigenous knowledge and customary methods of forest management (Bassett, Blanc-Pamard, and Boutrais 2007; Bassett and Crummey 2003). Like "Sankofa"—the legendary Akan bird whose long neck doubles back so the head looks over the tail at what lies behind—recent policy discourses frequently invoke the enduring value of the past as a source of inspiration and practical advice for the future.

Claims that the path to a better future lies through the past are not limited to the environment. The growing neoliberal emphasis on democracy and decentralization as conditions for sustainable development has led to calls for grassroots participation in development planning as well as practice, and an active search for African partners willing and able to make it happen. From the World Bank to World Vision, donors emphasize the progressive possibilities of community and custom as *local* institutions and forms of knowledge, better suited than exotic models for engaging ordinary people in governance and self-help. In a striking replay of the logic of indirect rule, the World Bank has made particular efforts to enlist African chiefs and elders as partners in local development and community mobilization. During an impromptu conversation, the chief of A__ described a recent World Bank-sponsored trip to South Africa, where he and other West African chiefs met with their counterparts in the former homelands to exchange ideas about the roles traditional authorities might play in development and governance today.

Asante occupies a pivotal place in the convergence of "traditional" and "modern" in contemporary Ghana. Dismantled in the late 1890s by British invaders, who packed the Asantehene (king) off to the Seychelles, along with an entourage of royal relatives and officials, the monarchy was slowly reassembled in the early 1930s, as the colonial government formalized indirect rule. Formally restored to the Golden Stool in 1935, King Prempeh II and his successors have been active in politics and governance ever since. Championing Asante farmers' and traders' demands for a larger share of cocoa export proceeds, the Asantehene and other senior chiefs backed the opposition party (National Liberation Movement) that challenged Kwame Nkrumah's rise to power during the 1950s. Defeated at the polls, the NLM disbanded after independence, while Nkrumah stripped the Asante chiefs of most of their former prerogatives, and sent his most outspoken opponents to prison.

After Nkrumah's government was overthrown in 1966, the imprisoned chiefs were released and restored to their traditional regalia, but traditional authorities remained outside the formal institutions of the state. Their influence has expanded nonetheless under the relatively weak military regimes that followed, and the civilian rulers of the Fourth Republic. Noting the preponderance of professionals among senior Asante chiefs enstooled between 1970 and 1995, Kwame Arhin asks how, in view of "the impoverishment of traditional rulers and the progressive minimization of their significance" under Nkrumah, does "one account for the tendency of well-educated, professional men and apparently prosperous businessmen to contend . . . and sometimes pay heavily for succession to" traditional office? (See Arhin 2001, 74; see also Allman 1997; Dunn and Robertson 1973; Rathbone 2000.) Arhin limits himself to the observation that chieftaincy has become "an alternative means of securing political influence by those unable or unwilling to engage in national party politics" (2001, 74). Defenders of the institution argue that electoral politics are incompatible with the sacred character of chiefly authority: were chiefs to start losing elections, the dignity of the institution would be compromised. Critics point out that by foregoing electoral politics, chiefs avoid accountability for their actions, especially their use of money received for "allocating" stool lands (see, e.g., Amanor 1999 and Ubink 2008). I return to this point later.

The active, if largely informal, role of chiefly authority in contemporary economic and political affairs evokes ambivalent reactions among the citizenry. Ghanaians decry the ability of unelected traditional authorities to appropriate resources in the name of customary prerogative without being held accountable for the way they use them, but value chieftaincy as an authentic symbol of their rich cultural heritage. As the biggest and arguably most powerful of Ghana's precolonial states, Asante looms large in contemporary historical imagination, and the Asantehene enjoys an almost iconic status among contemporary exemplars of past accomplishments and power. A multimillion dollar grant from the World Bank to the Asantehene's Education Fund received mixed reviews in the press.[4] In a survey of 242 residents of eight peri-urban communities near Kumasi, however, Ubink found widespread dissatisfaction with the performance of village chiefs, but near universal approval of the Asantehene (Ubink 2008; compare Crook et al. 2005). A symbol of the old Asante empire, who is also a successful businessman and an astute practitioner of cultural politics, the Asantehene bespeaks Asantes' enduring fascination with "the very latest tradition" (Clark 1999, 82).

The reemergence of traditional rulers as significant players in contemporary Ghanaian society rests on their authority over land. Ratified, as we have seen, in the constitutions of Ghana's Second, Third, and Fourth Republics, authority over stool land has provided chiefs with a fulcrum from which to exert leverage in other domains of economic and political life, including management and exploitation of the forests. For much of the twentieth century, expanding cocoa production generated an ever-increasing demand for forest land, enabling chiefs to parlay traditional claims to tribute into a substantial share of "forest rent" derived from cocoa grown on freshly cleared forest land.[5] Following a series of cocoa "hold-ups" by farmers and local traders protesting expatriate export firms' monopsony over the cocoa market, in 1939, the state abandoned its long-standing commitment to free trade and took direct control of export crop marketing, establishing a state-controlled marketing board as sole buyer of cocoa and other major export crops. Channeling a large slice of export earnings into the coffers of the state, the marketing board became a fixture in government and the export sector. Modified but not dismantled under structural adjustment, "Cocobod" remains in place today.

In the case of timber, the state moved soon after independence to preempt chiefly authority by vesting all naturally occurring timber species in the office of the president, regardless of who "owned" or controlled the land they stood on.[6] As we have seen, subsequent civilian regimes reaffirmed stools' allodial title to both land and timber, while the current constitution divides timber royalties between the chiefs and the local District Assemblies. In recent years, chiefs have taken advantage of the current political climate to apply for concessions to establish timber plantations inside Forest Reserves, and sometimes assumed the role of informal regulators in the timber market as well. In 1994, I happened on a pile of logs that had been impounded "by the chief's forest guards" because the logs were being moved without the proper government permit. I suspect that this was not an isolated case (Berry 2001).

Ancestors, Citizens, and the Division of "Forest Rent"

Debates over governance of both land and people have a long history in Asante. British officers who occupied Asante in the late 1890s found themselves nominally in charge of a complex mosaic of disparate, shifting, and overlapping claims to authority over land and people in which social and territorial jurisdictions did not necessarily coincide. Chiefs, sub-chiefs, village heads, families, individuals, and custodians of shrines exercised various rights of land use, allocation, and transfer over the same or overlapping territories, giving rise to recurring debates over whose claims should take precedence, especially when possibilities arose for using land in new ways. During the first half of the twentieth century, the rapid growth of cocoa exports produced a moving frontier of freshly cleared forest land where struggles between farmers and landlords over the division of forest rent turned on contested issues of jurisdiction and belonging, based on tradition and historical precedent. Called upon to settle disputes and regulate access to forest reserves, colonial officials and the courts had little choice but to engage, one way or another, in debates over historical knowledge and the authenticity of custom (Boni 2005, 2006; Berry 2001; Crook 1986; Dunn and Robertson 1973).

Debate turned not only on the authenticity of customary laws, but also on questions of to whom they applied. Seeking to systematize the practice of indirect rule, British officials worked on the assumption that every Asante owed allegiance to a particular chief, and that this fact of traditional governance determined individuals' privileges and obligations under customary law. Access to land and liability for chiefly tribute depended on whether one was a native or a stranger—categories defined in terms of ancestral origin. Ancestral origin is not simply a matter of biological descent, however, but a "constant and strategic process of forging one's ancestry and its deeds"—in particular, the history of one's ancestors' relations to the stool (Boni 2006, 170; Berry 2001). Now known as citizens rather than natives, members of local communities are customarily entitled to use portions of stool land without paying tribute to the chief. In theory, stool lands are inalienable; in practice, both chiefs and citizens have sold land for building or commercial use, giving rise to bitter disputes over the division of the proceeds between the chief, the community, families, and individual citizens (Kasanga and Woodman 2004; Ubink 2008). Since stool lands serve as territorial jurisdictions as well as de facto properties of the stool, many chiefs have used their authority over land to extend their influence in other directions, creating what amounts to a system of informal governance that parallels the District Assemblies and executives established in the late 1980s, during Ghana's last transition from military to civilian rule.[7]

In recent years, chiefly influence was reinforced by international donors' efforts to clarify ownership, increase security of tenure, and privatize state services and assets. The drive toward privatization has prompted challenges to former land acquisitions carried out by the state during the early years of independence. Made in haste, many of these acquisitions were never properly executed under the terms of the law or used for any public purpose. Under the aegis of privatization, pressure has mounted on government to return these lands to their original owners. In several recent cases, the courts ruled in favor of litigants who challenged the legitimacy of state land acquisitions, and the Ministry of Lands has endorsed the policy of returning state lands that were never formally acquired or put to public use. In the process, old disputes have been reopened and new claimants have come forward, reviving or reinterpreting custom and historical precedents to support their claims to "original ownership" (Lund 2008; Ghana 1999).

Encouraged by donors and recognized in the courts, the ratification of customary rights has strengthened the hands of traditional authorities, not only in Asante and other areas with long histories of chieftaincy, but also in parts of the country where chieftaincy is relatively recent. Rather than clarify rights of ownership and authority, these policies have tended to proliferate claims and debates over historical precedents, contributing to a legacy of ambiguity that has allowed traditional authorities to reinforce their influence in natural resource management, as well as land markets and de facto governance (Amanor 2007; Lund 2008).

Timber: A New Cash Crop?

As Ghana's forests filled up with farms and villages grew into towns, pressure on the Forest Reserves increased. Farmers extended their cultivated plots inside reserve boundaries, supplementing their income with gathered plant materials and trees illegally felled for firewood, building materials, or sale (Kotey et al. 1998). Restricted or prohibited by law, access to land inside Forest Reserves is often negotiated informally among farmers, customary landowners, earthpriests, chiefs and District Assemblymen, allowing officials to reap what Lund aptly calls "the rent of non-enforcement." "[R]ather than fining and evicting people [and] risking confrontation with the assemblyman and the political party in power, forestry officials would turn a blind eye [to infractions] while lining their pockets" (Lund 2008, 149). As policies shift in favor of privatization, the social dynamics of access complicate efforts to clarify the ownership of forest resources, not to mention possibilities for outsourcing forest management and recovery to informal authorities and private actors. Debates about ownership, in particular, promote reexamination of histories of past land acquisition, boundary demarcation, and shifting jurisdictions of state and customary authorities.

In addition to demarcating Forest Reserves and regulating the terms on which people may legally enter and/or use them, both colonial and postcolonial officials made numerous attempts to regenerate degraded portions of the reserves and promote the cultivation of timber species outside their borders.[8] As early as the 1920s, colonial foresters introduced a form of sharecropping, known as *taungya*, under which farmers were allowed to cultivate land inside the Forest Reserves if they agreed to interplant timber species among their annual crops. In theory, the trees would benefit from labor and other inputs that farmers applied to their annual crops. In practice, farmers concentrated their efforts on food and other short-term crops, often neglecting or even sabotaging timber saplings in order to protect their other crops from too much shade. Revived after independence, *taungya* continued to produce mixed results—a record that dovetailed neatly with the neoliberal critique of state ownership in general. Because the state retained ownership of the trees, it was argued, farmers had no incentive to take care of them year in and year out—let alone leave them standing for decades to replenish full forest cover and maintain or expand future supplies of timber. Despite these difficulties, forestry officials continued to experiment with *taungya* for years—ignoring farmers' own efforts to encourage regeneration of woody plants on fallow land (Amanor 1999).

Farmers' lukewarm reaction to *taungya* was in no way reflective of their interest in growing trees. Beginning the late nineteenth century, Ghanaian farmers grew cocoa on partially cleared forest land, establishing thousands of small-scale farms across the southern part of the country, and turning Ghana into the world's leading producer of cocoa for much of the twentieth century. In the late 1980s, as the national economy began to recover from a twenty-year decline, a few people began to experiment with teak and other hardwood species, planting small stands of trees on portions of inherited family land or plots they had acquired for the purpose. Inspired by a mix of environmental patriotism and hopes of future financial return, others followed suit. By the mid-1990s, small stands of teak were readily visible, clustered on the outskirts of towns and villages, or standing by themselves near individual houses along the highways and secondary roads. Faced with growing opposition to their efforts to protect forest reserves, officials in the Forestry Department began to look for ways to assist growers' efforts.

Officials' new interest in smallholder teak grew out of a general shift in forest conservation policy away from aggressive state policing of protected forest reserves, toward participatory programs designed to enlist local communities in managing forest resources. By the early 1990s, the need for such a change was acute. In 1984, following a near collapse of the economy, Ghana signed on to its first Structural Adjustment Loan, agreeing to relax state controls on domestic and foreign transactions in exchange for financial assistance from the World Bank and the International Monetary Fund. Output and income recovered, but the gains were concentrated in Ghana's traditional export sectors—cocoa, timber, and gold—accelerating the pace of natural resource depletion and creating new problems for resource users and regulators alike.

Dominated by the export of round logs, the timber industry depended almost entirely on the extraction of naturally occurring hardwood species that grew abundantly in Ghana's semitropical forests. From the mid-1980s, the industry's recovery touched off a series of increasingly bitter struggles between large timber companies, individual chainsaw operators, farmers, forestry officials, and the state agency in charge of granting timber concessions outside the forest reserves. Most concessions went to large timber companies, who hired private security guards to prevent chainsaw operators from encroaching on their terrain, and rarely paid compensation to farmers whose crops were damaged by newly felled trees. In frustration, farmers and chainsaw operators increased pressure on the forest reserves, planting crops and felling trees illegally inside reserve boundaries. Forestry officials responded by tightening controls—arresting illegal chainsaw operators, evicting unauthorized farmers, and sending armed guards to patrol reserve boundaries and apprehend encroachers.

Applied amid widespread uncertainty about the country's economic and political future, these aggressive tactics backfired. Forest guards reported being shot at when they tried to enforce the rules, and farmers confronted local administrators, demanding compensation for damaged crops and/or easier access to reserved forest land. Newly installed at the end of 1992, after decades of military rule, the government was anxious to defuse social tensions, especially in crucial export sectors. Following a full-scale review of forest and timber policies, the government changed its approach—placing a moratorium on the export of round logs, requiring timber companies to obtain farmers' written consent before issuing concessions, and establishing local forest committees of farmers, sawyers, timber company representatives, forestry officials, and chiefs to work collaboratively on improving forest management and protection. Initial results were encouraging. Between 1994 and 1997, illegal timber felling declined sharply, state revenues from timber royalties rose fourfold, and the Forestry Department anticipated that "people will start to plant trees like teak in the very near future . . . produc[ing] a new harvest of planted timber off-reserve after the present stock of naturally occurring trees is used up" (Kotey et al. 1998, 100).

To encourage private timber cultivation, the department established a Forest Plantation Development Centre outside Kumasi, charged with providing technical, material, and financial assistance to small-scale growers. Access to cash was particularly important: plots of young teak must be cleared and trees pruned every few weeks during the rainy season in order to produce marketable trees. Growers whom I interviewed in 2002 said that finding people to do the work was not difficult: "Our problem is money."

Apart from a single disbursement of loans to about 1,000 private growers in 2005, government has been unable to expand growers' access to credit. By the early 2000s, attention was shifting to large-scale plantations established by private concessionaires or the state in degraded Forest Reserves. Readily visible along the highways in 2008, the plantations appeared to be overgrown and the trees unpruned, leading one to wonder about their future market value. Rather than secure claims on land by taking out formal leases (a process that, as one grower put it, "will cost me plenty"), growers I interviewed in 2002 were planting trees to signify their ownership of the land, whether or not they could afford to maintain them.

"The Very Latest Tradition"

Over the course of the twentieth century, the social life of Asante forests reflected not only the power of market forces to remake natural environments, but also the myriad ways in which market transactions, political contests, and bureaucratic practices intersect with historical imagination. Drastically reduced in physical size, Asante's forests continue to loom large in social imaginaries, animating debates over sustainable development and the political and economic legacies of frontier expansion, export-led growth, and indirect rule. Ratified by the legislature and recognized in the courts, the rule that naturally occurring timber trees belong to the state operates against a shifting mosaic of individual, family, and customary claims that both reinforce and subvert the intent and sometimes the letter of the law. Official boundaries around protected forest reserves overlap (and often collide) with boundaries of farms, family lands, and chiefly jurisdictions that are no less real when stored in peoples' memories and imaginations than those found on official maps. By raising the stakes in making claims of original ownership, recent efforts to rehabilitate Ghana's forests by privatizing them run the risk of increasing opportunities for rent-seeking and social exclusion, at the expense of equitable access and sustainable management.

22 ✳ The Invisible Map: Community Tenure Rights

DEBORAH BARRY AND RUTH MEINZEN-DICK

Background

Over the last twenty years, a little-known trend of land tenure reforms has swept across the world's forests. In what has been classified as between 79 percent (White and Martin 2002) and 84 percent (Food and Agriculture Organization [FAO] 2006a) public property under the formal ownership of the state, we are now witnessing an official transfer of tenure rights to communities living in more than 250 million hectares of forestlands (White and Martin 2002; Sunderlin, Hatcher, and Liddle 2008). The result is that by 2001, 22 percent of all forests were owned (14 percent) or held in reserve (8 percent) for communities (White and Martin 2002). The trend continues, with the area of state ownership continuing to decline during the period from 2002 to 2008, and with corresponding increases in the area of forests designated for use by communities and indigenous peoples, individuals, and firms (Sunderlin, Hatcher, and Liddle 2008).

However, the process of this devolution is problematic, suffering from a tendency to overlook and/or exclude the rights and claims of local actors, often communities with existing systems of de facto or customary practices and self-regulations for access and use of resources (Sikor, 2006; Fitzpatrick 2006; Pacheco et al. 2008). This is particularly true as these are not only land-driven reforms, but rather resource-driven reforms in forest landscapes, where there are multiple existing resources and claimants. The predominantly indigenous land—and extractivist reserves, community and village forests, and community concessions being granted legal rights of tenure—have differing patterns of resource dependence, some well established over time, with rights, rules, and responsibilities for their use. Natural resource tenure is broader than land tenure.

When these local systems of resource rights are not considered, the reforms can be seeding conflict and laying the grounds for further disenfranchisement of the poor. Often unintentional, the process of forest tenure reform, land demarcation, and titling changes the access, use, and decision-making rights in ways that can profoundly affect people's livelihoods, governance structures, and quality of life. Local participation in these reforms is a fundamental step, often overlooked during the implementation of even the most well-intentioned efforts.

On the positive side, the legal devolution of forest rights to ancestral or new forest dwellers has spurred the emergence and growth of participatory community land-use mapping (PLUM), used by human rights activists, development practitioners, ethnographers, geographers, and even conservationists, helping to expand the opportunities for local communities themselves to participate (Herlihy and Knapp 2003).

When done well, the exercise gives communities a common *spatial* framework—a map—that strengthens their understanding of how physical, social, and economic factors interact (ILC 2008). Community participatory maps are used for a wide range of purposes, but today, most technically assisted projects are used to produce land-use maps as the basis for establishing external boundaries or perimeters and then deployed as an integral part of a legal procedure for acquiring land rights (ILC 2008; Van de Sandt and MacKinven 2007) or tenure mapping. This bridging of the technological gap between the state and the local claimants has allowed the latter to take a huge step forward, in presenting or disputing land claims based on their de facto, customary, or ancestral rights.

Land-use mapping reveals the bio-physical sphere, delineating distinct land and natural resource uses (agricultural, hunting, gathering, fishing, settlements, sometimes watersheds, tree groves, bird sanctuaries, etc.), often with cultural designations, such as burial grounds, sacred sites, or other ceremonial uses. The spatial rendition of these activities serves as the basis for the negotiation of the extension and perimeters of the land to be demarcated and eventually titled.

Some of the leading practitioners of participatory community mapping have taken the exercise several steps further, using it as a tool for strengthening collective identity and action (Chapin 2006; Colchester 2008; Di Gessa, Poole, and Bending 2008). Others have evolved the land-use mapping into an internal tool for communities to deploy in their land and resource management and negotiation of changing rights, to ameliorate internal conflict between community groups and higher level authorities, or for hazard mitigation (Participatory GIS [PGIS] 2008; Chapin, Lamb, and Threlkeld 2005).

It appears that most of the significant effort for developing these mapping practices has been at this interface of the applied technology and local knowledge for spatial renditions of land use (Chapin, Lamb, and Threlkeld 2005; Di Gessa, Poole, and Bending 2008; Poole 2006). However, when the effort is oriented toward establishing legal tenure claims, in most cases the underlying system of local rights remains are lost or rendered invisible in the negotiations. Spatial maps of boundaries are what are required to establish formal tenure rights: the perimeter. Spatial rendering of land use can help stake out the coordinates, but the system of rights behind them is forgotten when actual demarcation or titling takes place. Once demarcated or titled, externally crafted policies and regulations introduce new rules, often based on different tenure systems.

In many cases, tenure reforms are based on Northern concepts of "ownership" in which an individual or legal individual holds all bundles of rights over the land and other users are not acknowledged. On the other hand, most forest tenure reforms do not turn over the ultimate right of alienation from the state to communities or individuals, and the state may even create additional rules that can crowd out local tenure systems.

The consequences are many, including the undesired: unfolding confusion, illegality, conflict, or increasing poverty (Fox et al. 2008; Hale 2005). The local institutions that develop, support, and enforce the rules governing resource use and distribution may either be ignored and bypassed or made more rigid in order to be recognized by outsiders. Local disenfranchisement or increased tension is often the outcome, where empowerment or conflict resolution was the goal. Thus, the mapping of the systems of existing tenure rights becomes a key step in the process of conducting a reform (particularly a forest reform).

This chapter describes and explores the conceptual framework and potential uses of a practical tool for the mapping tenure rights, understood as a "bundle of rights" (Schlager and Ostrom 1992; see also Alchian and Demsetz 1973). The proposal is to develop a common conceptual framework and, through a guided process of shaping it in the field with communities themselves, produce a practical tool for understanding and mapping tenure rights. The effort should be focused on two basic purposes: (1) to help communities complement and go beyond the spatial renditions of their land uses, clarifying the internal system of rights, rules-in-use, and responsibilities, or noting where they do not exist; and (2) when they exist, give the systems visibility and credibility for negotiating with the state, any regulatory or normative framework, development project, or private investment being considered.

The conceptual framework that we call "Tenure Rights Mapping" needs to be honed through a process of practical application, where communities will be able to *map and present* the internal systems of rules and regulations that govern their rights in relation to their natural resources in an easily understood form. Mapping interaction with external change would follow. We argue that a more complete understanding of the nature of these rights of access, use, and decision making is necessary to take this process forward, and that if well designed, Tenure Rights Mapping could become highly com-

plementary to participatory community land-use mapping as a tool for negotiating tenure reforms. This effort could prove especially valuable as the
basis for the future of the tenure reforms in forests, where multiple forest
resources are used by many groups, and where real boundaries may not be
fluid and shifting, rather than fixed.

Community Land-Use Mapping and the Need for the Mapping of Community Rights

Although participatory community land-use mapping is not new, its deployment in the process of state recognition of tenure is growing but still
experimental. Our interest is in how these exercises can illuminate the local
knowledge and systems of rights that organize the rules for resource use
and bring them into the process of land and natural resource claims, demarcation, and titling, and in how such local knowledge and rights systems
are shaping the nature of the forest tenure reform and other land titling
events or creation of areas for conservation. We observe how their inclusion
allows the construction of the area (forestlands) under consideration to be
done on the basis of local knowledge and social criteria, building on these
institutions as a basis not only for titling, but also for the constant interface with the state. As most of these areas under the current forest reform
are collective lands, lack of understanding of the underlying institutions is
problematic in many ways. In the remainder of this section we discuss some
of the problems that arise when the richness of an expanded mapping exercise arrives at the demarcation and titling process, and how this interface
requires new approaches to transcend the spatial boundaries.

THE BOUNDARY ITSELF

The goal of demarcation and titling is to determine and focus on a series of
fixed points—now, geo-referenced—in order to delimit the perimeters of
the land to which rights will be transferred. An underlying premise is that
clear and bounded property rights will be more secure, reduce conflict, and
stimulate investment. Much of this stems from European-derived notions
of tenure, particularly freehold with title. Titling implies the imposition of
fixed points with complete and ultimate rights on each side of the perimeter. The modern system of land titling finds itself in a quandary of imposing
a rigid system of spatial data onto the fluid boundaries of operating tenure
systems. Even where private, public, and some forms of common or communal tenure are recognized, the implicit assumption is that there are clear
boundaries between land falling into each category, whereas in practice,

there are often overlapping claims by the state, community, and individuals on a given piece of land.

By definition, this process of focusing on the boundaries restricts attention to the underlying system for use of resources: how communities are organized to determine the rights, rules and responsibilities and use them. The principal recorded fact is the perimeter, staked out on a physical map. Land uses and the spatial distribution of their patterns can be recorded to substantiate the claim. But even these "facts" are soon lost in the process of registry. The intricate system of user rights and rights allocation, both inside and out of the perimeter, remains invisible. Visual maps of land uses make a major contribution to the understanding of the rationale for establishing the limits of land use, but do not provide a comprehensible rendition of who has what rights to these multiple spaces and resources, whether they are individuals, groups, or the collective as a whole, if the borders are fluid with mechanisms to negotiate passage.

In some setting of boundaries, such as demarcation for protected areas, the boundaries are a wholesale external imposition, based on incomplete information, often tainted by anti-anthropomorphic biases, where conservationists see primary forests, supposedly untouched by humans. Early delineation by international conservation organizations using highly sophisticated technology for demarcating large areas to set-aside were based on indicators of biodiversity. They mapped huge regions of forests, deserts, and coastal areas, assuming them to be "empty." Government agencies that need to delimit or demarcate for land transfers often use simple methods, such as transects running through large areas, based on past practice in the delineation of agricultural properties, or polygons with straight-running lines typical for demarcating industrial forest concessions. When these simple lines are applied to forest-based communities, it often provokes confusion, disruption of the patterns of local resource use, and even conflict by creating false barriers across contiguous resources spaces (pasturelands, brazil nut tree groves, watersheds, lakes). Locals who insist on maintaining previous resource use patterns are rendered illegal. Law enforcement may then be deployed, criminalizing and punishing the previously legitimate activity (Poole 2006; Van de Sandt and MacKinven 2007). These disruptions can have further negative consequences of eroding local governance structures for resource management, where rights and rights allocators have been overridden, and no alternative mechanisms for resolving disputes have been provided (Chapin 2006; Colchester 2008; Van de Sandt and MacKinven 2007).

In some cases, one can find a more strident contradiction with cultural perceptions of boundaries. What appears to be a relatively simple undertaking can in fact become a major process of internal negotiation. For example, many Miskitu Indians of Honduras do not have a notion of fixed boundaries between their territories that imply limits for passage. The concept of "pana-pana" is precisely the opposite, meaning "you pass-I pass" into each other's land. These rules of access by reciprocity make it difficult to demarcate the final perimeter of these large borderless territories (CCARC 2007).

OVERLAPPING RESOURCES AND RIGHTS HOLDERS

Although many Western-trained people think about property rights in a narrow sense as ownership—the right to completely and exclusively control a resource—property rights are better understood as overlapping bundles of rights, even a web of interest that connects different stakeholders (FAO 2002; Arnold 2002, cited in Hodgson, 2004; Meinzen-Dick and Mwangi 2009). This applies to all forms of land use (even for a private home or farm, the state or community has some rights to regulate what can be done on the land), but it is especially true of common pool resources such as forests. When land is titled, giving the rights of access to a single forest resource, such as timber, without an understanding of the implications on other resources and their users, negative outcomes can emerge, such as local conflict or loss of preciously needed income for forest-dependant locals.

In many cases the rights to timber extraction alone can seriously undermine the rights of those who depend on nontimber forest products. In the community forest concessions of Guatemala, rights were allocated to the forest for timber production, and new community organizations were established based on this productive activity. Gatherers of *xate*, a naturally occurring decorative palm found on the forest floor, were not allocated specific rights and not represented in the organizations. Management plans for timber harvest both limited access to these areas and in the end destroyed *xate* plants during the harvest process. The weekly trickle of income from the sale of *xate*, which provided crucial cash flows to poor households, was suddenly interrupted (Barry and Monterosso 2007). As women are the principal *xate* collectors and manage this income flow, tension between families—and sometimes within the same family (the women collect *xate*, the men cut trees)—strained household relations until local leaders embraced the issue and promoted changes in the regulations and in their own organization.

FLUIDITY

Whereas mapping gives fixed boundaries, in practice resource use and rights are often much more fluid as locally understood. The rights may vary by season (e.g., to harvest particular forest products, or to use land alternately for private cropping or for collective pastures). In a drought year people may go farther afield for critical resources, and reciprocal arrangements between local groups often accommodate this. Such changes in rights are adaptations to changing conditions of the resource itself (Meinzen-Dick and Pradhan 2002). Where they exist, customary processes may lead to periodic harmony, but clear property rights do not emerge because custom itself is inherently negotiated and contested (Fitzpatrick 2006).

Some changes in resource rights take place over even longer time periods and are even less predictable. A forest landscape in the highlands of Puebla, Mexico, supported intensive livelihoods from the ubiquitous amate tree through the 1980s. Today, the amate is a relic among agro-forestry systems of shade-grown coffee. Changes in local management capacity and intensity of resource use are especially important and can either increase or restrict the rights of local people and outsiders to use particular resources, as when new bylaws are passed that regulate harvesting of particular products. Projects associated with conservation or even land demarcation (indigenous lands) often emphasize destructive pressure on resources by outside actors, neglecting that the internal resource management practices of communities are under increasing pressure from change, such as changing settlement, demographic growth, and weakening of cultural values (van de Sandt and MacKinven 2007).

WHO SHOULD HOLD THE RIGHTS?

When titling or formal registration is the primary instrument of land tenure reform and is seen as necessary for tenure security, the issue of who should receive the title can become a challenge, especially in areas of overlapping land use. In many cases, governments have difficulties in determining to whom to title the land due to lack of clarity on whose rights to recognize. With no clear "mapping" of the rights holders—whether individual, family, community, or other collective—any kind of registration is likely to change the nature of their rights and who makes the decisions about those rights.

Representation of the collective becomes a central issue throughout the entire process. Where, when, and how the collective action of the communities shapes itself (through existing organizations, the creation of new ones) to interact with the external authorities, becomes the *quid* of the process. Simple mapping of rights will not be able to adequately address all the needs of this issue, but experiences with a range of strategies for recognition of customary rights in Africa may provide a starting point for this endeavor (Cousins 2000; Fitzpatrick 2005; Wily 2008).

Framework for Considering the Distribution of the "Bundle of Rights"

Property regimes are generally categorized as public, private, or of common property, defined in terms of who holds the rights: the state for public property, individuals (or legal individuals, such as corporations) for private property, and some form of defined group or community for common property (Bromley 1992; Feder and Feeny 1993). Open access is considered the absence of established property rights. A broader and more useful view of property rights differentiates rights further into various bundles of rights (Schlager and Ostrom 1992). There are many combinations of such rights, but they can be grouped into the following large categories:

· Use rights, such as the right to
  o access the resource (e.g., to walk across a field).
  o withdraw from a resource (pick some wild plants).
  o exploit a resource for economic benefit.
· Control or decision-making rights, such as the right to
  o manage (plant a crop, decide what tree to cut, where to graze).
  o exclude (prevent others from accessing the field or forest).
· Alienation, the right to
  o rent out.
  o sell or transfer the rights to others.

"Ownership" is often thought of as holding the complete bundle of rights over a particular resource, such as land. Accordingly, figure 22.1 illustrates an "ideal type" of distribution of the bundle in public, common, and private ownership regimes. Note that in public and private property, the state or individual is assumed to hold a complete bundle of rights, including alienation rights, but in most common property, the collective do not have alienation rights.

These ideal types are almost never found in practice. In most cases, there are overlapping sets of rights, underneath the general classifications. A national park area may be classified as public property, but individuals and groups are often allowed to use the park, either for access (e.g., bird-watching), withdrawal (taking a drink in the park), or even management (under comanagement arrangements or concessions). At the other end of the spectrum on individual private property, outsiders may have rights, for example, to cross the land with their animals (access) or to take drinking water or harvest particular products (withdrawal), or the right of the state to regulate (manage) land uses in most countries.

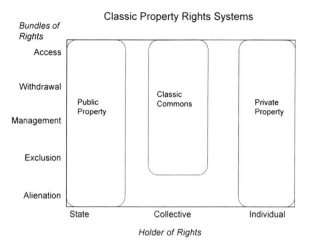

22.1. Classic property rights systems

The state (and even most analysts) cannot deal with the enormous range of such complex rules and negotiations for each tenure agreement. But there is a need for moving beyond the physical and conceptual "perimeter" defined on maps and conventional definitions of public, common, and private property, which have superimposed a rigid understanding of tenure when rights to land and resources demands flexibility, fluidity, renegotiation. This is particularly true of common pool resources like forests (Ankersen and Barnes 2005). We need conceptual building blocks to develop a practical tool that can play a role in redefining the nature of the relationship between property rights holders and the state.

In putting this into practice, the first step would be to identify the relevant manifestations of the state, collective, and individual. Rather than seeing these as three distinct categories, we may find more of a continuum: different central government agencies, municipalities in the state sector, chieftaincies, clans, social organizations, and smaller groups in the collective sector, blurring in to extended families, nuclear households, and single individuals. Which of these is relevant will depend on the local context. Then for a defined resource or set of resources, the next step would be to ask what rights each of these entities holds.

When attempting to actually map the different rights regimes, particularly with the full participation of communities, there are many dimensions that need addressing, all of which highlight the limitations of the classic system for understanding property rights. As mentioned before, forest tenure often involves rights to different resources on the same land. Where there are considerable seasonal differences in property rights, it may be necessary

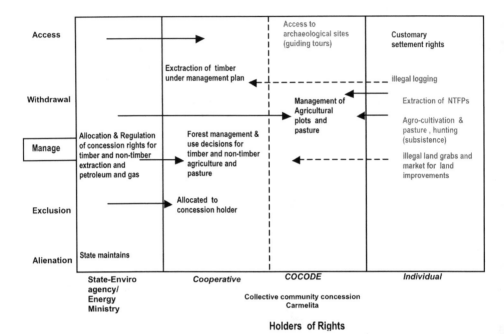

22.2. From individualization to the collective: The forest community concessions of Peten, Guatemala

to draw different figures for different seasons to include seasonal grazing, hunting, or gathering rights. Over time, there may be further changes in the institutions that allocate rights. Devolution processes involve the shifting of rights from the state to user groups (i.e., from the left of the framework figure to the middle). The most common shift is only in the use rights, not decision-making rights (Meinzen-Dick, Knox, and Di Gregorio 2001). The extent and seriousness of devolution can be indicated by how many of the bundles are transferred.

Barry and Monterroso's (2007) study shows how the establishment of community concessions created the collectivization of forest rights and established common property along 500,000 hectares of forestlands surrounding the Mayan Biosphere Reserve. Negotiations led to the establishment of the novel form of community forest concession contracts for a period of twenty-five years, renewable. Nearly the full set of tenure rights— access, use, withdrawal management, and exclusion—were transferred from the state to the newly established collective entities representing resident and nearby communities, as illustrated by the arrows moving from the left to the center of the diagram in figure 22.2. The arrows from the right to the center of the diagram represent the informal strengthening of the collective entity to govern the common resource base. The diagram shows the significant expansion of management rights, though highly regulated through certification schemes.

Developing a Tool for Community-Based Tenure Rights Mapping: Potential Applications of Tenure Rights Mapping

Although Tenure Rights Mapping builds on academic concepts, it also has practical applications. The first use lies in the exercise of identifying right-holders and rights or claims. When land or other resources are classified and demarcated as public, common, or private, it closes our eyes to the range of rights-holders, and the claims they may have. This may exclude many who depend on a resource for their livelihoods or who play an important role in the condition of the resource base. Consciously thinking beyond the normative categories helps to make these other uses and users visible, which may in turn strengthen their legal claims on the resource.

Unlike maps that focus on physical boundaries, the process of identifying the state, collective, and individual right-holders draws attention to the underlying system for use of resources: how communities are organized to determine the rights, rules, and responsibilities over their resources and how they use them. Initial exercises attempting to apply Tenure Rights Mapping in the field, particularly in communities where tenure reforms are under way, presented the challenge of capturing the organizational changes occurring as a result of the reforms themselves.

A second practical application of Tenure Rights Mapping is to help communities deal more effectively with claimants to their resources, and as a tool for deliberation for reaching decisions among themselves on the allocation of access and use of resources. Even laying out the option of nonexclusive use rights can help diffuse tensions over competing claims. If allowing one group to use a resource is seen as ceding all rights over the resource to them, it is more likely to cause conflict than if more limited rights of many different claimants can be recognized. At the same time, Tenure Rights Mapping can also help highlight where there is likely to be the most competition, and where to assuage looming conflict.

A third application of Tenure Rights Mapping is as a means of gaining recognition for a range of overlapping rights. While there is always a risk that codification will reify rights and make them less adaptable to changing situations, at least if more of the complexity of property rights are recorded, it can provide some protection for those who depend on a resource, but cannot claim the underlying ownership of it. Instead of strengthening exclusion rights through conventional titling or boundary mapping, this approach can lead to more inclusive rights, such as for groups dependent on nontimber forest products or grazing lands.

Conclusion

Land-use mapping has made important strides in identifying *where* rights are exercised. Tenure Rights Mapping offers an important overlay of *who* holds *what* bundles of rights. Together, these tools could help to shape the tenure reforms or conflictive claims in many rural—particularly forested—landscapes in the southern countries of the world. Greater tenure security with flexibility would emerge.

Tenure Rights Mapping provides a conceptual guideline for organizing the system of rights holders and transfers and could produce a visual representation of the system of rights, responsibilities, and rules that govern land and resources.

This could be done similar to the way in which PLUM has evolved, particularly as an instrument for legalizing land or resource claims. Tenure Rights Mapping can be seen as having enormous potential in both helping communities to design their proposal for tenure and regulatory reforms, as well as adjust their internal distribution of rights as their social, environmental, or market conditions change.

Putting Tenure Rights Mapping in the hands of the communities (where there is sufficient organization to appropriate it) can allow them to proactively propose the elements they want to be considered. How much visibility to give to what right, to what resources, and who holds them will depend on their interest and capacity to render them visible.

23 ∗ Re-Greening the Sahel: Linking Adaptation to Climate Change, Poverty Reduction, and Sustainable Development in Drylands

CHRIS REIJ

Desertification is the degradation of land in arid, semi-arid, and dry sub-humid areas. It is caused primarily by human activities and climatic variations. Desertification does not refer to the expansion of existing deserts. It occurs because dryland ecosystems, which cover over one third of the world's land area, are extremely vulnerable to over-exploitation and inappropriate land use. Poverty, political instability, deforestation, overgrazing, and unsustainable irrigation practices can all undermine the land's fertility.

Food and Agriculture Organization (FAO) of the United Nations

The tree-based farming that Sawadogo and hundreds of thousands of other poor farmers in the Sahel have adopted could help millions of their counterparts around the world cope with climate change. Already these practices have spread across vast portions of Burkina Faso and neighboring Niger and Mali, turning millions of acres of what had become semi-desert in the 1980s into more productive land. The transformation is so pervasive that the new greenery is visible from space via satellite pictures. With climate change, much more of the planet's land will be hot and arid like the Sahel. It only makes sense, then, to learn from the quiet green miracle unfolding there. Hertsgaard (2009)

Introduction

The Sahel region of Africa, the broad area of seasonally dry open forests and savannas bordering the Sahara desert, stretches from the Atlantic Ocean to the Red Sea and forms part of the countries of Niger, Mali, Senegal, Mauritania, Burkina Faso, Chad, northern Nigeria, Somalia, and the Sudan (figure 23.1). This region has been seen as a place of deforestation and desertification, processes generally laid at the feet of its human inhabitants, and used as a cautionary lesson about Malthusian problems exacerbated by climate change (Newman 1975; Glantz 1976; Franke and Chasin 1980). The imagery of African megafauna, including hippos and crocodiles, in Saharan caves

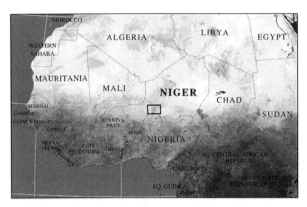

23.1. Map of the Sahel

has contributed to the perception that the region has experienced continu-ous degradation as a function of human practices and the failure of African populations to address environmental change. The environmental history of the Sahel is complex, with a record of climate fluctuations and a great deal of human-driven change (Bassett and Crummey 2003; Maranz 2009), but the "Deserts on the March" version of human engagement with the Sahel may need to be revised. Moreover, many ecological studies now highlight the resilience of the Sahelian vegetation, so that even when ecosystems are buffeted by drought, they have capacities for recovery (Vincke, Diedhiou, and Grouzis 2010; Turner 2004; Hiernaux, Ayantunde et al. 2009; Hier-naux, Diarra et al. 2009; Maranz 2009).

Niger's is a complicated story, one in which macro-level gloom often ob-scures positive micro-level change (Bassett and Crummey 2003; Reij and Smaling 2008). This chapter focuses on forest recovery in Niger, one of the poorest countries in Africa, and one that has mostly Sahelian vegeta-tion types. Despite the common perception of Niger as environmentally degraded and in permanent crisis, there has been large-scale landscape re-covery, even in the absence of significant government or foreign aid inter-ventions, at population densities higher than those that prevailed during the very intense drought years of the 1970s.

Over the last twenty years, Niger has experienced a net gain in tree cover, despite its harsh environment. A recent study on long-term trends in agri-culture and environment in Niger found large-scale on-farm re-greening, especially in regions with dense populations, a pattern reflected in some other Sahelian studies (Reij and Smaling 2008; Hiernaux, Diarra et al. 2009). In itself, this is not surprising, as increasing population densities can induce farmers to intensify agriculture, and trees are part of the production system. It is the scale of this farmer-managed re-greening that is unexpected and significant. Using traditional soil conservation techniques based on local knowledge, particularly "zai holes" or planting pits to restore fertility to degraded land, as well as farmer-managed natural regeneration, some 5 million hectares have now been transformed into a new landscape of dense agricultural parklands with the relatively dense tree cover on farmers' fields resembling an open forest. Niger offers a striking example of how positive trends have remained unobserved.

Fighting the Desert: Niger Farmer-Managed Re-Greening on a Spectacular Scale

Niger was struck by severe drought in the mid-1980s. Farmers have noted that they began protecting and managing spontaneously regenerating trees ever since this time. According to them, the main reason for this transition is that they had to "fight the Sahara." By this, they do not mean fighting encroaching sand dunes, but fighting the impacts of the dust, dessicating wind, and sand storms. The failure of formal reforestation efforts, the mobilization of their local knowledge of resource management, and experience from other areas (both as migrants and by sponsored exchanges with other Sahelian farmers in other regions) also helped inform parts of the farm strategies and practices that led to on-farm re-greening (Reij, Tappan, and Belemvire 2005).

Re-greening has led to more complex and more productive farming systems, improved household food security, changes in local climate, increased drought resilience, local increases in biodiversity, improved soil fertility management, and a reduction in time women need for the collection of firewood. This does not necessarily mean that all impacts occur everywhere, but most of them do. An adequate quantification of the different impacts remains an important challenge. These positive outcomes often remain invisible because of the way national statistics are gathered and fed into the statistics that are compiled by the Food and Agriculture Organization (FAO). Local populations often benefit from surpluses, which they can sell locally, a transaction that does not show up in national commercial statistics. Because of this, productivity can easily be underestimated and benefits to women (firewood, herbs and oils) may not be counted (Nielsen and Reenberg 2009; Van Haaften and Van de Vijver 2003). There are often many positive benefits, with "non-target" results that may be equally or more important to farmers than the targeted outcome, like yields. Thus, from one perspective a program may look problematic, but practices may spread because they respond to a need that can go undetected. These systems also include a great deal of farmer innovation and institutional change, which has helped the system scale up.

In this chapter I (1) highlight the unprecedented farmer-managed re-greening in Niger over the last twenty years; (2) indicate some of the beneficial impacts of this on-farm re-greening; and (3) offer a way forward for publicizing and replicating the Niger success story in other Sahel countries. On-farm agroforestry mosaics are a successful, low-cost, grassroots strategy for addressing food security. It is not a quick and simple technical solution, but a long-term process of social and environmental capital building, requiring changes in policy, legislation, and practices that support long-term resilience.

Across much of south-central Niger, farmers began protecting and managing on-farm trees in the middle of the 1980s. Remote sensing images and field visits show that this farmer-managed re-greening stretches across about 5 million hectares, an area the size of Costa Rica. This means an average increase in on-farm trees of 250,000 ha/year over a period of twenty years, an outcome never achieved by any formal tree-planting projects in Africa. Indeed, of the approximately 60 million trees planted in reforestation projects in Niger, only about 12 million have survived. In contrast, with an average of forty trees per hectare over 5 million hectares, this means that there are now some 200 million new trees established by on-farm protection. If each tree produces an average annual value of one euro per tree (firewood, fodder, fruits, honey, oils, medicinal products, etc.), this amounts to an annual production value of 200 million euros, even without considering the value of the standing tree stock (asset building), or the environmental services of soil building and protection, micro-climate mediation, carbon sequestration, pollination, and wildlife habitat, as well as non-economic symbolic and spiritual values.

Although in some places organized projects have played a key role in stimulating farmers to protect and manage on-farm trees, these practices subsequently spread spontaneously in a process of farmer-managed natural regeneration (FMNR). FMNR is based on local knowledge and practices that support the natural emergence of native seedlings, recolonization, and protection of existing trees: it is low cost and has produced multiple benefits, both for humans and the natural environment.

WHAT TRIGGERED FARMERS TO PROTECT AND MANAGE ON-FARM NATURAL REGENERATION IN NIGER?

Tree Tenure

The droughts of the 1970s and 1980s triggered major production problems; these were not helped by government policies that discouraged farmers from protecting naturally regenerating trees and favoring introduced species like eucalyptus.[1] Across the Sahel, national governments traditionally claimed ownership of all trees, sanctioning farmers caught pruning or cutting trees with various fines. Moreover, planted trees often displaced native trees with more traditional uses. Farmers, whose relation to these planted trees was mainly labor, commonly uprooted seedlings in order to avoid future prosecution, a strategy that led to a reduction in woody cover. They often removed the eucalyptus, seeing it as an intruder in the agro-forest landscape, knowing that its demand for water can deplete the water table, a serious problem in the Sahel. Single-species reforestation was a widespread but unpopular policy among the local populations, whose agriculture and tree farming were negatively affected by these policies.

A significant factor promoting on-farm re-greening has been the perceived shift in rights to trees from state-owned to private ownership. These policies were enacted in the mid-1990s. Once farmers recognized that they held exclusive rights to the trees in their fields, the stage was set for the revival of several traditional modes of water conservation that have proven beneficial to natural regeneration and management of indigenous multi-use trees. The difference between countries that recognize private ownership of on-farm trees and those that do not is striking. Systematic analysis of satellite imagery from twelve *terroirs* used for comparison on both sides of the Niger–Nigeria border, for example, show a marked difference in woody cover, with the government-owned trees of Nigeria present much sparser cover than the farmer-owned trees on the Niger side (figure 23.2).

Clearly, farmers will invest in on-farm trees when they have exclusive rights to those trees or where regulations against tree-cutting are not enforced, though there is also a vital contribution to be made by outside organizations and governments, which can lower barriers to investment in trees through national legislation, policies, and incentives.

HOW HAS RE-GREENING TAKEN PLACE?

For the most part, forest regeneration has taken place as a consequence of accelerated protection of spontaneous sapling growth rather than deliberate planting. Sahelian farmers have long dug *zai* or *tassa*, small shallow pits that concentrate rainfall around the base of a crop plant. Starting in the 1990s, farmers began to enlarge these pits and even fill them with manure. While these practices improved productivity of such rainfed crops as sorghum and millets, they also encouraged sapling growth, encouraging the expansion of agroforestry across the Sahel, especially where property rights in trees were recognized. Small-scale water harvesting, involving small

Comparative Overview of Terroirs on Opposite Sides
of the Niger-Nigeria Border

Niger

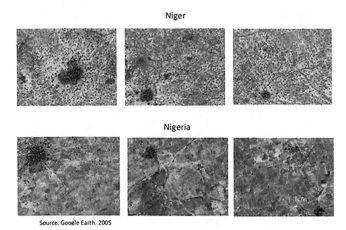

Nigeria

Source: Google Earth, 2005

23.2. Comparative woodland cover Nigeria-Niger

stone bunds as well as pits, led to tree regrowth, a process that accelerated
as the seed stocks represented by new trees continued to expand (Reij, Tappan, and Smale 2009). Combining trees and annual crops into agricultural
regimes is not new in Africa, where wood still serves as the primary source
of fuel, and agroforestry initiatives clearly build on this long tradition as
re-greening practices spread from farmer to farmer across vast areas of
Burkina Faso, Mali, Senegal, and Niger.

WHAT ARE SOME OF THE MEASURED OR PERCEIVED IMPACTS OF FARMER-MANAGED RE-GREENING?

One significant effect of the expansion of tree cover has been higher crop
yields and improved household food security. As noted, before the re-
greening, farmers had to sow two to four times before the crops succeeded,
as the strong winds covered the crops with sand or damaged the young
plants. Now they only sow once, which increases the length of the growing season. In general, farming systems have become more complex—with
a mosaic of open agro-forests, grasslands, and crops—and more productive, which leads to a reduction in rural poverty and increases in household
food security. Trees produce fodder, which allows farmers to keep more
livestock. More livestock means more manure, which is no longer used as
a source of household energy, but can be totally allocated to the fields to
enhance soil fertility. Partial shading by trees enhances the micro-climate,
resulting in reduced soil temperature, protecting crops against dessication.

Another change has been a reduction in the amount of time women spend on the collection of firewood, from about 2.5 hours per day to 0.5 hours per day. In addition to this increased efficiency, wives of farmer-innovators have informed us that because the men work in their *zai* fields, sandy soils not suited to *zai* agroforestry have been allocated to the women, allowing them to grow groundnuts for the market (Reij, Tappan, and Smale 2009, 20). Trees also produce marketable products such as fodder, medicines, and construction wood.

There is some evidence that the local climate has changed, as the direct effects of sun and wind are mitigated. Rainfall studies have shown that large-scale re-greening also leads to locally higher rainfall (an increase of about 30 percent) (Nicholson, Tucker, and Ba 1998). Re-greening has also reduced vulnerability to drought. During the 2005 famine, child mortality in villages with significant on-farm tree cover was much lower than in villages without. In drought years, poor families can survive on tree products, and there is also evidence that tensions between farmers and herders have been reduced by the overall expansion of the resource pie.

The environmental effects are also significant. Trees, of course, sequester carbon, but they also provide other services, such as animal habitat. Although some regions of FMNR are dominated by *Faidherbia albida*, elsewhere tree biodiversity is increasing.

The economic benefits to farmers of investing in the protection and management of on-farm natural regeneration are high. A study by economists, which is based on a review of some of the benefits, shows an internal rate of return of 31 percent. One farmer interviewed by Hertsgaard (2009, 3) noted, "Twenty years ago, after the drought, our situation here was quite desperate, but now we live much better. Before, most families only had one granary each. Now they have three or four though their land has not increased. We have more livestock as well."

HOW TO PROMOTE RE-GREENING OF THE SAHEL BASED ON FARMER-MANAGED NATURAL REGENERATION?

The lessons of recent woodland resurgence in the Sahel show that it is vital that national legislation support investments by farmers in trees. A key policy measure is that farmers be granted exclusive rights to the trees on their fields. Tenurial regimes in African contexts are notoriously complicated, but local control over and access to forest products makes a difference (Larson, Barry, and Dahal 2010; Andrade 1980; Mwangi and Dohrn 2008; Coulibaly-Lingani et al. 2009; Kant 2009). The contrast between Niger and Nigeria, where such rights are not recognized, is an example in point.

It is important to identify and analyze existing grass-roots success stories in FMNR in the Sahel and use these success stories as a basis for organizing farmer study visits, a proven tool for spreading good practices. There are numerous cases of farmer-managed natural regeneration in the other Sahel countries, and these all provide a basis for future action. This vernacular knowledge is key in understanding eco-social resilience in dynamic environments (Van Haaften and Van de Vijver 2003).

It will be critical to use mass media to inform the farmers and the widest possible audience about success stories and the impacts of farmer study visits, as well as about forestry legislation. Unless farmers are aware of their rights, they will have no incentive to protect on-farm saplings.

In order to promote re-greening, in each participating Sahel country (at present Niger, Burkina Faso, Mali, and Senegal) a national alliance of non-governmental organizations (NGOs) and other partners have been created. All partners in the national alliances will jointly promote the protection and management of on-farm natural regeneration by farmers. This alliance is linked to an international alliance of NGOs and research institutions created to support them.

Conclusion

Promoting FMNR in the Sahel is complex, but it contributes to realizing some of the Millennium Development Goals as well as the international environmental conventions on climate change, desertification, and biodiversity. It also offers a relatively low-cost, rapid and sustainable means of fulfilling the African heads of state desire to combat desertification through the Great Green Wall initiative, while avoiding weaknesses of previous major tree planting projects. This is an opportunity for developing substantive civil society action on the ground, which builds on existing grass-roots success stories.

PART V
The Urban Matrix

Urban Ecologies

CHRISTINE PADOCH

Sometime in 2008, the world's population passed an important milestone: for the first time in history, more people found themselves living in urban areas than in rural areas (Martine and Marshall 2007). All signs suggest that cities will become demographically even more dominant in the future. The towns and cities that now house the majority of humans are a very diverse lot, as are the urbanites themselves. But city dwellers, whether new or experienced, have not cut their ties to forests; they continue to participate in the constantly evolving social life of forests.

The ties between cities and forests are many. In their various forms and from their various locations, forests provide products and services that cities and city folk need or desire. These include food and fiber, clean water supplies, sinks to absorb wastewaters, a place to find solace from urban anxieties, and frequently a place to relocate, to invest, and ultimately to transform into yet another urban space. Cities, in turn, have offered—or at least have promised—forest and farm people an opportunity to find employment, to get an education, to improve their health, to taste foreign and rare forms of excitement, and to share in the power, sophistication, and anonymity that resides in urban spaces. But cities' promises of a better life often falter. Globally, urbanites have repeatedly returned to forests or forest products for many essential supporting services, and even in the world's largest and densest cities have created, reconstructed, rediscovered, and increasingly prized the forests within their concrete landscapes. The urban–rural dichotomy as we imagine it in the temperate zone—a model derived from a history that empties out rural zones—does not appear adequate for understanding the complex linkages, processes, and shifting strategies in the way that the urban and the rural are deployed in the world of burgeoning cities.

In this section we bring together five explorations of the complex linkages between forests and cities. The two geographic areas examined appear to be polar opposites: the great, industrialized metropolis of Chicago, and the seemingly empty green expanse of Amazonia. As the authors of these chapters and others point out, however, when connections between forests and cities are actually examined, many opposites or dichotomies disappear, and the continuous, shifting, and nuanced reality of what we once thought were contrasted categories emerges. Chicago and the Amazon are both locations of complex, historically important urban-forest interactions, and the forests they hold however different in scale and structure, dynamics, and development, bear the imprints of centuries-long histories of shifting urban demands and values. But perhaps more to the point, they reveal possibilities of coexistence in ways that may seem quite foreign to those accustomed to thinking of cities as mostly human and not more generally biotic spaces. This question becomes ever more important as we take on the idea that all cities need to be carrying out environmental services, as well as maintaining their primacy as locations for economic processes. Thus, creating habitats—new kinds of forests—also involves creating new kinds of people and ideologies about the place of cities in "nature," not only as metropolis of consumption, but as sites of biotic sustenance, resilience, and creation.

Each chapter in this section tests and questions accepted truths about forests, cities, global markets, and forest resurgence. Perhaps the most compelling and surprising of the four Amazon-focused contributions is chapter 24 by Heckenberger et al. The major discoveries that it reports from Brazil's Xingu River basin show that centuries ago Amazonian ancestors may have created cities that today's progressive urban planners only dream about: a networked constellation of garden cities that coexisted with productive rural landscapes. These discoveries may even indicate pathways to the sustainability of tropical urban spaces that their descendants today might successfully follow.

The three chapters on recent trends in Amazonia, by Brondizio, Padoch et al., and Sears and Pinedo-Vasquez all point to the emergence of new forms of households, social networks, and markets linking rural and urban spaces. They show that explosive growth in global, urban-based demands for exotic foods and tropical timbers need not lead to unrelenting deforestation, but can actually drive a local resurgence of forest cover. They also indicate that local patterns of afforestation may be linked to accelerating migration from the hinterlands to Amazonia's cities, but not through the mechanism predicted; the forested sites that send urban-ward migrants are not simply emptied of their peasant populations. The contrast between urban and rural peoples, jobs, and futures in Amazonia has become so indistinct, that it has virtually disappeared. Finally, the chapter by Peter Crane and colleagues shows that even in Chicago, one of the planet's most highly developed urban places, the recovery of forest, even of a "wilderness," is not only possible, but can be ever more important to its resident communities.

24 ∗ Amazonia 1492: Pristine Forest or Cultural Parkland?

MICHAEL J. HECKENBERGER, AFUKAKA KUIKURO,
URISSAP'A TABATA KUIKURO, J. CHRISTIAN RUSSELL,
MORGAN SCHMIDT, CARLOS FAUSTO, AND
BRUNA FRANCHETTO

Was the Amazon a natural forest in 1492, sparsely populated and essentially pristine, as has been traditionally thought? Or, instead, were parts of it densely settled and better viewed as cultural forests, including large agricultural areas, open parklands, and working forests associated with large, regional polities (Balée 1989; Denevan 1992a; Lathrap 1970). Despite growing popularity for the latter view (Mann 2000, 2002),[1] entrenched debates regarding pre-Columbian cultural and ecological variation in the region remain unresolved due to a lack of well-documented case studies (Heckenberger, Petersen, and Neves 2001; Roosevelt 1999). Here, we present clear evidence of large, regional social formations (circa 1250 to 1600 AD) and their substantial influence on the landscape, where they have altered much of the local forest cover. Specifically, archaeological research in the Upper Xingu (Mato Grosso, Brazil), including detailed mapping and excavations of extensive earthen features (such as moats, roads, and bridges) in and around ancient settlements, reveals unexpectedly complex regional settlement patterns that created areas of acute forest alteration.

The Upper Xingu is unique in the southern peripheries of the Amazon as the largest contiguous tract of tropical forest still under indigenous resource management (Parque Indígena do Xingu [PIX]). It remains little affected by twentieth century mechanized development (fig. 24.1). The Upper Xingu is a long-standing case study of indigenous Amazonian agriculture and ecology (Carneiro 1957, 1983), and one of the few places where contemporary observations about indigenous agriculture, land use, and settlement pattern can be systematically linked with archaeological and oral historical evidence (Heckenberger 1998). Long-term in situ cultural development of Xinguano peoples over more than 1,000 years[2] is clearly documented by continuity in (1) utilitarian ceramics used to process and cook staple foods[3]; (2) settlement placement (at forest/wetland transitions) and local land use, marked by substantial forest and wetland alterations; and (3) settlement form, notably circular plazas with radial roads.[4]

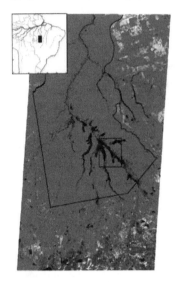

24.1. Upper Xingu region image [Landsat 7 Enhanced Thematic Mapper, path 225, rows 68 to 69; 12 August 1999; image is a composite false-color infrared with bands assigned as 5-4-3 (red-green-blue) to provide the appearance of natural vegetation] with the approximate boundaries of the PIX and the Kuikuro study area (inset of Upper Xingu, Brazil). The ecological transition between the closed forests of southern Amazonia and the more open savanna/gallery forest of the Brazilian Planalto Central is shown (lower right).

In the Kuikuro study area (fig. 24.2),[5] nineteen major pre-Columbian settlements have been identified, generally separated by 3 to 5 km and linked by a system of broad straightroads.[6] Recognition and mapping of major earthworks at these sites reveal their articulation in a remarkably elaborate regional plan. The earthworks include (1) excavated ditches in and around ancient settlements (up to 2.5 km long and 5 m deep); (2) linear mounds or "curbs" positioned at the margins of major roads and circular plazas (averaging about 0.5 to 1.0 m in height); and (3) a variety of wetland features, such as bridges, artificial river obstructions and ponds, raised causeways, canals, and other structures, many of which are still in use today. Similar constructed features (such as settlements, roads, weirs, and ponds) are known from culturally related peoples (principally Arawak speakers) across the southern Amazonian periphery (Erickson 2000a; Heckenberger 2002; Métraux 1948).

The integrated settlement configuration was in place by circa 1250 to 1400 A.D., based on radiocarbon dates from stratified deposits at X6, X11, and X13 (table 24.1). Major curbed roads (10 to 50 m wide) articulate with plazas, ditches, and partition space within villages and across the broader landscape, notably linking settlements into "galactic" clusters across the region (Tambiah 1985).[7] The Ipatse cluster includes four major residential settlements (X6, X17, X18, and X22), linked to a fifth unfortified hub site (X13), with only limited residential occupation (fig. 24.3, panel A). Another cluster of sites, centered on X11, shows a similar pattern of a large (50 ha) fortified

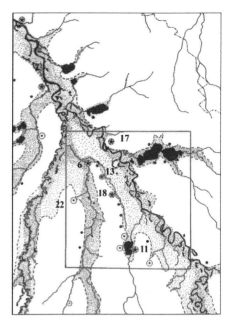

24.2. Kuikuro study area showing the distribution of major ditched plaza centers (stars in circles), major plaza centers (open circles), and small plaza and non-plaza villages (black dots).

settlement connected to other smaller but still elaborate settlements. In the case of the Kuhikugu (X11) cluster, the largest residential center is the hub (fig. 24.3, panel B). The primary nodes of each cluster are also linked by roads to smaller plaza settlements. On the basis of artifact and dark-earth distributions, it has been estimated that large sites, such as X6 and X11, had 15 to 25 ha of residential space, medium sites (X17, X18, X22) had 5 to 10 ha, and small sites had 2 to 5 ha.[8] Thus, the actual residential area of a cluster like X6 or X11 was at least 40 to 80 ha in an area of about 400 km², with an estimated population of between 2,500 and 5,000 persons (or about 6 to 12.5 persons per km² in the study area).

Areas within each galactic cluster can be characterized as saturated anthropogenic landscapes, because virtually the entire area in and between major settlements, although not entirely cultural in origin, was carefully engineered and managed. Indeed, the road networks, oriented by the same system of cardinality that characterizes plaza spatial organization, partitioned the landscape into a gridlike or lattice-like organization of nodes (plazas) and connecting thoroughfares, although patches and corridors of secondary and perhaps managed forests were likely common, as they are today. Mapped archaeological features correspond to patches of acutely modified secondary growth, distinctive from surrounding forest and easy to recognize in satellite images. These patches or islands are identified in the

Table 24.1. Radiocarbon dates from Nokugu (X6) and other sites in the Kuikuro study area, southern PIX

Lab no.	Site/unit	Conventional radiocarbon	2Σ-calibrated age range	Provenance
		Historical Xinguano (1700 A.D.–present)		
Beta 176142	X6/ET2	20 ± 50	modern*	Ditch 3 (S), upper ditch infill
Beta 72260	X6/ET1	180 ± 60	1520–1940 A.D.	Ditch 2 (S), upper ditch infill
		Terminal "galactic" period (1400–1700 A.D.)		
Beta 176137	X6/ET10	350 ± 60	1460–1640 A.D.	Ditch 2 (N), upper ditch infill
Beta 1301	X6/ET1	360 ± 70	1420–1640 A.D.	Ditch 2 (S), mid-ditch infill
Beta 176135	X6/ET3	440 ± 60	1420–1480 A.D.	Small plaza, subcurb intact
Beta 72262	X11/EU1	440 ± 70	1400–1650 A.D.	North road, intact/curb interface
Beta 176140	X6/ET3	530 ± 60	1400–1430 A.D.	Small plaza, subcurb intact
		Initial "galactic" period (1250–1400 A.D.)		
Beta 176139†	X6/ET2	590 ± 60	1300–1420 A.D.	Ditch 3 (N), basal fill
Bea 177724†	X6/ET2	670 ± 60	1260–1410 A.D.	Ditch 3 (N), basal fill
Beta 88362	X13/EU1	690 ± 60	1260–1300 A.D.	Central plaza, subcurb intact
Beta 78979	X6/ET1	700 ± 70	1230–1410 A.D.	Ditch 2 (S), sub-berm intact
Beta 176136	X6/ET4	710 ± 50	1270–1300 A.D.	Ditch 1 (S), basal fill
		Late developmental (900–1250 A.D.)		
Beta 72263	X11/EU1	900 ± 60	1000–1250 A.D.	North road, subcurb, basal intact
Beta 88363	X13/EU1	910 ± 80	1040–1250 A.D.	Central plaza, subcurb base intact
Beta 72261	X6/ET1	1000 ± 70	950–1210 A.D.	Ditch 2 (S), sub-berm, base intact
Beta 176141	X6/ET5 1	30 ± 60	980–1030 A.D.	Central plaza, subcurb base intact
		Initial Xinguano (pre-900 A.D.)		
Beta 176143	X6/ET2	1370 ± 60	640–690 A.D.	Mid-ditch 3 (N), mid-ditch
Beta 176138	X6/ET10	2110 ± 40	190–60 B.C.	Ditch (N), basal fill

Notes: Beta 176135 to Beta 176144 are reported here for the first time. Calibrated age ranges for samples dated in 2003 (numbered 176135 and up) were reported by Beta-Analytic laboratory; previous (1994) dates were calibrated using CALIB 4.0 (Stuiver and Pearson 1993). ET, excavation trench of 1.0 by 10.0 m or more; EU, excavation unit of 1.0 m 2; S, southern side of plaza or site; N, northern side; Ditch 1, outermost; Ditch 2, middle; Ditch 3, innermost ditch in all sites.

*Two additional modern dates, Beta 98978 and Beta 176144, are considered invalid. The former was redated with a sample slightly higher in profile (Beta 81301). †Beta 176139 is inversed with 176142 and was redated by 177724. Beta 176138 also comes from stratified but mixed context of ditch infill, and both likely represent earlier materials that are mixed in ditch construction.

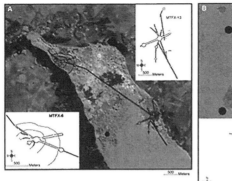
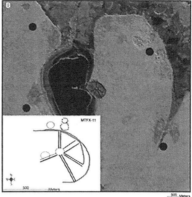

24.3. Satellite image [Landsat 4 Thematic Mapper, path 225, row 69; 21 June 1992; bands were assigned as 5(red)-4(green)-3(blue)] with global positioning system-mapped Ipatse cluster sites X6 and X13 (A, insets) linked by the "north-south road" and transit-mapped X11 (B, inset). Ditches are colored in red; road and plaza curbs are black. Fieldwork in 2003 demonstrates that roads extend fully from X13 to X18, and continue on to X19, X20, and beyond along the north-south road, as well as across high ground to X17 and X22 (fig. S1); X11 roads also connect it to the four satellites. MTFX, Mato Grosso (the state), Formadores do Xingu (the archaeological region). The number refers to the site number.

indigenous knowledge systems, including diverse species whose distributions are generally restricted to anthrosols (dark earth)—called *egepe* by the Kuikuro—associated with ancient settlements (Petersen, Neves, and Heckenberger 2001; Woods and McCann 1999).[9] Some areas related to major pre-Columbian settlements, such as X6, X11, and X13, have not returned to high forest after about 400 years of abandonment, although other areas (such as roads, hamlets, gardens, fields, and parklands) show a highly patchy forest regrowth typical of postabandonment (after 1600 to 1750 AD) succession.

Recognition of the massive forest alterations associated with pre-Columbian occupations requires an understanding of local biodiversity in the context of the complex cultural history of the area. The composition of forest and wetland habitats reflects long-term cumulative changes, given that the settlement areas were occupied more or less continuously over many generations, as well as the large-scale alteration and management of local environments by dense late-prehistoric occupations (c. 1250 to 1650 A.D.). Present soil and biotic distributions, often isomorphic with the distribution of archaeological features, notably plazas, residential areas, roads, and roadside hamlets, are in large part the result of pre-Columbian land-management strategies. After circa 1600 to 1700 AD, catastrophic de-

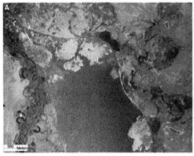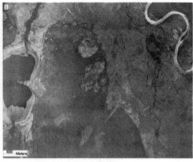

24.4. Aerial photographs (US Air Force/Força Áerea Brasileira 1967; 1:60,000) of Lake Ipatse (A) and Lake Kuhikugu (B) showing land-cover change during six-year period (1961 to 1967) after abandonment (Kuhikugu) and reoccupation (Ipatse). Comparison with fig. 25.3 (1992) of same areas shows a ~thirty-year period. Graphic testimony of post-1492 decline in village size is shown by comparing the size of X11 with the four sequential Kuikuro villages (c. 1870s to 1961) in and north of the ancient site and by comparing X6 with the active village in 1967 and 1992.

population (Heckenberger 2001) led to the abandonment of these works and many settlements, resulting in extensive reforestation in many areas. The scale of the prehistoric settlements, including exterior constructions, such as roads, hamlets, wetland structures, and cultivation areas, suggests that agricultural and parkland landscapes, rather than high forest, character-ized the broad landscapes around ancient villages, as is true in contempo-rary villages. Metal technology, however, has increased the speed at which forests can be converted into mosaic parklands of dispersed manioc gar-dens, sapé grass fields, piqui groves, and secondary forest (Denevan 1992b; fig. 24.4).

The Upper Xingu is a unique Amazonian example of a tropical forest way of life that supported large, densely settled, and integrated regional populations over the past 1,000 years. Local ecology reflects the dynamic interaction between the natural environment, the influence of fairly large, settled human populations, and the legacy of Euro-American colonialism over the past 500 years. Evidence of large, well-engineered public works (such as plazas, roads, moats, and bridges) in and between pre-Columbian settlements suggests a highly elaborate built environment, rivaling that of many contemporary complex societies of the Americas and elsewhere.[10] To suggest that Xinguano lands were intensively managed and developed circa 1492 AD, however, does not imply that indigenous land-use strategies, based on patchy development within long rotational cycles, are comparable to modern nonindigenous clear-cutting strategies (Denevan 2000; Smith et al. 1995). Xinguano cultivation and land management, indeed, provides a viable alternative. The present research emphasizes the critical importance of collaborative research strategies, including archaeological and ethnographic fieldwork, remote-sensed data analysis and geographic information systems, and most important, indigenous participation, to understand the complex interplay of ecological, historical, and political conditions in Amazonia before and after 1492.

25 ✳ Urban Residence, Rural Employment, and the Future of Amazonian Forests

CHRISTINE PADOCH, ANGELA STEWARD, MIGUEL
PINEDO-VASQUEZ, LOUIS PUTZEL, AND MEDARDO
MIRANDA RUIZ

Rural-urban migration and the growth of urban areas have historically had important impacts on rural environments, including forests. With the world's population having very recently become predominantly urban (Martine and Marshall 2007) and urban areas in many tropical areas growing at record levels, interest in how rural–urban shifts specifically affect tropical forests is receiving considerable attention (DeFries et al. 2010; Rudel et al. 2009; Lambin et al. 2001; Aide and Grau 2004; Grau and Aide 2008; Wright and Muller-Landau 2006; Garcia, Soares-Filho, and Sawyer 2007; Padoch et al. 2008). The urbanization of tropical populations could lead to a halt in ongoing deforestation and to the eventual afforestation of substantial areas if emigration from the countryside were to drain rural zones of farmers and return their abandoned fields and pastures to forest. Just such a trend, which characterized regions of northern Europe and North America, as well as some areas of South America (Baptista 2008; Baptista and Rudel 2006; Grau et al. 2008; Aide and Grau 2004; Wright and Muller-Landau 2006), is credited with leading to the forest transitions that occurred in nineteenth-century northern Europe and more recently in parts of the Western Hemisphere. The applicability of these experiences to much of today's forested tropics, and especially to the Amazon, is being discussed and widely questioned (Parry et al. 2010; DeFries et al. 2010; Rudel et al. 2009; Rudel, Bates, and Machinguiashi 2002; Padoch et al. 2008; Fearnside 2008; Hecht, this volume, and others).

Researchers have shown that despite sizable urban growth and rapid rural–urban migration, rural areas in the humid tropics are far from empty and fields and pastures are rarely abandoned to regrow into unmanaged and unused forests. In some cases, rural emigrants are merely replaced by new immigrant farmers (Fearnside 2008), in others by industrial-scale agricultural enterprises fueled by growing regional urban and global markets; in either situation, few lands apart from the most remote and marginally arable return to forest (DeFries et al. 2010). Other studies have examined patterns of urban-ward migration and find that many emigrant families actually are engaged in circular, rather than unidirectional, rural to urban movement (Tacoli 2009; Padoch et al. 2008; Pinedo-Vasquez and Padoch 2009) and again, rural sites are not abandoned. Migration in the tropics, as in much of the world today, appears to be far more complex than national censuses and general overviews suggest. Some researchers propose that not only has contemporary rural-urban migration been largely misunderstood, but that no meaningful distinction can be made between what and who is rural or urban (Tacoli 2002; Smit 1998; Rigg 2003).

In this chapter we contribute to the examination of the complex impacts of urban-ward migration on rural environments, by focusing specifically on employment patterns among the urban poor, many of whom are recent migrants. Employment is key to understanding urbanization impacts on forests. The switch in employment opportunities from predominantly farm work to overwhelmingly urban industrial, as well as informal labor, is believed to have spurred the nineteenth-century rural–urban migrations mentioned before and the subsequent regrowth of northern European forests. Using a variety of field-derived data from in and around the western Amazonian city of Pucallpa, we show that, in contrast to these historical patterns, many Pucallpa households, though residents of the city, rely on rural-based employment, notably logging and other timbering activities, rather than on city-based jobs to support themselves and their families. We also find that timber extraction in the region, now dominated by urban-based loggers, is a fundamentally different activity—with different impacts on rural environments—than logging as carried out by farmers.

We suggest that urban and rural areas including forests, as well as rural and urban peoples, can no longer be characterized by distinctly rural or urban occupations or behaviors, nor are boundaries between rural and urban places and peoples easily distinguished or clear-cut. Finally, we argue that as urban and rural peoples, resources, and places become increasingly interconnected in complex and dynamic ways, the planet's greatest forest is experiencing new, and as yet little-understood patterns of change.

The Complexities of Rural and Urban Employment: "De-Agrarianization" and "Ruralization"

In recent years, scholars have paid increasing attention to an important change in rural and urban employment patterns. Since the mid-1990s researchers have documented a global trend of de-agrarianization, or livelihood diversification in the world's rural areas. Peasants are abandoning full-time agriculture and other essentially rural activities to engage in nonfarm public sector and informal service work (Bryceson 1996, 1999; Bryceson, Mooij, and Kay 2000; Bebbington 1999; De Janvry and Sadoulet 2001; Finan, Sadoulet, and De Janvry 2005). In many parts of the world, including Latin America, rural nonfarm work has been estimated to account for up to 80 percent of total household income (Reardon and Escobar 2001). Broad livelihood shifts away from completely or even largely rural-based activities are attributed to a variety of factors, some of which are problems that push, and others opportunities that pull farmers to nonfarm jobs. Some push factors include obstacles that have arisen from globalization processes that have made it difficult for smallholders to survive through agriculture alone. In particular, structural adjustment programs (SAPs) following the debt crises of the late 1980s called for trade liberalization or the opening up of international markets across the globe (Jansen 2000; Zoomers 2010). As a result, agricultural markets and small farmers have become integrated into global networks of trade in agricultural products. The rise of agro-industry in many countries such as Brazil has deprived peasants of opportunities to market their produce. Structural adjustment also called for decentralization and decreased government spending, and thus in many countries, market changes coincided with the loss of agricultural subsidies and rural extension programs targeting small farmers (Bryceson 1996; Ellis 1998; McMichael 2006; Pacheco 2006; Caldas et al. 2007). In some areas, globalization, structural adjustment, and decentralization have benefited the rural poor by bringing industrial development, wage labor, and government income in the form of pensions and salaries to the countryside (Jansen 2000; Rigg and Nattapoolwat 2001; Steward 2008). In those areas where economic development initiatives have failed and off-farm employment is rare, however, many households have seen some or all of their members leave the countryside for the city (Bryceson 1996; Ellis 1998; Rudel 2002; Barbieri and Carr 2005; Portes, Escobar, and Radford 2007; Carr 2009; Sanderson and Kentor 2009).

Far less frequently discussed than rural patterns of de-agrarianization— although widely documented over the years in many parts of the world (Tacoli 2002; Baker 1995; Kamete 1998)—is a parallel process taking place in urban areas. Increasing household income diversification is an urban phenomenon as well. Millions of urban residents—among them many new immigrants who recently left the countryside—rely not only on urban-derived earnings, but also on rural resource management and natural resource extraction to feed themselves and to generate a substantial part of household incomes (Satterthwaite and Tacoli 2002; Smit 1998). The growing need to diversify livelihoods, ostensibly derived from living in the city, reflects many of the same types of economic problems (and occasional opportunities) that rural folk have faced, as the lives of families in poor countries have become increasingly precarious. In many regions, job opportunities and incomes have declined in recent decades following downturns in employment in the public sector as well as in wage levels. These declines have affected both formal sector workers and those in the informal service sector who must largely depend on the spending of the cities' formally employed and more securely employed. Widespread cuts in public spending, especially in the areas of education, public health, and infrastructure development, coupled with rises in food and fuel prices, have particularly affected the urban poor, especially the newly urban and shanty-town residents. Researchers have noted these effects in changing patterns and livelihood strategies such as significant increases "in mobility accompanied by strong social and economic links with home areas; and higher levels of multi-activity, especially among younger generations" (Tacoli 2002, i).

Such ruralization of urban household incomes takes a variety of forms depending on the actual places of residence and of employment of household members. In one widely discussed pattern, urban residents, especially newly urban and peri-urban residents, create and cultivate farms, pastures, or orchards located in urban (Del Castillo 2003; Drechsel and Dongus 2010; Long and Nair 1999; Vicentini 2004; Stoian 2005; Stark and Ossa 2007) or more frequently in peri-urban zones (also termed the urban fringe, transition zone, or urban–rural interface) where they produce goods for their own use and the market (Simon 2008). In addition to urban or peri-urban farming, many of the poor residing in cities are increasingly turning to fishing in regional lakes and rivers for household consumption and sale, or raising small livestock in city gardens or peri-urban sites. As many researchers note, these patterns are not new, but have recently been rediscovered by planners and researchers as environmental and sustainability concerns have come to the fore.

A second common way the incomes of urban, especially newly urban, households are being diversified involves long-term multisitedness or dispersion of household members among urban and rural homes, combined often with circular migration of individuals between these urban and rural sites. The family members residing in the countryside usually continue to maintain fields, fallows, and forests in production, while their urban counterparts seek employment in the city. The rural folk may then send produce urban-ward to feed their city relatives and to have them sell farm items in urban markets; the city dwellers in turn remit cash to the family members in the countryside. City and rural members may, and often do, switch places of residence and of employment. The continuing or even increasing economic importance of maintaining such dispersed economic units and kin-based networks in what WinklerPrins and de Souza (2005) have termed an "economy of affection" has been documented around the globe. Many studies have particularly focused on the crucial economic and in some instances environmental roles played by remittances both internationally (from more to less industrialized countries), and internally (from cities to rural area; Hecht 2009).

In a significant and probably growing percentage of cases, however, the urban poor fail to find remunerative employment in regional cities and rural relatives or peri-urban gardens cannot satisfy urban needs. This situation leads to yet a third distinct urban–rural configuration of residence and employment, one that appears to reverse not only the usual patterns of labor migration, but of income flows as well. Unemployed urban household members frequently leave for rural areas where they work for wages, and remittances flow from work done in the countryside to the city. Informal employment in forestry and mining feeds significant numbers of urban families, especially the newly urban; most frequently these activities are carried out far from the city, often in remote areas (Stoian 2005, 2006).

The pattern of leaving the city to work in rural zones may be less unusual than it first appears since, despite the common assumption that rural-urban migration is invariably driven by the search for better incomes and labor opportunities, in many regions initial decisions to migrate from farming villages to cities are made for reasons not directly related to employment (Stoian 2005, 2006; Padoch et al. 2008.). The desire or need for urban educational and health services is often an initial factor that pulls rural folk to cities or pushes them out of the countryside. Access to medical attention, educational opportunities, entertainment, and other cultural resources is often radically different in rural and urban settings, with the urban areas offering far better services. Thus enhanced employment opportunities in cities may not even be considered when rural folk first make the move to the city. In any case, when the search for urban employment does become important, even relatively insecure and low-paying jobs are often unavailable. Work in rural areas diversifies household income streams, and allows poor households throughout the world to tap into a variety of labor opportunities, income sources, and resources, thus increasing incomes while reducing uncertainty. Rural-based jobs are, of course, not only the choice of the urban unemployed and desperate; some families of means, including large landowners, find that rural assets are a way of adding to incomes that they already enjoy.

Both de-agrarianization of rural areas and ruralization of cities have various implications for the ways in which we perceive not only rural and urban populations but also rural and urban spaces and rural, including forest, resources. Urban and rural areas are no longer characterized by distinct land-use patterns and resource use behavior, nor are boundaries between rural and urban spaces and places easily distinguished or delineated. As much recent literature suggests, we can no longer assume that residence necessarily defines livelihood structures, activities, or forms of employment. While the influence of urban economic and social patterns is doubtless increasing in rural areas—leading, for example, to "de-peasantization" and dissolution of more traditional village structures—a parallel trend of ruralization of cities and towns has also been noted. This trend results in rural occupations and broader rural lifestyles, attitudes, and consumption patterns characterizing even large urban places (Krüger 1998; Padoch et al 2008). This complex of new and newly important urban–rural relations is and will continue to have a variety of impacts on the extent, configuration, and use of forests and forest resources. Little of this complexity is reflected in most of the debate about urbanization and the future of tropical forests.

Urban and Rural in Amazonia

In this chapter, we look at rural–urban interactions in western Amazon, with attention to the work and employment patterns of the urban poor in the Amazonian city of Pucallpa, Peru. We focus specifically on households located in the rapidly growing informal settlements or *asentamientos humanos*, which constitute the great majority of that city's residents, and on the participation of these urban households in forestry, particularly in logging, an activity that is usually viewed as eminently rural. We then discuss the importance of logging as a source of employment and income for urban households, and the possible impacts of changing logging patterns on the future of the forests of Peruvian Amazonia.

The majority of the residents of the Amazon Basin have for decades been urbanites, although regionally the level of urbanization varies. In 2000, the national census indicated the population of Brazil's "Amazônia Legal" (i.e., the states of Amazonas, Para, Acre, Tocantins, Mato Grosso, Maranhão, Amapá, Rondônia, and Roraima) was already about 70 percent urbanized; in the Peruvian national census of 2005, the enormous Amazonian Region of Ucayali, where Pucallpa is located, was 72 percent urban, (Instituto Nacional de Estadística e Informática [INEI] 2007). On the other hand, the Amazonian areas of Ecuador and Guyana are still predominantly rural (GEO Amazonia 2009). The Brazilian urban geographer Bertha Becker (2005) noted the need to understand urbanization processes in Amazonia in a broader context and suggested that mere numbers showing the growing populations in cities and towns provide an inadequate picture of the extent and meaning of urbanization in the region. The transformation of the values of Amazon society is another, if less easily calculated, measure of urbanization. Becker famously described all Amazonia as an "urbanized forest."

The urban demographic predominance in the western Amazonian Region of Ucayali largely reflects the dramatic growth of Pucallpa, its largest city and capital, beginning in the 1960s. Between 1961 and 1993 the population of the city, which lies at the terminus of a highway linking the main stem of the Amazon River to Peru's highlands and coast, increased more than sixfold (Santos-Granero and Barclay 2000, 286). We must reiterate, however, that no matter how dramatic these numbers and how real the patterns of change, these census data present a deceptively simple picture of urban and rural shifts in the region. While people in Amazonia, as elsewhere, are classified as residing in either urban or rural places, and while there has definitely been an important change in both residence and occupations in the region, as we noted earlier, the urban–rural distinction is more often than not, difficult to make. Urban and rural people and places are linked in numerous ways (Nugent 1993; Wagley 1953; WinklerPrins 2002; WinklerPrins and de Souza 2005; Siqueira 2006; Raffles 2002). As noted before, a great many newly urban households are multisited, multilocal, or dispersed, maintaining houses and as we will show, economic activities in rural areas as well as in the city (Padoch et al. 2008; Pinedo-Vasquez and Padoch 2009). Such multisited households are not new to Amazonia, although recent changes in communications and transportation, markets, and labor opportunities have greatly increased their incidence (Stearman 1985; Nugent 1993; WinklerPrins 2002; WinklerPrins and de Souza 2005). It is important to note that migration, including large and frequent displacements in search of resources, is also not new. As research on both indigenous and nonindigenous Amazonian groups has shown, the history of the region is one of very frequent, complex, and often quite massive movement, some forced, some voluntary (Alexiades 2009; Little 2001; Hecht 2011). Demographic exchanges between rural and urban areas have often occurred in the contexts of regional economic booms and busts, including the famous Rubber Boom that spanned the end of nineteenth and beginning of the twentieth centuries. Recent decades have seen shifts temporarily reversing urban-ward trends in many Amazonian regions in response to, for example, opportunities to participate in Bolivia's seasonal Brazil nut harvest (Stoian 2005, 2006), government settlement programs in Brazil (Paganoto 2007), and oil exploration and development in Peru (Padoch et al. 2008). A program to formalize tenure of informally held land parcels in Peru may now be encouraging urban residents to occupy and use rural lands in the hopes of gaining permanent ownership (Putzel 2010, and see, e.g., COFOPRI 2007).

Rural people from Peru's lowland Amazon, foothill areas, as well as the highlands and the coast, have come over the last several decades to Pucallpa, Peru's second largest Amazonian city, for a variety of reasons: for access to secondary and university education, to seek medical attention, to escape the violence of the drug trade and conflict between Marxist insurgents and the military, to improve their economic lives, to escape destructive floods and other disasters, and to share in the modern urban life that small communities do not offer. Since the 1970s Pucallpa has accommodated these many new migrants in a large number of informal settlements (*asentamientos humanos* or *AA.HH.*) located throughout the city and in the spaces between Pucallpa and neighboring Yarinacocha. In mid-2007 these AA.HH. were said to number 579 separate communities (Padoch et al. 2008). Pucallpa is the administrative hub for the Ucayali Region as well as Peruvian Amazonia's most active center of logging and timber milling. Pucallpa is also important as a port where goods imported from Peru's highlands and coast are distributed throughout much of the lowland Amazon, including Iquitos, Peru's largest Amazonian city. Pucallpa has grown in population not only with the arrival of farmers from neighboring villages and towns, but also with immigrants from many of the strikingly different environments that Peru comprises, particularly from the Andes.

We have been carrying out research on resource use in and around Pucallpa since 2003, focusing on issues of urbanization, forests, forest policy, markets, and trade. The case study in this chapter is based on three specific sources of field data: (1) a survey of households in ten of Pucallpa's AA.HH.; (2) a survey of timber camps located around the river towns of Contamana, Orellana, and Pampahermosa on the Ucayali River north of Pucallpa; and (3) a series of in-depth interviews with residents and former residents of the area of the lower Shesha River (a tributary of the Abujao, itself a tributary of the Ucayali). All three of these data-gathering activities focused on unraveling the complex relationships between rural–urban migration, residence, and employment in logging or other aspects of the timber industry. The survey of the AA.HH. were carried out between August 2008 and February 2010; the timber camp surveys cover approximately the last ten years, and the Shesha interviews were carried out in February 2010.

Urban Residence and Rural Employment in Pucallpa

Although official statistics derived from the 2007 National Census state that only 4 percent of the economically active population (*población eco-*

nomicamente activa) of Pucallpa is unemployed, interviews, observation, and anecdotal evidence suggest that a more accurate unemployment number may be about 20 percent. When the underemployed in Pucallpa's AA.HH. are added, the percentage probably reaches at least 50 percent. Of those who are formally employed in stable and/or salaried jobs, few live in the informal settlements that are home to virtually all recent migrants from rural zones. Employment opportunities that do exist in Pucallpa itself tend to be concentrated in the timber and wood industries, including jobs in the many sawmills, the river port and other wood transport and shipping facilities, in the city's highly polluting charcoal-making facilities, and the plywood factories that dominate the economic life of the city (Putzel 2010; Putzel, Padoch, and Pinedo-Vasquez 2008).

Many of the poor, however, fail to find adequate casual employment in the city and turn to work in the countryside to support themselves and their families. In 2008 and 2009 we conducted 207 interviews with randomly selected households residing in ten informal settlements in Pucallpa. In each household an adult member was interviewed about the employment status of all the adults in the family who identified themselves as urban residents and who had contributed to the income of the urban household in the preceding two years. It should be noted that some of the households interviewed were distinctly multisited with at least two houses (usually one rural and one urban), among which resources and sometimes people circulated. For this survey only the employment of the family members who were considered members of the urban household were registered. Those who may have contributed resources but owned and lived in the rural home were excluded from the sample. Households differed considerably in size, with some having only one potentially economically active member and others having as many as six.

Table 25.1 summarizes the results of the interviews. We found that 199 of the total 207 households sampled (96 percent) included at least one member who had engaged in at least one rural-based economic activity over the preceding two years.

While a large majority of households had participated in some form of urban employment as well, only 6 percent of the total sample reported having a household member with a stable salaried city job, while 72 percent listed occasional or informal jobs in the city. The majority of the latter were poorly paid day-labor jobs that included activities such as unloading cargo from riverboats, casual construction or house repair, or driving a leased *motocarro* (three-wheeled motorcycle taxi). Women in the sample who reported urban employment often sold produce or other goods on the street

Table 25.1. Number of households with urban, rural, and rural/
urban employment from a sample of 207 urban households

Category	Number	% of Total
Households with any rural employment	199	96
Households with only urban employment	8	4
Households with only rural employment	41	20
Households with both rural and urban employment	158	76
Households employed in timber	148	72

or just outside one of the municipal markets. Many also sold meals cooked in their homes.

Rural-based employment included work in agriculture, fishing, cattle raising, coca production, fishing, and mining, as well as timber activities. The timber activities were highly important even though they commonly required workers to travel to areas far from the city and spend months away from their families working as loggers, woodsmen (*materos*), equipment operators, boatmen, drivers, and cooks. Some informants in this category also worked in timber processing, either in the field or in or around the city. Urban-based loggers included some who were self-employed, others were financed or otherwise enabled (*habilitado*) or working on contract to deliver specified volumes and types of timber to mill owners and timber merchants. The majority, however, worked as short-term wageworkers. Logging along the upper Amazon and its tributaries tends to be seasonal, the harvest coinciding with high water levels that permit access to small streams and allow logs to be floated out. Logging trips reportedly can extend up to seven months, during which time workers remain in often isolated timber camps while their families collect their wages in the city. Although smallholder farmers and other rural actors also engage in logging, timber extraction in Ucayali is very largely an urban-based activity. Logging crews are hired in the city, equipment is brought from the city and returned to it, and workers need to be in the city in order to get paid.

Table 25.2 summarizes eight years of data on the place of origin of loggers working in forests along the Ucayali River and its tributaries north of the city of Pucallpa. These data were collected from 2000 to 2005 by the provincial government centered in the town of Contamana; in 2006 and 2007 we collected the same information. These data show clearly that urban-based workers dominate what is ostensibly the most rural of occupations: logging mature forests. As we mentioned earlier, these data should

Table 25.2. Logging camp workers and their areas of origin in the
Provincia Ucayali, Región Loreto

Year	No. of logging camps	No. of workers	No. of workers from Pucallpa	No. of workers from Contamana	No. of workers from rural areas
2000	32	254	237	5	12
2001	47	470	438	15	17
2002	21	189	167	18	4
2003	38	304	283	14	7
2004	17	170	158	7	5
2005	22	154	137	9	8
2006	14	140	133	3	4
2007	26	208	188	11	9

not be taken to suggest that rural Amazonians do not harvest timber or en-gage in other forestry activities.

As we have argued elsewhere, smallholder farmers in Amazonia have long included management, harvesting, and sale of timber and other forest products as an integral part of their production and livelihood systems (Putzel 2010; Putzel, Padoch, and Pinedo-Vasquez 2008; Pinedo-Vasquez et al. 2001; Padoch and Pinedo-Vasquez 1996; Sears, Padoch, and Pinedo-Vasquez 2007). Most, if not all, Pucallpa loggers learned the techniques of the trade prior to migrating to the city. We observed, however, that while the knowledge and skills that enable city folk to successfully log Amazonia's forests reflect their rural upbringing, present-day practices of logging by both rural residents and by recent urbanites differ in several significant ways. We explored these differences further through interviews with five families in Pucallpa and in several communities along the Shesha River. We specifically inquired about the informants' knowledge and experience with the logging of forests before and after their families migrated from their rural homes to Pucallpa.

Life and Logging along the Rio Shesha

The five informants and their families first settled in the sites, later incorpo-rated as the villages of Las Mercedes and Nueva Vida along the lower Shesha, in the late 1970s to mid-1980s; here, they found a few families that had lived there since the late 1960s. Like their predecessors, the settlers established themselves as mixed production farmers who maintained swiddens for small-scale production of a range of agricultural staples, some small plots of coca, and a few cattle and other livestock. At this time, local logging was limited to the extraction of timber from nearby forest stands to meet household needs and for occasional regional sale of valuable timber species. In the early 1990s, these settlers and their entire households left their farms due to increasingly threatening confrontations with Marxist insurgents and the fear that their children would be recruited or forced to join these groups in the neighboring forests. After a decade of living in Pucallpa all the families had essentially become urban. Several of the original Shesha families, however, have begun returning to their former farms in the last few years, but they no longer engage in small-scale diverse farming. The majority of those who have returned to their original lands are the sons of former farmers who have gone back to extract timber from the forests along the Shesha. We interviewed five of the original farmers to understand how the scale, focus, and character of logging in the two periods differed. We found notable contrasts between the logging activities of the five families as a group before they decamped for the city and after they returned. These differences principally included shifts in the volume and species of logs extracted, in the tools used to log, as well as in the environmental effects of logging as described by the five informants.

When residents lived along the Shesha River, logging was important to all households as a source of wood for building and repairing their houses and as a source of occasional cash. In each case, however, logging was only one activity of a large repertoire of integrated activities, each demanding labor, skills, tools, and capital of limited availability in the households and community. During this time period, in all cases but one, logging was highly selective and focused on a few valuable species, which were extracted from nearby forests with axes and chainsaws. One informant reported that while living along the Shesha his household harvested an estimated 400 board feet of timber from three species in a typical year; another listed 500 board feet of four species per year, and yet another recalled that he sold eighteen standing trees of three species to local loggers. Among the informants who recalled the quantities of timber they felled, several reported noteworthy variations from year to year: one informant reported that he took out no more than 500 board feet in most years, but in 1989 he sold 1200 board feet to loggers. The one case from this period that contrasts sharply with all the others is that of a farmer whose forest was heavily damaged by an urban-based logger without the farmer's prior knowledge or permission; 20,000 board feet of timber were felled and exported by the logger to Pucallpa, and only some small compensation was paid.

Before the move to Pucallpa, logging on Shesha properties generally took place in the same general environment where all other economic activities of the household occurred. Again, except for the one case of surreptitious logging, which according to the informant, resulted in "destruction of my holding, deep tractor tracks, and destruction of the stream," informants report that environmental damage was limited to the local extraction of timber itself. The one farmer who reported selling standing timber to loggers did mention that they left "tractor tracks"; others cited "no damage."

The residents who recently returned to the lower Shesha River after living in Pucallpa now engage almost exclusively in timber extraction; current logging practices are far less selective, volumes extracted are immeasurably higher, and the environmental damage reported is far greater. The household that reported harvesting at most 400 board feet of three species annually in the 1980s reports having recently taken out 50,000 board feet of timber of at least nine species. The family that had sold only a few standing timbers reports recent sales of at least 40,000 board feet of eight traditional hardwood species and an additional one thousand trees of *bolaina*, a fast-growing hardwood that has just recently found a very active market. The family that reportedly harvested the least among our informant families still took out 30,000 board feet of timber from Shesha River forests.[1] The amount of money that the families now realize from timber sales is much higher than what it was in the 1980s.[2] However, it should be pointed out that because of the present industrialization of logging, with large machines replacing what was once largely an artisanal activity, we estimate today's urban-based families generally see only about 13 percent of the local market value of their timber, while the farmer-foresters of the 1980s often realized 83 percent of that value.

Informants' reports of recent environmental damage due to logging focus on dammed streams, destroyed secondary forest resources, and networks of logging roads throughout the forests. Returning to their native communities as urban residents, families seek quick economic profit from their holdings; they no longer depend on the forest's diverse resources nor broader environmental services. The shift from diversified and more sustainable use of a variety of resources to intensive extraction of timber has obvious environmental impacts. While the older residents sincerely lament the destruction of their waterways, wildlife, and a broad range of nontimber products, the lower Shesha is no longer the loggers' home, and they demand little more than a supply of saleable timber from the forest.

Conclusions

Employing three different sets of field data, we explored the issues of urban residence, rural-based employment, and the state of forests of western

Amazonia. Our results point to three basic conclusions: (1) rural employment is extremely important for poor urban households in the Peruvian Amazon, and work in timber—frequently logging in distant forests—is a very frequent and important source of employment; (2) members of active logging crews in the region are overwhelmingly urban residents; and (3) the way logging is done by rural residents differs from that carried out by urban residents in a number of significant ways, including the impacts it has on the logged forests.

Our first two findings provide concrete examples of a process of ruralization of urban livelihoods in western Amazonia and strengthen the arguments of previous researchers who described a blurring of the categories of urban and rural (Stoian 2005, 2006; Padoch et al. 2008). We clearly see that in and around Pucallpa, where one resides is rarely a predictor of the types of employment upon which one depends.

Furthermore, our conclusions contribute to ongoing debates about the effects that rural–urban migration has and will continue to have on tropical forests. Confirming previous research conducted on a variety of spatial and temporal scales, our data indicate that despite demographic flows out of forested rural areas and into cities, forests are neither abandoned nor left to regenerate; people living in both cities and rural areas continue to use and change forest resources for a variety of economic reasons. We show that forests provide the urban unemployed and poor with incomes. We also find that the rural and urban poor engage differently with rural economic activities. While broader in scope, management by rural dwellers may preserve some environmental features, including ecosystem services (such as water and regeneration of secondary forests), which urban dwellers neither need nor value. These broader forest qualities suffer when rural folk leave and are replaced by urban dwellers whose demands are narrow, but whose relationship to the forest is profoundly different. While the forests are no less economically valuable, they are no longer home. The effects of this demographic shift on the persistence and management of Amazonia's forests has yet to be understood.

Finally, we suggest that close attention be paid to the issue of underemployment in urban areas, a factor that our data show may result in the observed patterns of forest destruction and unsustainable use. The combined effects of the processes of urban-ward migration, underemployment, and change in natural resource use in rural areas by the urban poor may subvert efforts to conserve and sustainably use forest resources in an era of urban expansion and change.

26 ✳ From Fallow Timber to Urban Housing: Family Forestry and Tablilla Production in Peru

ROBIN R. SEARS AND MIGUEL PINEDO-VASQUEZ

Forests as Development Space

Studies indicate that forest cover has been increasing in some areas of the Amazon during the past decade, due in part to transition from open land to secondary forest cover. In some cases this results from the abandonment of agriculture or pasture (Zarin et al. 2001), but also to the decision by rural smallholders to practice secondary forest management (Arce-Nazario 2007; Brondizio, chapter 27, this volume, and 2008). Rudel, Bates, and Machinguiashi (2002) have even shown that forest transition, in this case reforestation, is occurring alongside roads in the Ecuadorian Amazon, a trend spurred by a reversion from pasture to short-cycle shifting cultivation to feed nearby markets. In other parts of the Neotropics, such as El Salvador, forest transition (Hecht et al. 2006) has occurred because of a reduction in land use, in this case because of remittances sent to rural households from abroad. At the same time, deforestation remains a threat to biodiversity and to rural livelihood in some regions of Amazonia (De Sherbinin et al. 2008).

With the mounting concerns about the state of the world's forests, there is a corresponding suite of proposed solutions to degradation and effective methods for restoration of valuable forest ecosystems. Over the past several decades many researchers and agencies have attempted to examine and reconcile the apparent dichotomy of ensuring the conservation of forest ecosystems and improving livelihoods of forest-dependent people (e.g., Ebeling and Yasué 2008; Sunderlin et al. 2005; Wunder 2001). Codifying the dependency between forest ecosystems and rural development, and indicating hope in the reconciliation, Sunderlin et al. (2005) at the Center for International Forestry Research (CIFOR) highlight options for "forest-based poverty alleviation," whereby the use of forest resources leads to "lessening deprivation of well-being" (1386). They recognize that forest-based enterprise may be one component of a diverse portfolio of the rural household, which we find is very much the case in Amazonia (Coomes 1992; Netting 1993; Pinedo-Vasquez et al. 2002).

While professionals in development and conservation sectors are busy redefining the forest as a development space (Salafsky and Wollenberg 2000; Wunder 2001), farmers continue to toil on the land and in the forests. They continue to use both local and outside knowledge to innovate in production practices and strategies for engaging in markets (Pinedo-Vasquez et al. 2002; Sears, Padoch, and Pinedo-Vasquez 2007). Attempts by external actors to help rural producers improve production in agriculture or engage more equitably in markets in Peru have not been overly successful (Coomes 1996; Pinedo-Vasquez et al. 2005). For example, the state agricultural credit program in Peru in the mid-1980s was a major attempt to improve agricultural production and livelihoods in rural areas throughout the country. In Amazonia the program was widely adopted, but some argue that it had quite the opposite of the intended effect, in fact, pushing control of land out of the hands of capital-poor farmers and into those of the middle class (Coomes 1996).

Citing the widespread failure of externally driven initiatives to create markets for forest products in Amazonia, Marianne Schmink (2004) poses the question, "Under what conditions can communities engage with the market on their own terms?" (125). We address the question in this chapter through a case study of the production and processing of a fast-growing timber species in the Peruvian Amazon by smallholder farmers. While significant gains could be made in rural livelihood, obstacles remain to their legal and profitable access to the market. These obstacles include complex regulations, cumbersome bureaucracy, and a culture of cheating in forestry transactions. We discuss the ecological, economic, social, and policy preconditions for successful engagement of smallholder farmers in regional timber markets, with a specific focus on fast-growing tree species. We argue that there is little justification for regulatory control over timber production on the small private landholdings of rural farmers without distinguishing between produced and extracted timber.

We have been studying the systems of smallholder forestry in Brazil and Peru during the past decade to first understand the legal, economic, and social conditions under which implementation of the new forestry laws and regulations in both these countries do or do not work, and second, to strategize with district and local politicians and natural resource authorities, as well as timber producers, on how to facilitate smallholder participation in the sector. We have found that most rural smallholders produce some form of wood products on their landholdings, some with a more active management strategy than others, some for subsistence and barter, and others for sale (Pinedo-Vasquez and Rabelo 2002). The smallholder timber management systems that we observed in both countries also produce a diversity of ecosystem products and services, including pollination services, wildlife habitat, micro-environment cooling, and soil moisture retention. These systems yield both direct and indirect benefits for livelihood, environmental management, and biodiversity conservation.

What we have also found in Peru is that virtually no smallholder farmer producing timber has developed an annual operating plan, and thus none has obtained the proper permits for harvest and sale of timber from their landholdings. Therefore, most of their forestry activities are carried out beyond the purview of regulatory agents. This is largely because the regulatory system makes it virtually impossible for smallholders to comply with policies and regulations that simply do not apply to them. In the next section we will briefly describe the smallholder timber production, processing, and marketing systems that are in action on the ground. Then we will discuss the conditions for their engagement and the window of opportunity that current political and economic conditions provide to integrate smallholders into the forestry sector more effectively and equitably.

Production, Transformation, and Marketing of Smallholder Timber

THE PRODUCTION SYSTEM

Small-scale landowners in the Peruvian Amazon practice shifting cultivation on both uplands and low-lying floodplains (Coomes 1992; De Jong 2001; Denevan 1984; Denevan and Padoch 1987; Hiraoka 1989). On their landholdings, rural farmers manage several production spaces concurrently, each with a distinct management objective or set of objectives (Padoch and Pinedo-Vasquez 2006). Under this system smallholders start managing forests in their fields by encouraging and managing natural regeneration as well as enriching their fields with desirable timber species. Farmers depend largely on natural regeneration for timber production in their agricultural systems, from stem coppicing, the soil seed bank, or through dispersal into the fallow area. In addition, farmers may enrich their fields by planting seed or seedlings of important tree species, therefore increasing the ecological and commercial value of the landholding (Montagnini and Mendelsohn 1997). In enrichment plantings in the region of study, farmers emphasize tropical cedar (*Cedrela odorata*, Meliaceae) and mahogany (*Swietentia macrophylla*, Meliaceae), among others. What is clear from our studies is that almost all smallholder farmers maintain some important wood species and stocks of commercial timber in their landholdings.

Smallholders in both the Brazilian and Peruvian sites produce and conserve timber species by maintaining a very diverse and patchy landscape.

Land surveys conducted in a sample of twenty-one landholdings revealed most farmers had more than one fallow and the average size was 6.0 ha (thirty-nine fallows total). Half of the landholdings in this community had forest areas at an average size of 7.0 ha. These figures indicate that fallows are the dominant cover type on this landscape. In the same survey, thirteen timber species were found, five fruit tree species, and six other useful tree species. Among those, the fast-growing timber species locally known as *bolaina* (*Guazuma crinita*, Sterculiaceae) and *capirona* (*Calycophyllum spruceanum*, Rubiaceae) were dominant: bolaina was found in thirty-eight of the thirty-nine fallows and in eighteen of twenty-three fields, and capirona was found in twenty fallows.

Our studies in floodplain fallows show that bolaina grows very fast, at an average diameter increment rate of 4.8 cm per year. Thus, farmers can produce poles in two years and trees of 15 to 20 cm diameter at breast height (DBH) in four to five years. Capirona grows more slowly at an average rate of 1.5 cm in diameter per year in managed fallows, but produces a much denser wood (Sears 2003). Smallholders tend to harvest capirona pole trees (7 to 9 cm DBH) in four to five years and small-diameter sawtimber (15 to 20 cm DBH) in eleven to twelve years. Both poles and small-diameter timber are highly marketable in the local and regional markets in both the Peruvian and Brazilian Amazon.

Fallows, considered in the paradigm of "domestic forest" (Michon et al. 2007), present an ideal place for a short-cycle timber production system. Fast-growing species that occur there, such as bolaina and capirona, are suited for small-dimension lumber, such as the *tablilla*. As such, the system of smallholder timber production is dynamically coupled with local, regional, and national timber markets. It is feeding the process of rapid urbanization in municipal and urban centers in the Peruvian Amazon (Padoch et al. 2008; Padoch et al., chapter 25, this volume). Of interest in this chapter is not the management system per se, which is well-documented elsewhere (Padoch and Pinedo-Vasquez 2006; Sears, Padoch, and Pinedo-Vazquez 2007), but the product from the system (small-diameter timber transformed locally into small boards), the increasing demand in regional and national markets, and the regulatory framework that purportedly applies to these production systems and the timber.

THE LUMBER

Timber in Amazonia is traded and transformed into a dizzying array of shapes and sizes and products, from charcoal to tongue-and-groove boards to high-value veneer. A Peruvian forestry group (Palomino 2006) reports

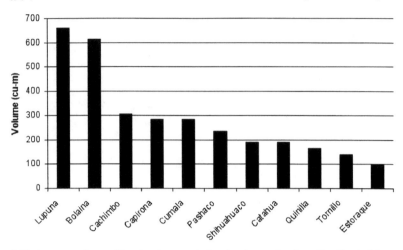

26.1. Average volume of lumber transported on trucks per day from Pucallpa to Lima, 2007. Source: Forestry control checkpoint records.

timber production volume in twenty-three categories. A recently emerging category is the *tablilla*, a unit of lumber of small dimension. A *tablilla*, sometimes called a *ripa*, is a small board of 2.5 m by 10 cm by 1 cm. They are used primarily in house construction in rural and peri-urban areas for walls and flooring. House construction in the rural areas of the Amazon Peru has traditionally been with round logs and rough-hewn timber. In the rapidly expanding peri-urban construction, and for those who can afford it in the village, people prefer processed lumber such as *tablillas* for both house framing and enclosure materials over the rustic split-palm trunk or hand-hewn boards (Padoch et al., chapter 25, this volume). The next best house material would be cement blocks and stucco.

In the past couple of years the market for *tablillas* has surged in Lima as workshops for prefabricated houses have been established to serve the poorer neighborhoods on the outskirts of Lima and elsewhere. Rural assistance programs are reportedly purchasing this housing material for highland community development projects. Bolaina was the second-highest volume of commercial timber passing through the road checkpoint from Pucallpa to Lima, with an average of over 600 cubic meters transported on twenty trucks per day, superseded only by shipments of lupuna (used for plywood) at 660 cubic meters (fig. 26.1).

Tablillas can be produced from any timber species, but the majority are produced from bolaina. A survey over two years of permanent small-scale millers in Contamana revealed that six tree species are represented in their *tablilla* production, but by far the majority of boards produced were of bo-

Table 26.1. Quantity of *tablillas* produced per
month by small-scale millers in Contamana

Species	Average number of *tablillas* produced per month
Bolaina	14,886
Cedro*	471
Capirona*	343
Utucuro	243
Copaiba*	143
Lagarto*	29
Total	16,114

*Species also sold as round logs.

laina (table 26.1). The species of valuable timber (cedro, lagarto, copaiba, capirona) were likely processed opportunistically, since these species fetch a higher price as larger dimension sawnwood.

MARKETING TABLILLAS

The sawtimber for producing *tablillas* from bolaina and capirona comes from both managed and unmanaged fallows on both the floodplains and uplands. On the floodplain, both species occur in monodominant natural stands in disturbed areas. Sometimes the natural stands have an apparent owner (an individual or, more often, a community) with whom the *tablillero* must negotiate. The timber is also sourced from fallows of smallholder farmers. Bolaina is not a concession timber; since the species is a late pioneer it does not appear in mature forest and concessions are primarily given on upland mature forests. Records from the *Instituto Nacional de Recursos Naturales* (INRENA), the Peruvian forest authority, indicate that the harvest and transport of this species is requested on permits issued for extraction on private landholdings and authorizations for riverside extraction, not from forestry concessions. Locally based forestry technicians working for INRENA recognize that bolaina is a produced and not extracted timber, but the forest law does not recognize the difference.

The producer is a rural farmer engaged in agricultural and/or livestock production on small plots of land ranging from 5 to 50 ha. The farmer sometimes has the title to the land but more commonly has only a Certificate of Possession, which allows them to live on and work the land, but not to engage in more intensive activities, such as extractive or plantation forestry.

Those activities would require a permit from the natural resource authority, INRENA. This is the first obstacle that impedes the development of small-holder forestry, or family forestry, which is emerging as the main supplier of small diameter sawtimber to local and regional markets.

The actor who produces *tablillas* from small-diameter timber is the *tablillero*, who possesses the machinery; these can be sessile or mobile mechanical units, usually employing a table with a recessed-disc sawblade and a diesel motor. *Tablilleros* may be residents of small villages who process their own timber and that of their neighbors as a cottage industry. Others may come from urban areas (e.g., Pucallpa) with transport and machinery, often financed by larger operators. They travel throughout the region, temporarily setting up on river banks and in towns, sourcing trees from natural stands with no apparent owner and negotiating the acquisition of timber from apparent owners of trees. Finally, a *tablillero* may have a relatively permanent operation in a rural regional center (e.g., Orellana, Contamana) and process logs delivered to them by producers and itinerant loggers. These mills operate with two to four workers with an average daily output of 1,000 to 1,500 *tablillas*.

Smallholder producers currently have several options for marketing their wood. They may sell standing stock or add value to the timber. Occasionally, there are straightforward transactions, but often producers and buyers negotiate a deal that includes some cash, some in-kind service, some product, and maybe some credit. Different marketing options are described here and summarized in table 26.2. The figures presented are based on a hypothetical 1-hectare, four-year-old mixed species fallow with 200 bolaina trees of 25 cm diameter; these are conservative figures based on field data.

The easiest option for farmers to sell timber is to invite a *tablillero* to the stand with a portable saw. The *tablillero* will either offer a single purchase price for all the standing timber stock in the stand (as low as S./100 per ha) or pays the landowner per *carga* (a bundle of 10,000 processed *tablillas*, which usually requires 2 ha or more). In 2007 the amount offered to producers per carga was from S./200 to S./300. The producer loses the majority of the value of this timber, since the seller of the *tablilla* will receive S./120 to 180 in Pucallpa for 100 *tablillas*, which translates to a gross income of S./12,000 to S./18,000 per carga. Expenses for the *tablillero* vary, but could run up to S./3000 per carga.

In a second marketing scenario, woodlot owners may sell their standing trees to a *tablillero* at a rate of S./2 per tree. Alternatively, the owner may offer the trees *a medias*, where the cutter gets half the processed *tablillas* and the owner gets half, with no cash transaction. The owner may then

Table 26.2. Estimated expense, revenue, and profit from a stand of 1 ha with 200 *bolaina* trees at a diameter of 25 cm using four different marketing options

Unit	Quantity	Expenses (S./)	Price/unit (S./)	Revenue (S./)	Profit (S./)	Profit (USD*)
Area (ha)	1	0	100	100	100	35
Tree	200	0	2	400	400	140
Log (3 per tree)	600	143	2	1,200	1,058	372
Tablillas (8 per log)	4,800	1,590	1.80	6,900	5,310	1,870

*Exchange rate used, May 2007: US$1 = S./2.84.

choose to sell a portion of the *tablillas* back to the *tablillero*, who is usually willing to buy and in a better position to transport the lumber to the urban market.

A third option is for producers to sell the standard unit of sawn logs (2 m in length) to a *tablillero* at the mill, who will process the logs and sell the lumber. A tree of bolaina yields anywhere from two to five logs. Generally, the owner transports the logs to the miller, who pays S./2 per log. This requires some investment by the producer to fell the trees, process the logs, and transport them to the mill by land or water. Depending on the distance and terrain between stand and mill, the expenses can run from S./150 to S./500 per ha. Net revenue to the producer for one hectare of bolaina under this arrangement can be as high as S./5000. The miller then sells to middlemen, receiving from S./70 to S./120 per 100 *tablillas* locally, depending on the quality and finish of the lumber.

In a fourth scenario, the woodlot owner may have his or her own machinery for processing logs to *tablillas*. In 2006, these local *tablilleros* received on average S./70 per 100 *tablillas*, either from a buyer at farm gate or slightly more at the market in Pucallpa. These local *tablilleros* face more risk in the sale of lumber because they do not have guaranteed buyers, transport, or storage facilities. Often the lumber molds and then is either not sellable or of lower quality and fetches a lower price. They also face risk transporting their lumber, since they rarely have the money or connections to obtain even a falsified transport permit.

The most profitable option for farmers, of course, is to process and sell their own lumber. Fourteen thousand *tablillas* are transported in one truckload. If a conservative estimate of yield is 24 *tablillas* per tree of 25 cm DBH, then one truckload accounts for the processing of 583 trees. Those trees could come from a single fallow of 3 ha, which is less than the average size of fallows on smallholder farms. The farmer, if she produced her own *tablillas*, could stand to net just under US$5,600 with 3 ha of trees. Even if she sold the logs instead of sawn lumber, she could still net US$1,100. If she had four fallows of different ages she could rotate harvest annually in a system of continuous production and enjoy a secure annual income of more than one thousand dollars from timber production, enough to pay school fees and more. But until this farmer can legally transport the logs or *tablillas* from her farm to the mill or port, she stands to gain little (other than rebuilding soil nutrients for agricultural production) from maintaining tree cover on her land. The one remaining obstacle, however, is regulations.

Opportunities for Engagement

Byron (2001) has identified four necessary conditions for smallholders to be able and willing to engage in commercial tree farming on their small land-holdings. First, land and resource rights must be secure. Second, a viable production technology and all necessary inputs must be available, including credit. Third, farmers must have reasonable confidence that they can protect their trees from risks until maturity. Fourth, there must be a market demand and a market structure in which small producers are permitted to participate. There also must be physical access to that market.

Byron (2001) has identified a consistent bias against smallholder forestry in most countries, which we recognize in Peru: inadequate technical support, market structures, and government policies. In Peru the small-scale producer is, at best, ignored in the regulatory framework and, at worst, regulated out of the market. In the first case, existing regulations simply do not apply to the smallholder situation. The documentation necessary and the costs of obtaining permits to harvest and transport fallow timber are out of reach to many rural smallholder producers. Regulations require management and annual operating plans, plans require that the producer hold title to the land and resource, and implementation of the plans requires permits. Smallholders are also sometimes paralyzed by lack of access to credit and capital to add value to the timber in the field. As indicated earlier, farmers often resort to selling standing stock to loggers for a very low price. If they decide to process the timber themselves, without access to credit options, good market information, and reliable buyers, smallholders often become indebted to private money lenders who give loans under very unfavorable terms.

Nevertheless, marketing small-diameter timber is more viable and favorable for smallholders than marketing larger and more valuable timber for several reasons. First, the growth and harvest cycle is rapid (two to four years for bolaina, six to twelve for capirona), and with several stands of different ages, producers are guaranteed some income each year. Second, with small-diameter logs there is less room for buyers to cheat the producer out of a fair price. Third, investments in labor and machinery to harvest, process, and transport these products are small, since the wood is of small dimension and relatively light weight. Fourth, because of the expanding market for *tablillas* there are multiple options for selling the product. All of these factors lower the investment risks to the producer.

We have identified several enabling conditions in Peru that currently create a window of opportunity for validating and including smallholder forestry in the nation's forest sector to achieve sustainable development goals of biodiversity conservation and poverty reduction. The first is a series of new legal frameworks that integrate principles of sustainable development, specifically sustainable forest management and broad participation of actors in the sector. The second is a renewed focus on poverty alleviation programs by the federal government, spurred by the requirement of the World Bank lending policies for borrowers to demonstrate commitment to poverty alleviation and environmental sustainability. The third enabling condition is the decentralization of the Peruvian government, which should put more decision-making power into the hands of regional and local governments, who, presumably, understand the needs and opportunities of their constituents better than do government offices in the capital.

First, new approaches and concepts of forestry are being integrated into policies and laws in Peru and other Amazonian countries. Specifically, the principles of sustainable forest management, biodiversity conservation, and poverty alleviation are integrated with production goals in the new legal framework for the management of forest and fauna in Peru (Law 27308). The new law provides at least the legal framework for the participation of small-scale timber producers in forestry, specifically indigenous communities, producer associations, and small private landowners. The precise regulations relevant to fallow forestry by smallholders were in development at the time of writing (and have been for years).

Second, with the support of external donors, the Peruvian government is developing programs for poverty alleviation with a focus on improving rural livelihoods. Smallholder forestry fits squarely in this realm, and because there is a high demand for the product and the production systems are already in place, little needs to be done to make the system work for the people. Poverty alleviation programs for promoting alternatives to coca production are ongoing in the Peruvian Amazon, largely supported by the United States Agency for International Development (USAID). Because these projects are often developed by external agents with little knowledge of the ecological and social landscapes in rural areas, they are met with varying degrees of success (Dávalos 2001). We are working to ensure that family forestry is recognized as a viable development opportunity.

Third, political decentralization in Peru provides opportunities for locally-defined development strategies. The entrepreneurial spirit is alive in villages across Peru, and decentralization should result in more supportive regulations for business transactions. It should help with credit access, better access to information, and oversight of activities on the ground. It is recommended here and elsewhere that smallholder producers form local associations with administrative capacity and linkages to local and regional political structures (Muñoz, Paredes, and Thorp 2007). Market information about nontraditional timber is available through informal communication networks, which helps producers make decisions about management strategies, harvest schedules, and more importantly, acquisition of credit. There is a certain amount of risk associated with implementing decisions based on information that carries no guarantee. Therefore, some measure of price guarantees and favorable credit opportunities would help rural producers gain advantage. The formation of a Committee of Fallow Producers and an Office of Family Forestry in municipal capitals would help to dissolve those impediments.

Conclusion: Translating Policy to Action, Action to Policy

As currently configured, the Peruvian forest law falls short of its goal of ensuring the sustainable use of Peruvian timber resources (Oliveira et al. 2007), and especially short of its goal of promoting rural livelihood improvement through engagement in forestry activities. Until the natural resource authorities recognize the positive role small-scale farmers have in the timber sector they will continue to exclude or ignore them from regulatory and market developments. The family-forestry system described in this case study is one that can help to change the Amazonian forestry paradigm from an extractive to a productive activity. Plantation forestry in the Peruvian Amazon is still not well-developed, but productive forestry is commonly practiced by rural smallholders.

Family forestry, timber production in smallholder fallow forests, is a viable endeavor that presents an economic opportunity to the rural agricultural households and that carries with it a high degree of conservation value. Family forestry helps to maintain a diverse forest cover where pressure on forests comes from agricultural policies for monoculture. The system should be supported by public policy as a strategy for poverty alleviation, reduction of deforestation and forest degradation, and securing timber supply to expanding markets. One option that we support is to deregulate the production, harvest, and transport of fast-growing timber in small private landholdings, or at least establish regulations that are favorable to the poor rural producers. Fay and Michon (2003) also make a historical case for deregulation of timber produced on small, private landholdings, where public goods and public welfare are not influenced by private action. Sunderlin et al. (2005) also call for the deregulation of timber from smallholder production systems. Our recommendation has precedent elsewhere, including some Brazilian Amazon states, where timber products from rural smallholders are treated as agricultural products, and as such, fall under agricultural regulatory framework. In Brazil, the state of Amapa instituted a family forestry management plan, which was developed together with the state natural resource agency (Embrapa) and a workers' union (Sindicato de Trabalhadores Rurais do Amapá). Through this plan, municipalities adopted rational regulations that allow smallholders to produce and sell timber under agricultural regulations, presenting fewer regulatory obstacles than if it were controlled under normal forestry regulations. We think this would work in Peru.

27 ✳ Forest Resources, City Services: Globalization, Household Networks, and Urbanization in the Amazon Estuary

EDUARDO S. BRONDIZIO, ANDREA D. SIQUEIRA, AND NATHAN VOGT

Introduction

One of the remarkable sides of the fast and intense transformation of the Amazon in recent decades has been the coupled process of urbanization and forest resurgence taking place in the Amazon estuary. Since the mid-1970s, urbanization and a forest-based economy have emerged side by side, set in motion by regional, national, and global demographic and economic forces and shaped by the responses of local populations and institutions to these processes. As such, people–forest interactions in this region have become predicated on the complex nature of these relationships. To borrow a term from Tsing's (2005) work on globalization and the Kalimantan forests, people and forests are part of the *frictions* created by the interdependencies between global and local, rural and urban, culture and economy, and the (often unexpected) outcomes they produce. This chapter examines one dimension of these processes: the intrinsic connections between the emergence of a forest economy, the reorganization of households and social networks, and urbanization in the Amazon estuary.

The economic and political gateway to the Amazon basin since the sixteenth century, the Amazon estuary is not a new front to capitalism or globalization; this is perhaps why it expresses so well their contradictions still today: wealth and poverty, forest and pollution, inequality and social conviviality coexist as in few other places. The pace of change, demographic and economic, during the past decades adds to the region's complexity. Not surprisingly, understanding people–forest interaction in this context poses many challenges to conceptual models, which are built from a either structural or processual perspective and/or models that are level-specific or strictly disciplinary (VanWey, Ostrom, and Meretsky 2005). Furthermore, such complexity renders concepts like rural and urban, household and family, or native and domesticated forests more useful as heuristic tools to describe a continuum than to represent actual categories of reality. These are some of the challenges embedded in understanding the social life of forests[1] in this region, and in this sense, Tsing's (2005) model offers, if not a solution, at least a provocative starting point.

Not always passive subjects or economic actors engulfed within regional transformation, the regional population has fashioned, on their own terms, ways to negotiate the economic and demographic forces shaping the region's society and forests. Again, for the lack of a better term, this represents a kind of "glocalization" (Comaroff and Comaroff 2000) Amazonian style. One can see signs of the often unanticipated outcomes of global–local intersections in this region from many angles: for instance, in the context of Latin American neoliberal multiculturalism (Hale 2002), regional ethnic and class identities (e.g., the use of the terms *caboclo*, peasant, extractivist, and small farmer) have become increasingly situational and manipulated in the face of regional, national, and international expectations and opportunities. In the context of Amazonian urbanization (Browder and Godfrey 1997) households have increased their multisitedness and further blurred difference between rural and urban spaces (Padoch et al. 2008; Pinedo-Vasquez and Padoch 2009). Also, in the context of global markets expansion (Brondizio 2008), new forms of local economic relations have emerged that combine colonial and new practices (Brondizio 2011).

These abstract frictions offer a point of departure for the analysis that follows on the intersection between a rising forest economy, the reorganization of households, and urbanization in the Amazon estuary. In pragmatic terms, we argue that the social life of forests in the Amazon estuary is predicated on two interrelated processes.[2] First, the current forest landscape of the estuary results from increasing market demand for forest products created by urban areas in the region and elsewhere and accentuated by decreasing economic return from competing land uses for annual agriculture and cattle ranching. The region has seen a forest transition and the emergence of a forest-based economy embedded within multiple market chains and characterized by social and economic interdependency between rural and urban spaces. Second, households have developed new forms of multisited organization to capture control and value of forest resources as well as enhance their ability to access urban markets and services through their social networks. These arrangements mediate the flow of resources between forests and markets as well as access to services and economic opportunities in urban areas. These relationships do not occur in a historical vacuum; conversely they are linked and shaped by the social and political milieu that underlies economic relations and differential and unequal access to resources and infrastructure among the regional population.

The concept of a multisited household (Padoch et al. 2008; Pinedo-Vasquez and Padoch 2009) presented here is twofold. It places a unit of organization that is place-based (i.e., a house inclusive of nuclear and extended families) within a larger unit that is multisited (i.e., a network of houses interlinked by recognized kinship, and/or social and economic relations). Multisited households operate as a social and economic unit with a shared goal of facilitating the flow of people, resources, and services through chains of reciprocity and economy. The existence of multisited households thus depends on the maintenance of shared interests and obligations. As such, related families may organize themselves as part of a multisited household in order to link the production of resources in the forest, the control the multiple paths of commercialization to regional and global markets, and to access urban services and opportunities. This concept makes evident the paradox of defining rural and urban spaces. It calls attention to the rural and urban as a continuum, although representing different spaces of production and consumption of resources, thus acknowledging some heuristic value to each term. In the absence of alternatives, the categories of rural and urban are necessary to interpret census data, where each space is clearly differentiated for demographic purposes, and also to recognize the different nature of resources offered within each: forest products on the one side and city services on the other.

What follows is a brief examination of the two points outlined earlier: an analysis of the region's forest-based economy of relevance to rural and urban residents, and the emergence of new forms of household and social networks linking rural and urban spaces in the Amazon estuary. These analyses benefit from long-term ethnographic work in the Amazon estuary, particularly the municipality of Ponta de Pedras (Pará State), primary and secondary remote sensing of forest change, archival and census data from 1950 to 2000, and recent household surveys among seven rural communities (264 households and 2,168 individuals) and three urban areas (100 urban households and 1,063 individuals). We conclude this article by reflecting on the implications of these processes to the understanding of forests, livelihoods, and urbanization in the Amazon.

The Emergence of a Forest-Based Economy: Its Opportunities and Limitations

Contrary to deforestation trends elsewhere in the Amazon, the estuary has seen an increase in forest cover and an economy closely linked to forest resources. This is a trajectory documented by regional scale assessment (Instituto National de Pesquisas Espaciais, Projeto Desmatamento [INPE-PRODES] 2008) and census of forest products (Instituto Brasileiro de Geografia e Estatística [IBGE] 1950–2006), as well as locally through ethnographic and land-use assessments in different parts of the region (Brondizio 2008; Pinedo-Vasquez et al. 2001). These trajectories result from several processes interconnected in different degrees. On the one hand, the progressive influx of agricultural products from other parts of Brazil, such as rice, beans, and corn have slowly created a difficult economic environment for estuarine farmers to compete. Products arriving at cheaper prices and in large quantities in the markets of Belém also make their way to smaller cities where farmers face limited access to technology, storage, and transportation. Even products with a wide production basis (e.g., manioc flour) arrive from other parts of the Amazon, such as the Bragantina and Santarém regions, at cheaper prices. Low market prices, land tenure constraints, the high cost of labor, government subsidies such as "bolsa familia," and the hardship of manufacturing manioc flour have discouraged most families from continuing its production not only commercially, but also for consumption. Although they still may be inquisitive about it, local residents of small estuarine towns have become used to observing tons of manioc flour arriving by boat every day. In general, census data show a decline in annual crops among estuarine municipalities not only in total production volume, but also in the number of establishments dedicated to production (figure 27.1; IBGE 1950–2006).

Parallel and closely connected to these trends, the estuary has experienced an increasingly stronger forest resource economy. Commercial and continuous logging in the floodplain forests of the region (Anderson, Mousasticoshvily, and Macedo 1993; Barros and Uhl 1995) dates back at least to the 1950s and has moved to different phases and emphasis of exploitation. The so-called post-boom logging documented by Pinedo-Vasquez et al. (2001) illustrates the resilience of both forest and farmers to overcome forest overexploitation while finding new markets, such as the booming demand for low-cost, fast-growing timber used for houses in urban areas throughout the region. Most illustrative however, has been the boom of the açaí fruit economy since the 1970s, a process widely documented by one of the authors (Brondizio and Siqueira 1997; Brondizio 2008) and others (Hiraoka 1994; Anderson 1990; Mourão, Jardim, and Grossmann 2004). The açaí palm economy has first seen a rise as part of a growing demand for heart of palm, a process that, at least during the 1970s and 1980s, was mostly characterized by its predatory nature (Pollak, Mattos, and Uhl 1997) but progressively found increasing synchrony with açaí fruit production. Long ingrained in regional life and culture, the rise of the açaí fruit economy starts during the early 1970s as demand for the fruit, consumed regionally as a staple food, increases in urban areas experiencing fast expansion with the arrival of rural migrants. Local producers seize the opportunity of an emerging market and regional demand for a preferred food source through the intensification of forest management and agroforestry techniques, lead-

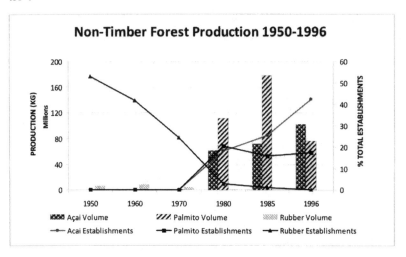

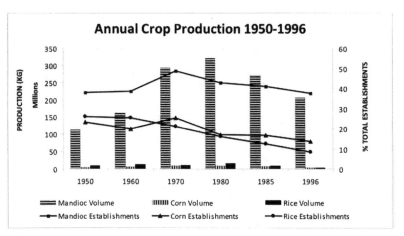

27.1. Five decades of production changes for municipalities of the Amazon estuary: Non-timber forest products and annual crops (IBGE published agro-pastoral census and archival data 1950–1996)

ing the region through an agroforestry economy as pervasive as the tidal waters of its rivers. As a species already widely occurring in the floodplain forests, açaí palm responded well to the management largely based on local knowledge of forest and species ecology and technologies of intercropping of great, albeit invisible, agronomic sophistication (Brondizio and Siqueira 1997). The "açaízation" of the Amazon estuary (Hiraoka 1994) illustrates and defines the importance of açaí to a regional forest transition. To make a long story short, from its rural and later regional urban expansion as a staple food, açaí fruit moves into national markets during the 1990s, fast reaching a phase of industrialization and internationalization that one of us has defined elsewhere as representing its rise as a fashion, vis-à-vis a staple food (Brondizio 2004). The consequences of these processes, social and environmental, are multiple and illustrate both positive and negative dimensions of the interplay between a globalizing economy and colonial social structures of land ownership and access to markets (Brondizio 2008; 2011).

Furthermore, the emergence of the açaí fruit economy has reaffirmed the intrinsic and central place of forests within estuarine regional culture, which, it is fair to say, distinguishes this part of the Amazon from other nonindigenous areas of the Amazon. Forests define most of what is rural and caboclo in the Amazon estuary. Living at the intersection of water and forest, with houses framed almost idyllically within palm groves, adults and children alike are familiar with its shadows, comfortable walking barefoot, and knowledgeable about forest foods, delicacies, and the raw materials it offers. Although it sounds romanticized, this is a description not too far from reality. However, denizens share similar familiarity with life in a boat or at the many peripheries of regional cities. In this sense, açaí palm and its material and symbolic consumption also helps to define the urban in the Amazon estuary. Its insertion into local, national, and global urban markets associated with its intrinsic regional origin has brought a sense of appreciation of its symbolic and aesthetic value as well as contributing to a sense of place and identity for people of the region; it is also an icon manipulated as part of the politics of regionalism. At once, it symbolizes culture, environment, and economy of the rural and urban areas of the region.

The wide participation of families in the açaí fruit economy and its potential to offer a so-called sustainable land-use solution for the region have become its most recognized strengths. Among rural communities the vast majority or the totality of households have been able to engage in production and/or commercialization of açaí fruit. Around and in urban areas, it has offered thousands of jobs, albeit informal, in processing, commerce, transport, and a variety of other sectors. At the same time, its globalization has created new challenges for estuarine families. As market demand increased, production and competition from other areas did as well. Price fluctuations during different parts of the season and high transportation costs have offered opportunities, but also much greater risks to producers. Furthermore, it has become increasingly gendered, where men tend to control and work on management, production, and harvesting, as well as commercialization and merchant trading during the off-season. Women's increased participation has occurred in cases where they share land ownership with husband or children, or at best, in processing stalls. It has also proved that it is no cure for regional problems or even for substantial economic improvements of families. The lack of transformation industries adding value locally and offering a tax and employment base for municipalities where the product is produced has limited its contribution to improve economic development in the region.

A recent survey of rural households of Ponta de Pedras (Brondizio 2008), one of the main producing municipalities of the estuary actively engaged in the production of açaí fruit, shows that only 20 percent of households have açaí fruit as their first source of income, and a similar percentage have it as a second source. In general, rural households depend on multiple sources, and particularly on retirement and different forms of government aid and subsidies. While such diverse portfolio of economic strategies has long being defined as a trademark of the Amazonian caboclo, today it includes not only the traditional array of agriculture, fishing, and extraction of resources, but also wage labor and particularly government aid, various forms of merchant and petit commerce, and public employment. Furthermore, rural families have found a variety of niches within the expanding economy of açaí fruit. Most households have moved from passive subjects of merchants to dispatching their own middleman to trade during the off-season in distant regions or as brokers of various levels in the larger markets in Belém and Macapá. One of us has argued elsewhere that while sizable and complex, the açaí fruit economy is largely controlled by family networks no matter their economic status and ownership of resources (Brondizio 2011). Resources circulate and are traded largely within family networks or family-controlled networks that connect land and resource ownership, work arrangements in management and harvesting, and commercialization from the edge of the forest to a variety of market niches in urban areas, including how açaí fruit is negotiated with larger corporations.

This is the context within which the region has seen significant transformation of its households and forms of social organization. While shaped by the forces of a globalized economic system (e.g., the need to insert a family member as market broker or middleman or even commercializing açaí juice direct to consumer in urban peripheries), individuals and households have responded to much wider processes, not least of which are equally changed expectations and lifestyles at the interface of rural and urban areas. Increasingly, families have actively sought to educate their children in the city, often investing in urban houses and/or sending members of their households to live with relatives. Not surprisingly, the urban population of most small towns in the estuary has, on average, doubled between the 1980s and 2007 (Costa and Brondizio 2011). However, it is important to understand these processes as facilitated by a variety of other changes, particularly the availability of electricity, faster transportation, and communication.

Household Economy and Livelihood Nested within Social Networks

While we tend to assume that rural (or urban) households are always linked through complex social networks of various kinds (e.g., reciprocity and mutual support, political ties, etc.) one cannot ignore how technological changes have impacted social life and household organization in the Amazon estuary during the past decades. For instance, electricity, significant reduction in travel time between capital and interior, significant increase in frequency of transportation, and changing the ability to communicate from one central phone booth to cell phones have greatly influenced how individuals and households relate to each other, make choices, and seek opportunities within the region. One cannot assume that households today have the same playing field as just twenty years ago, and thus the nature of social networks and demographic mobility put in place are significantly different. These changes have reduced the distance between what is considered a rural forest space and urban life. Because of the bewildering distances peculiar to the Amazon, as social spaces, forest and city are one and the same for most towns around the Amazon estuary.

Until the early 1990s the availability of electricity around most estuarine towns was restricted to a few hours a day and few days a week, thus limiting the number of events and the kind of public social life available in town, as well as the availability of products (like a cold beer) and different kinds of stores. Progressively, investment by the state has allowed electricity to become available 24/7, creating new economic, social, and cultural opportunities previously unfamiliar to rural and urban residents alike. Most rural areas, however, remained in the dark, except for the state of Amapá where significant investment was made to provide electricity to households and communities along the floodplains. More availability of night schools, public shows and parties, as well as church activities in town, have increasingly attracted rural residents to urban areas. More years of education, more exposure to media, and increasing commerce of imported goods have also changed the expectation of youth seeking to keep an economic base in the rural area, while maintaining an urban lifestyle.

Transportation has equally changed the possibility for interaction between rural and urban areas in the interior and the capital city of Belém for instance. Until the early 1990s, one could find one or two boats a week connecting Belém to many small towns, such as Ponta de Pedras, where we have been working since 1989. A trip, often at night when the waters of Marajó Bay are even more dangerous, would take five to six hours, if not eight, depending on the boat. A visitor to the same town today can choose from a variety of boats departing every day, some of which are air-conditioned, very comfortable, and take less than two hours between stops. The frequency of visits has increased proportionally. Likewise, the city has put in place boat transportation for school children, a facility absent a decade ago. Communities located in the uplands have today daily buses, albeit precarious, an unimaginable condition in recent years past. Similarly, decentralization of phone services from a central station to houses, and later the availability of cell phones, has also changed dramatically not only how açaí fruit commerce has developed, but also how household members communicate with each other and maintain a relationship. In other words, these transformations have created a new environment for household organization and social networks to form, issues that are explored in more detail in the following.

COMMUNITY-LEVEL NETWORKS

The doubling of urban residents in estuarine municipalities since 1990 has happened largely with the arrival of residents from the same or surrounding municipalities. Yet, there has not been a proportional decrease of rural residents, in some cases, there has actually been an increase motivated by an active açaí fruit economy. What underlies this process is the simultaneous presence of households in rural and urban areas, a process referred elsewhere as the formation of multisited households (Padoch et al. 2008; Pinedo-Vasquez and Padoch 2009). Before looking at household-level networks, an analysis of community-level networks illustrates the wider nature of these processes. Data on social networks from seven communities considered to be rural in the municipalities of Ponta de Pedras illustrate similarities in the nature and density of social networks within and between communities and urban areas (e.g., the local town, Ponta de Pedras, and the state capital, Belém). Figure 27.2 illustrates the density and distribution of networks of kin-households.[3] While important differences exist, as expected, one can observe across communities that households have a larger number of kin-households within the community where they control land and/or work arrangements and form close networks of reciprocity and support. The percentage of kin-household within the same community ranges from more than 30 to 60 percent. Although connections to other rural communities exist, on aggregate households have stronger connections with kin-households in the local city, Ponta de Pedras. Most communities have more households connected to relatives in the state capital in Belém than connected to other neighboring rural communities. In other words, networks of kin-households are based on a strong intracommunity system coupled with strong presence in the local town and equally important con-

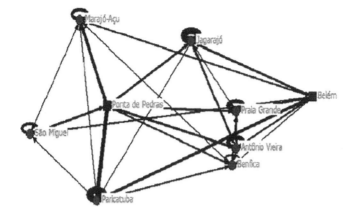

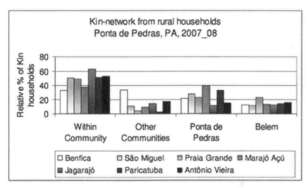

27.2. Relative percentage of recognized kin-household distributed within and between rural communities and urban areas (household-level data aggregated at the community level), Ponta de Pedras, PA, 2007–2008

nections to the state capital Belém, and these connections are, in most cases, stronger than their connections to other rural or neighboring communities.

Ethnographic data suggest that these connections serve different purposes and function in a variety of ways. Labor flow between rural and urban households is intense during parts of the year, such as during açaí fruit harvesting or periods of agroforestry preparation and maintenance. Economic connections are equally strong during the peak of açaí fruit season when product moves from rural to urban markets for commercialization. The mobility of household members between rural and urban range from daily and weekly visits to seasonal or even permanent movement. People visit towns for services such as banks, commerce, or doctor visits, or they stay weekly for schooling. Conversely, people go to rural areas to work on açaí harvesting and commerce and to access other resources and land for cultivation. As a whole, they form a wide safety net both within the rural space and in-between urban areas, each playing different roles in the provision of resources and services. This is particularly important in a region where economic opportunities are informal and based on family networks and where services are precarious and dispersed.

MULTISITED HOUSEHOLDS

Understanding the underlying structure and processes of these networks, however, requires attention to household-level organization. One can understand the concept of multisited households when considering where members who are considered to be part of a household are located, as further illustrated by figure 27.3.

During the above-mentioned survey, we asked household heads to list those considered to be members of the rural household (i.e., those who are part of the household and who maintain connections through social interaction, joint economic activities, or ownership of land resources, as well as some level of participation in decision making). Not surprisingly, in aggregate terms (all households in all communities), the number of members living (includes seasonal and permanent) in urban areas correspond to about 40 percent of all members. Also not surprisingly, of those members living in urban areas, close to 60 percent are women. The gendered nature of the açaí economy has progressively led female members to seek opportunities, including study and work, in urban areas. There is also a clear pattern of rural–urban connection, in part explained by the geography of the region and the string effect of household expansion through previous networks. Families have members predominantly in the town of Ponta de Pedras (more than 60 percent) followed by Belém (close to 30 percent). The presence of family members in Barcarena, for instance, is a relatively new phenomenon, although prone to increase. Barcarena is one of the few cities in the region where employment opportunities exist because of a large aluminum plant functioning in the region. Recent employment success of one individual has already led other members of the same community to have a supporting base there when searching for opportunities. This connection evolves as word of mouth spreads and people become more familiar with the area and understand the risks and opportunities available "across the bay."

The frequency of visits between rural and urban areas is equally intense. Our survey showed that more than 95 percent of rural household members around Ponta de Pedras visit town at least once a week. Improvements in transportation have allowed people to make daily trips, even to Belém. As

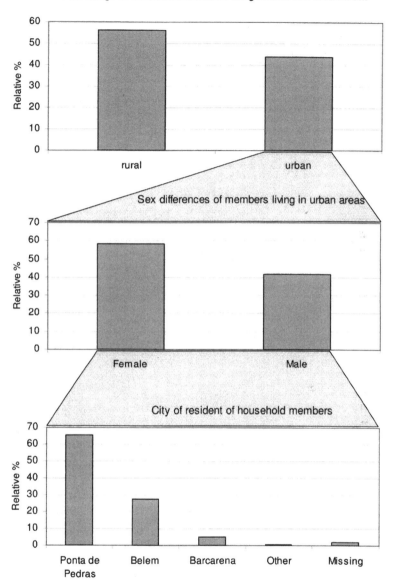

Percentage of Household Members living in rural and urban areas

27.3. Multisited households: Household members in rural and urban areas

Eduardo S. Brondizio, Andrea D. Siqueira, and Nathan Vogt

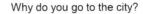

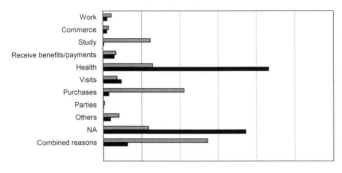

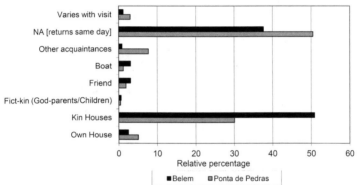

27.4. Reasons for city visits and lodging while away among rural members of households, Ponta de Pedras, PA, 2007–2008

figure 27.4 indicates, a large number of people go to Ponta de Pedras and Belém without needing to seek lodging. When they do so, they predominantly stay in the houses of relatives, particularly when traveling to Belém (50 percent of the time) but not as much in Ponta de Pedras (30 percent). While the frequency of visits to the state capital varies greatly between households and communities, thus defying any average, they are equally important. Reasons for visiting Belém differ from those for visiting Ponta de Pedras, particularly in terms of the nature of service sought. More than 40 percent of people visiting Belém do so because of health services unavailable in Ponta de Pedras. In Ponta de Pedras they tend to seek study, purchase basic consumption goods, or particularly to combine several activities when visiting town (e.g., shopping, visiting, and receiving retirement pensions and government aid). While it appears that visits for the purpose of commerce represent a small percentage (relative to number of individuals), at the household level they are very important, but not captured by this figure. Most households deploy a member to Belém and/or Ponta de Pedras in charge of commercializing açaí fruit or other products usually involving day trips several times a week during the main harvesting season.

Cities and Forests as Complex Social–Ecological Systems in the Amazon Estuary

In this chapter we have highlighted the close connections between a forest-based economy, the reorganization of households, and urbanization in the Amazon estuary. We have stressed the multiple ways in which individual and households reorganize their lives in the context of larger transformations and forces affecting the region during recent decades. Elsewhere (Brondizio 2008, 2011), one of us has stressed the role of recurrent structural problems undermining the wider participation and economic benefits for those farming the forests of the region and for estuarine municipalities where forest resources are produced. The ingrained structural problems, dating back to colonial times, and recent phases of globalization and technological change appear to us as two sides of the same coin. While the people–forest relationship is ingrained in regional culture and subjected to unequal forms of economic exchanges, it is constantly rearticulated from the bottom up. In the process, people have crafted new forms of household organization, narrowing the gap between rural and urban, developed new expectations and ideas of well-being, and fostered ways to access city services and forest resources. These are important dimensions shaping, today and in the near future, the social life of forests in the Amazon estuary.

28 ✳ Chicago Wilderness: Integrating Biological and Social Diversity in the Urban Garden

PETER CRANE, LIAM HENEGHAN, FRANCIE MURASKI-
STOTZ, MELINDA PRUETT-JONES, LAUREL ROSS, ALAKA
WALI, AND LYNNE WESTPHAL

In 1912 Henry Chandler Cowles, professor of botany at the University of Chicago, and one of the most influential early American ecologists, wrote, "Many might think it strange that forest conservation should have a place of importance in the 'Prairie State.'" Cowles, part of a long-established conservation movement in Illinois, argued for a balanced approach to the stewardship of Illinois forests, to combat what he called "the undue cupidity of the lumber interests" (Cowles [1912] 2007). With others, he urged the Illinois state legislature to establish state parks and protect forests, and as part of the Conservation Council of Chicago, he helped defend the newly designated forest preserves of Cook County against the grazing of sheep, the poaching of native orchids, and many other assaults. But Cowles is important not just because he added his scientific credibility to the cry for a balanced approach to the management of Illinois forests, but because his research in the Indiana Dunes lead him to the idea of ecological succession (Cowles 1901), the principle that underpins both natural and human-induced ecological change, and the common theoretical framework in which both can be understood (Hobbs and Huenneke 1992). Although some aspects of Cowles' ideas have been questioned (Whittaker 1957), and others have been refined (e.g., Odum 1969; Connell and Slatyer 1977), his key insight remains. Vegetation in any setting is dynamic and must be understood as a result of constant change. While some processes may be deterministic and predictable (Connell and Slatyer 1977; Inouye et al. 1987), they are also contingent: their outcomes are constrained by preexisting conditions. Both the starting point and the ways in which successional processes are manifested (Chazdon et al. 2007) may be influenced, in both subtle and obvious ways, by disturbance (Pickett et al. 1989; Pickett, Cadenasso, and Bartha 2001), including disturbance by people.

In Illinois, the interface between forest and prairie itself is, in part, a matter of succession influenced by human intervention. Fire has been a key natural process in ecosystems for millions of years, but it is also a potent tool through which people have long sought to manage their environment. Fire certainly was used by Native Americans, but because ecological outcomes differ depending on the type of fire, the season, and the local climatic context (Collins 1992; Howe 1994, 1995), disentangling the relative importance of the different natural and human-induced processes that influenced the forest–prairie transition is not likely to be straightforward.

Urban Forests

Many of the chapters in this volume focus on the biologically diverse forests of tropics, but temperate forests add a further important dimension (Neumann, chapter 3, this volume; Foster 2008). To the degree that social and economic trajectories in the tropics ultimately follow those that have played out in North America and Europe, temperate forests may provide insights into the shape of things to come elsewhere in the world. In this chapter, we focus on temperate forest communities in Illinois—especially the Chicago area—adding what we hope is another illuminating dichotomy: between forests in urban versus rural settings.

Definitions of urban forests vary from those focused on the relationship between plants and people—"ecosystems characterized by the presence of trees and other vegetation in association with people and their developments" (Dwyer et al. 2000, 19)—to others that take a broader ecological view—"a collection of living organic matter (plants, animals, people, insects, microbes, etc.) and dead organic matter (lawn clippings, leaf-fall, branches) on a soil . . . through which there is cycling of chemicals and water and flow of energy" (Duryea 2000, 2). But no matter how one defines them, urban forests are of disproportionate and increasing importance. Globally, more people live in urban areas than outside them, and around the world urban populations are on the rise. Virtually all population growth between 2000 and 2030—an estimated 2 billion people—will be in urban areas (United Nations 1999). Increasingly, as is already the case in the United States, most of the world's population will have much of their direct contact with nature in an urban setting.

Urban forests enrich our lives and provide habitat for diverse plants and animals, but their importance goes beyond aesthetics and biodiversity. Trees help improve air quality and mitigate some of the effects of air pollution (Nowak, Crane, and Stevens 2006). A growing body of research also points to the positive impacts of trees and vegetation on people in cities. Increased green space can speed healing, foster a greater sense of community, lower levels of aggression, and reduce symptoms of attention-deficit/hyperactivity disorder (Ulrich 1984; Kuo 2003; Kuo and Sullivan 2001; Westphal

2003; Kuo and Faber Taylor 2004). Urban forests and green space are no panacea for all urban ills, but they are more than just amenities; they have broad importance for human well-being.

The spectrum of Chicago's urban forests extends beyond trees in backyards, streets, and parks to include rich remnants of landscapes that existed before human settlement, all now merely fragments embedded within and around the metropolitan area, and all impacted to varying extents by people. These forests provide important ecological services, for example, floodwater storage and the potential for recharging of underground aquifers (Dwyer et al. 1992). Additionally, they are important refuges for biodiversity and, in some cases, are the best remaining seminatural communities of their kind (White and Madany 1978). But they are also instructive for the future of biodiversity conservation (Heneghan et al. 2009). Relatively pristine protected areas are currently about 14 percent of global land surface, but most are under some level of human influence (Kareiva et al. 2007). Around the world, as around Chicago, most biodiversity conservation will take place in new kinds of ecosystems (Hobbs et al. 2006) in greatly modified landscapes dominated by people. For this and other reasons, urban ecology and urban conservation are no longer peripheral interests within their broader disciplines.

Forests of the Chicago Area

The patchwork of remnant local ecosystems in the Chicago area, of which forests are just a part, owes its existence to a particular history. Geologic and climatic factors of the last glacial advance left behind a mosaic of diverse habitats that were colonized by species from the east, south, and west to create a palimpsest of diverse ecological communities on plains, moraines, kettleholes, wetlands, river bottom lands, and beaches (Greenberg 2002). The result is an exceptionally diverse biota (Greenberg 2002). Birds, amphibians and reptiles, plants, and many other organisms show impressively high levels of diversity in northeastern Illinois. The greater Chicago area is a biodiversity hotspot in the state as a whole. Almost 2,700 species of plants alone are found in the Chicago Wilderness region, of which many are threatened or endangered. Belgium—which is similar in area (roughly 30,000 square kilometers)—is home to about half that number (The Vplants Project, n.d.).

Within the Chicago region are nearly a dozen types of woodland communities, including savannas, oak-hickory woodlands, and upland forests (Chicago Wilderness 1997). But these once-rich forest resources have diminished with expanded human settlement. Native Americans undoubtedly influenced this landscape for thousands of years, but through the nineteenth century, the magnitude and scope of human impact increased dramatically. In the early 1800s, almost 40 percent of Illinois was forested. But with the arrival of European settlers, demands for land and wood increased (Iverson 1991). Timber use for houses, fences, and fuel rose with growing populations; over 300,000 people settled in Illinois in the 1830s alone. By 1923, only 22,000 of the original estimated 13.8 million acres of forest remained (Iverson 1991). Chicago grew, in large part, due to the rich natural resources of its hinterland, and especially its forest and deep fertile prairie soils (Cronon 1991).

As early as 1837, Chicago took for itself the motto "Urbs in Horto," the City in a Garden. But it quickly became apparent that the "Garden in the City" was also important. In 1869 Dr. John H. Rauch, author of an early study on the "moral and physical effects" of public parks on city inhabitants (Rauch 1869), formally proposed forming a Chicago Park District to secure lands for park purposes. As a result, lands were designated for Humboldt, Jackson, Washington, and other parks, which are still major components of Chicago's urban forest. Also pivotal was a revolutionary report, led by architects Jens Jensen and Dwight H. Perkins, calling for the designation of land for recreation, and for preserving some public lands in their "natural" states (Perkins 1905). Daniel Burnham, the chief designer of the Columbian Exposition, and Edward H. Bennett went still further in their visionary *Plan of Chicago* (Commercial Club of Chicago 1909/1970). They pictured Chicago not just as a "City Beautiful," with parks and broad avenues echoing those of the grand cities of Europe, but they also advanced the vision—first put forward by Dwight Perkins in 1904—of a city set within an "outer belt" of parks throughout the region (Chicago Historical Society 2005). Among the *Plan*'s many supporters was Aaron Montgomery Ward, who passionately insisted that Grant Park, created at the heart of Chicago's lakefront, remain *"for ever open and free"* (Morgan 1935). Today it remains an important refuge, not just for people, but also for migrating birds (Brawn and Stotz 2001). The Burnham Plan also furthered the creation of the Forest Preserve District of Cook County in 1914, whose holdings include more than 68,000 acres of open lands, about 11 percent of the county's land area.

The foresight of Burnham and others served Chicago well. But today this vast, sprawling conurbation is home to more than nine million people, more than triple the population at the time of Burnham and Bennett's Plan (US Bureau of the Census 1913). Here, as much as anywhere else in the nation, there is the jarring juxtaposition of past and present heavy industry and rich biological diversity that survives in an archipelago of remnant, semi-natural prairie, woodland, dune, and wetland fragments. This proximity raises many questions regarding the long-term viability and management of these ecological communities and the organisms that comprise them, but it also creates opportunities to realize the full benefits they bring to people.

The Late Twentieth Century

Of all the land under the stewardship of the Forest Preserve District of Cook County, roughly one-third—21,000 acres—is ecologically significant as the last vestiges of former Illinois landscapes. These areas contain most of the more than 1,000 native plant species found in the Forest Preserve District. But succession and other processes ensure that these habitats are not static; 76 percent of these lands show signs of decline (Antlitz 2005). They face a variety of threats, from invasive species and overgrazing by deer, to diffuse pollution and dramatic changes in hydrology. As in Illinois as a whole (Illinois Natural History Survey 1994), the predominant trend is toward simpler ecological systems. Over time, rich woodlands, prairies, and wetlands become degraded and species poor. Complex rivers and streams are replaced by simpler canals, levees, and dammed watercourses. And with increasing simplicity of habitats, animals and plants that are ecological specialists disappear. In and around Chicago, just as over much of the state, a "generic" biota, dominated by carp, starlings, deer, and generalist weeds, continues to advance.

There are also specific problems on the edges of the metropolis. As urbanization competes with agriculture for available land, habitat fragmentation continues, threatening watersheds and those habitats that do remain. In northeastern Illinois, these problems are particularly acute, with the accelerating movement of the population into what the United States Census calls "urbanized areas" on the fringes of cities. Between 1970 and 1990, the Chicago region's population grew by an average of 0.2 percent annually, while the quantity of urbanized land grew at more than eight times that rate: 1.85 percent per year. Satellite images of the Chicago region for the period 1972 to 1997 confirm urban development expanding from 880 square miles to about 1,270 square miles, with combined forest and grassland areas shrinking by 160 square miles (Chicago Metropolis 2020 2001).

It was against this background that the concept of a more interventionist approach to ecosystem management gained momentum in and around Chicago. One key early initiative was the re-creation of prairie at Fermilab, a pioneering effort in ecological restoration (Wiggers 2000). Another was the Prairie Project, later called the North Branch Restoration Project, which began in 1977. A group of enthusiastic citizens volunteered to help rescue several small native prairie remnants along the North Branch of the Chicago

River, and over time transformed them into some of the finest "natural" areas in Cook County (Stevens 1995). Today, still working closely with the Forest Preserve District of Cook County, the group provides stewardship for more than a dozen such sites.

Key in both these cases was the energy and labors of dedicated volunteers who restored prairie, savanna, and woodlands. Cutting the dense brush that choked savanna and woodlands, reintroducing key prairie species, and reinstating fire produced spectacular results. "Gardening" on a grand scale created a resurgence of wildflowers, grasses, and sedges. And with the plants came birds, insects, and other animals that had managed to persist in small refugia within areas that were already protected, but largely unmanaged.

Chicago Wilderness

The success of early ecological restoration and other projects in the Chicago region created momentum. Recognizing the unique ecological communities that needed attention, and also the power of engaging volunteers in their preservation, leaders from thirty-four public and private organizations united to work more strategically to conserve and manage significant natural communities and watersheds across the Chicago region. In 1996, with significant support from the US Forest Service and US Fish and Wildlife Service, the Chicago Wilderness initiative was launched. Its aim was to create a metropolitan area in which nature thrives in harmony with people. This unconventional regional conservation effort now covers more than thirty counties and 481 municipalities in the four-state Chicago metropolitan area (fig. 28.1). Its overall goal is to protect the natural communities of the Chicago region and restore them to long-term viability, in order to enrich the quality of life of its citizens and contribute to the preservation of global biodiversity.

Since its inception, the alliance has grown to more than 260 public and private member organizations, including federal, state, county, and local agencies, municipalities, conservation organizations, universities, park districts, homeowners associations, faith-based organizations, schools, and a thirty-member Corporate Council. Especially at the outset, the alliance faced many obstacles in transcending political and institutional boundaries, and achieving successful collaborations. Each member organization has its own mission and priorities, which must be respected. Complexities of budgets, governance, and data-sharing require patience and openness. But nevertheless, through dialogue and engagement, Chicago Wilderness has been successful in defining common goals and in moving diverse groups toward the same broad objectives. The alliance is entirely self-governed. Representatives of its member organizations volunteer their time and ex-

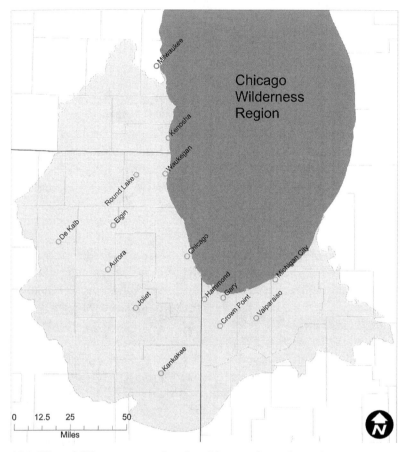

28.1. Chicago Wilderness covers a broad, multistate region at the southern end of Lake Michigan where globally significant natural communities coexist with millions of people

pertise, as well as contribute organizational resources and funds, to support the work of the alliance and its small staff.

DOCUMENTING CHICAGO WILDERNESS

An early objective of Chicago Wilderness was to document the biological wealth of the region. This culminated in *Chicago Wilderness: An Atlas of Biodiversity* (1997), which drew attention to the natural heritage and resources of the Chicago region. It also highlights the global significance of local communities, such as prairies, fens, and sedge meadows, and the rare and endangered species they support (Chicago Wilderness 1997). Among the tree-dominated communities are oak savannas, woodlands, and upland

and floodplain forests, all of which are crucial habitat for birds, herbaceous plants, and a variety of other species.

A BLUEPRINT FOR RECOVERY

After the *Atlas*, the alliance shaped a plan to restore and maintain the region's habitats and thereby protect native plants and animals, including many that were under threat. Three years of work with an array of private and public stakeholders resulted in the Chicago Wilderness *Biodiversity Recovery Plan* (Chicago Wilderness 1999). It outlines the steps necessary to achieve the overall goal of the Chicago Wilderness collaboration and prioritizes the region's ecological communities in terms of their regional and global significance and the degree to which native species are being lost. The *Recovery Plan* serves as the alliance's roadmap for recovery, providing a guide for where and how member organizations can restore and protect the region's ecosystems. It also sought to strengthen the scientific basis of ecological management. Lack of management—largely lack of management of ecological succession—together with poorly planned developments, overgrazing by deer, and the spread of invasive species, were identified as the key factors responsible for the continuing decline of environments and species. Appropriate actions to address these factors were established.

FOSTERING STEWARDSHIP

The *Biodiversity Recovery Plan* addresses not just the region's nature, but the connections between nature and people, particularly the importance of engaging with and involving citizens, organizations, and agencies in regional biodiversity conservation efforts. Especially critical to the success of Chicago Wilderness is engaging and training volunteers who help restore habitat, for example, through invasive species removal, planting native shrubs and flowers, and who assist with controlled burns. Many are trained as "citizen scientists" to help monitor a wide range of organisms from plants to dragonflies. Volunteers engage in Chicago Wilderness projects largely through individual member organizations. Five to ten thousand volunteers are engaged in active stewardship in the immediate six-county Chicago area alone. Volunteerism has grown in numbers, and also has spread to all types of communities, from those in outlying areas to the heart of Chicago's neighborhoods. Children—initially not part of volunteer efforts—now are engaged by the thousands through programs such as *Mighty Acorns*, which introduces fourth through sixth graders to nature and conservation stewardship through education, restoration, and exploration activities.

ENGAGING WITH STAKEHOLDERS

Beyond those volunteers who are active in land management, over the past decade Chicago Wilderness has built opportunities to engage and involve the broader residents of the greater Chicago region. Especially important was *Chicago Wilderness Magazine*, published from 1997 to 2009, with a circulation of over 13,000 and pass-along readership in excess of 25,000. This quarterly publication highlighted dozens of natural areas in the region, and featured many opportunities for people to participate in conservation activities. It also built awareness of local plants and animals with regional and global significance, as well as the work of local people who are making a difference.

Chicago Wilderness has sought to develop a new approach that integrates social science research with ecological research and conservation action. Psychological, anthropological, sociological, and communications approaches all have been engaged in an effort to understand residents' interests regarding nearby open space, how to effectively mobilize people against threats to those spaces, and how to broaden participation in management and other activities. This research has revealed untapped potential for diverse forms of engagement in different communities. Attachment to place—and the desire for healthy open or green space—spans class, ethnic, and racial boundaries. Indeed, residents in low-income communities that have often been deprived of access to green space, have a long tradition of fighting for parks, forests, and open space as part of their efforts to make their communities more livable (Washington 2005). Even in many neighborhoods otherwise characterized by blight or neglect and abandonment, very local efforts to manage green space persist, for example, by maintaining small gardens or caring for the waterways in which people fish (Field Museum, n.d.; Westphal et al. 2008).

MONITORING AND REPORTING

In April 2006, ten years after the alliance was formed, Chicago Wilderness released *The State of Our Chicago Wilderness: A Report Card on the Health of the Region's Ecosystems* (Chicago Wilderness 2006). Following up the *Biodiversity Recovery Plan*, this report found that 68 percent of the preserved land in the region was in poor condition, while very little was in good or excellent condition. The *Report Card* described some successes, particularly in places where active management and restoration was taking place, but overall, it underscored the poor health of many of the region's natural areas. It remains a crucial benchmark for communicating priorities for land protection, restoration, and outreach to community leaders.

Opportunities and Challenges

QUALITY OF EXISTING HABITAT

The *Chicago Wilderness Woods Audit*, initiated in 2001, was the first study to examine the region's remaining seminatural habitats in detail (Glennemeier 2004). Focused specifically on the state of the region's woodlands, it assessed quality through an index that included size classes of canopy trees, and frequency, coverage, and measures of invasive species. Findings for oak woods are representative: only 4 percent of oak wood plots were in excellent condition, 17 percent were rated as good, 38 percent fair, and 42 percent poor. The *Woods Audit* estimated that more than twenty-six million sapling-sized, individual invasive buckthorns (*Rhamnus cathartica* and *R. frangula*) occur in Chicago Wilderness woods, with the highest concentrations in Cook, DuPage, and Lake Counties. Buckthorn quickly dominates the shrub layer, shades out native wildflowers and other herbaceous species, and changes soil chemistry, reducing seedling recruitment of native trees. Early and prolific production of seeds, and long-distance seed dispersal by birds contribute to buckthorn's rapid spread (Wisconsin Department of Natural Resources, n.d.). Buckthorn is a classic example of the capacity of invasive species to change ecosystems.

Data from the *Woods Audit* highlight many aspects of the changing character of wood and forest communities. For example, black cherry—a common species invading bur and white oak plots—increases shade cover and creates a significant threat to oak reproduction. Native white-tailed deer are also a major problem. Early in the last century, white-tailed deer came close to local extinction around Chicago. But today, in the absence of predators and of hunting in suburban areas, their population has exploded. Development and farming have created more of the "edge" environments deer prefer. Their persistent browsing decimates wildflowers, oak seedlings, and other native plants, leaving a denuded understory (Spencer 2003).

LARGE-SCALE RESTORATION AND SMALLER SCALE MANAGEMENT

Chicago Wilderness has set an ambitious goal to have 70 percent of its wooded lands in healthy condition by 2025. The challenges are daunting: 29,802 acres of upland forest, woodland, and savanna in northern Illinois and the Indiana Dunes will need to be restored. Meeting this goal, and improving the *Report Card* next time around, will require more investment in managing

28.2. Trained ecologists initiate a controlled burn to clear weedy plants choking out native trees and wildflowers. Source: Photo, Carol Freeman.

28.3. Restoring the natural fire cycle brings back native species, including a diversity of spring wildflowers. Source: Photo, Carol Freeman.

succession. There is also a need to better understand the plant communities on a site and to develop more specific recovery goals, indicators, and monitoring protocols. To return wooded communities to sustainable good health, natural processes that have been disrupted or eliminated need to be restored (figs. 28.2 and 28.3). Historically, the frequency and intensity of fires has helped to determine the type of wooded community at a given site. Lack of fire has been a major factor in the development of an unnaturally high density of trees found in most wooded areas and the subsequent loss of habitat diversity. Removal or thinning of invasive exotic and native tree species is critical. Fire and many of the other processes that once sustained the plant and animal communities of the Chicago area no longer work in the ways that they once did, but there are significant challenges to building broader support for the hands-on management practices needed to replace them. For example, most people do not connect such efforts, including their own, to the larger idea of conservation. And some conservation agencies, especially government agencies, are not trusted. There is also much to be learned about which models of restoration planning will result in maximum participation of the diverse residents of the region and optimal outcomes for natural areas (Field Museum, n.d.; Heneghan et al. 2012).

PUBLIC DIALOGUE AND ENGAGEMENT

Practices, such as culling deer, burning prairies, and thinning woodlands, present both scientific and communication challenges. In 1996, controversy erupted surrounding ecological restoration practices—particularly controlled burns—leading to protests against their use from residents in several Chicago area counties. Scientists and restoration professionals in the field supported such ecological restoration, but the Boards of the Forest Preserve Districts of Cook, DuPage, and Lake Counties took action in response to concerns voiced by their citizens. DuPage and Lake held educational hearings on the benefits and consequences of restoration practices, made appropriate adjustments to practices to satisfy resident requests, and ultimately restoration programs continued with little to no interruption. Cook County, however, issued a moratorium on all management in most areas for ten years. During this period, much land became further degraded as it grew dense with buckthorn and other invasive species.

This and other experiences emphasize that communicating with the public is an essential part of land stewardship. Chicago Wilderness now offers communication tools on its website to help land stewards when using fire to manage land. These tools emphasize the need to engage the public about methods used, the ecological benefits of restoration, and the details about when and why particular burns will occur.

IMPROVED CONSERVATION PRACTICE

In addition to public support, there is also a need for better science to guide restoration, improve the quality of management, and inform better stewardship of existing protected areas. Through both experience and experimentation, we are still learning which management practices are most likely to be effective and efficient in a given situation. As a scientific discipline, restoration ecology is still in its infancy. Work in the Chicago Wilderness region provides opportunities to apply the insights of this new discipline and to investigate novel approaches to restoration management. The current research agenda, which incorporates ecological/conservation science and the social sciences, seeks to address questions such as what management procedures are most effective at restoring woodlands to healthy communities, which invasive species are of most concern and what are the best means of controlling them, and how do current groundwater, stormwater, and wetland laws at all levels of government impact plant and animal communities? Over 100 one-hectare plots have been identified as research sites in four Chicago Wilderness counties to evaluate the effectiveness of current management, as well as to test and foster new strategies.

LAND ACQUISITION

While habitat restoration and management is generally poorly funded and often carried out by volunteers, land acquisition often receives more direct support. Chicago area residents have approved close to $1 billion in bond funds for open space acquisition since 1988 (Vogel 2007). Together with citizen activism, creative thinking by conservation organizations, and the right combination of external factors, this has resulted in substantial land acquisitions over the last decade. Nevertheless, building on recent successes, land acquisition must be a continuing focus for Chicago Wilderness in the coming years.

One key success was the creation of the Midewin National Tallgrass Prairie on the site of the decommissioned Joliet Arsenal. So far, this has resulted in 15,454 acres—of 19,165 acres designated for conservation—being transferred from the Army to the US Forest Service. Midewin is now the largest parcel of conserved land within the Chicago Wilderness area and, as its prairies have been restored, many species have returned or increased in numbers, including rare grassland birds that do not thrive in smaller fragments.

A similar acquisition opportunity came with the closing of the Fort Sheridan Naval base. Less than half of the sixty miles of Illinois' Lake Michigan shoreline is accessible to the public. Openlands, a Chicago Wilderness member, saw the regional significance of this lakefront property and successfully advocated for the transfer of its northern end to the Lake County Forest Preserve District (Openlands, n.d.). The resulting preserve, which includes more than one mile of shoreline, is home to five plant species on the state's threatened and endangered list, and provides habitat for thousands of migrating birds each year.

REGIONAL PLANNING

In 2009 the Chicago region celebrated the centenary of the Burnham Plan, commissioned more than one hundred years ago by The Commercial Club of Chicago. In the late 1990s, The Commercial Club again helped stimulate a forward-thinking plan to help the Chicago region remain socially and economically vibrant. In 1999, through their report *Chicago Metropolis 2020: Preparing Metropolitan Chicago for the 21st Century* (Johnson 1999), they sought to evaluate different development options for northeastern Illinois and took steps to help ensure a viable future for the region's people and environments.

Central to delivering the aspirations of Chicago Metropolis 2020 will be the Chicago Metropolitan Area for Planning (CMAP). Formed in 2005, this new organization combines the previously separate Chicago Area Transportation Study (CATS) and the Northeastern Illinois Planning Commission (NIPC) into a single agency. The Chicago Metropolitan Area for Planning integrates planning for land use and transportation in a seven-county area of northeastern Illinois. By understanding how issues including land use, transportation, natural resources, and economic development are inter-related and shape our communities' futures, CMAP wants to change the way planning is conducted in northeastern Illinois. Chicago Wilderness is an active partner in these efforts. In particular Chicago Wilderness' Green Infrastructure Vision has become the foundation for guiding conservation and land-use planning in CMAP's (2010) *Go to 2040* plan. However, Illinois is only part of the Chicago Wilderness area. A key issue will be how CMAP and its equivalent bodies in Indiana (NIRPC) and Wisconsin (Southeastern Wisconsin Regional Planning Commission) can coordinate their activities around common aspirations and a shared vision.

The Green Infrastructure Vision

The Green Infrastructure Vision (GIV)—developed by Chicago Wilderness, and led by the NIPC (prior to the formation of CMAP), along with many partners—cuts across four states and all thirty-four counties of the Chicago Wilderness region. This vision is of an interconnected network of land and water that protects watersheds, enriches the lives of people, and provides resilient habitat for diverse communities of native flora and fauna at a regional scale. As currently formulated, the GIV encompasses roughly 360,000 acres of protected land including large areas of remnant woodlands, savannas, prairies, wetlands, lakes, stream corridors, and related natural communities.

The GIV also acknowledges that, even if these protected areas are perfectly managed, conservation attempts will fail if management is not co-ordinated with activities on adjacent lands where human impacts are even more intense. In this, the GIV is a regional manifestation of a global reality: protected-area approaches will never be enough without taking into account changes in adjacent areas, without striving for maximum connectivity among ecological communities, and without committing to ongoing monitoring and management.

The GIV is most easily visualized as a map (fig. 28.4) that identifies on-the-ground, regional opportunities for biodiversity protection. Within the more than seven million acres of the Chicago Wilderness area, GIV identifies 1.8 million acres (including the already-protected 360,000 acres) as "Resource Protection Areas" (NIPC 2004). The aim is to highlight macro-scale opportunities for innovative land-use planning that unite conservation-friendly

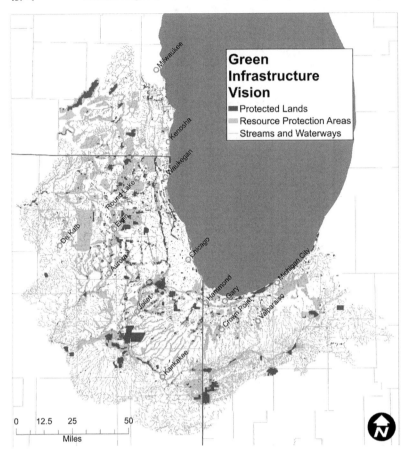

28.4. The Green Infrastructure Vision identifies opportunities for land acquisition, restoration, and conservation development to create a network of "Resource Protection Areas" that connect and buffer protected lands

land development with habitat conservation and restoration opportunities. The GIV's focus on the scale of watersheds is of particular importance for management of northern Illinois aquifers, which are under increased stress from substantial population and economic growth. The GIV will help facilitate improved water supply planning and management and will help ensure current and future water demands can be met.

The Future of Chicago's Wilderness

In just over a decade, Chicago Wilderness has made strides in developing a new and more inclusive model of regional conservation—one that has attracted attention in other urban areas, from Cleveland, Houston, and San Francisco, to the state of Hawaii. Chicago Wilderness has also attracted attention from outside the United States, for example, from Mexico City and the Curitiba metropolitan region of Brazil. Curitiba lies in the transition zone between Araucaria pine forest and the Serra do Mar tropical evergreen forest, both diverse and severely endangered ecological communities of the Brazilian Atlantic Forest. The Condomínio da Biodiversidade in Curitiba and Chicago Wilderness face similar challenges and opportunities and have committed to long-term collaboration. The work of Chicago Wilderness also has drawn the attention of the World Conservation Union (IUCN), with Chicago Wilderness' experiences helping to inform the development of a new code of ethics for biodiversity conservation (Mackey et al. 2008).

And yet even as Chicago Wilderness attracts interest from elsewhere, the work at home is really just beginning. In terms of science, there is a clear need to better understand the successional and other processes at work in Chicago-area forests and other habitats. And in terms of conservation practice, there is a clear need to understand the science underpinning restoration and management. But the GIV is of overriding importance as a powerful organizing vision for the future, comparable to the Burnham Plan of a century ago. In large part, the success of the Chicago Wilderness initiative will depend on the extent to which it is able to deliver that vision. This will require leadership from Chicago Wilderness and commitment from its member organizations, as well as individual efforts, community participation, and region-wide partnerships around a shared sense of purpose.

Beyond the science and conservation questions there is also much to be done to further engage stakeholders and better understand their aspirations and needs. Broader support is also necessary for the interventionist "gardening" practices needed to repair, restore, and keep healthy the forests and other plant communities of the Chicago area. This will require understanding the value of forests in all their many forms, increased public awareness of their importance, and better communication about why seemingly "natural" areas need to be managed. Through better engagement, more active planning, and informed management, Chicago can thrive in its human-dominated wilderness. Henry Chandler Cowles would have approved. His vision was of "forest tracts . . . numerous enough to be within easy reach of all of our people" and "made to serve many interests" (Cowles 1912, 259). Finding the right balance between these many interests will be crucial for the future of forests in some of the most dramatically human-modified landscapes anywhere on our planet.

NOTES

Chapter 2

1. This, like all subsequent quotations, have been translated from the original French by the authors.
2. More details of the following historical vegetation analysis are given in Leach and Fairhead (1994).
3. Such alternative social science analysis and its considerable evidence is documented fully elsewhere (Leach and Fairhead 1995).
4. For more on the management-use continuum (i.e., the way people use a resource in the way they manage it) see Roe and Fortmann (1982).
5. Further details concerning this case are given in Fairhead and Leach (1994).

Chapter 3

1. www.conservation.org/where/priority_areas/hotspots/africa/Guinean-forests-of-West -Africa/Pages/impacts.apx, accessed April 4, 2013.

Chapter 4

1. For the history of woodlands in the Hebrides, see Smout (2003, 41–43); on Walker see also Albritton Jonsson (2013, chapters 2, 3, and 4). Compare Henderson and Dickson (1994, 88, 154) and Sinclair (1791–99, vol. 18, 91).
2. Edinburgh University Library (EUL), Walker Papers, La III. 352/3/4; in Aeneas Macpherson's essay from 1793, the destruction of the Caledonian Forest marked the shift from the hunting stage to agriculture, though he did not discuss the cause. See EUL, La II. 412 Draft Essay, "Establishment of Inland Villages in the Highlands of Scotland," Aeneas Macpherson 1793, 14.
3. Obviously, we cannot infer from ownership that Smith read and understood every book in his possession, but there is sufficient reason here to assume a probability of selective interpretation on Smith's part.
4. On the early reception of Smith, see Teichgraeber (1987). For some recent literature on the importance of natural historians and botanic gardens within the British Empire from

1760 onwards, see, for example, Grove (1995), Gascoigne (1998), Drayton (2000), Webb (2002), and Arnold (2006).

Chapter 5

1. Including Malayic peoples, such as Javanese.
2. This approach is still in place; see Groninger and Ruffner (2010).
3. The latter occurred when the second president of Indonesia criminalized Communism and was hunting down both West Kalimantan and Sarawak members of the guerrilla forces trained by Sukarno and his army inside Indonesia (and by the Indonesian army). For a detailed account of these low-impact wars, see, for example, Mackie (1974), Coppel (1983), Dennis and Grey (1996), Poulgrein (1998), Davidson and Kammen (2002), Davidson (2002).
4. These, in fact, were places of considerable conflict, later recast as nature's empire. See Cosgrove and Daniels (1988); Bak and Hölbling (2003).
5. As has been the case in Vietnam and, more recently, in Laos.
6. Many spoke several dialects, in addition to other languages used regionally, on a daily basis.

Chapter 6

1. The US Department of Energy's (2003) Long-Term Stewardship Program mission statement, available from http://lts.apps.em.doe.gov/mission.asp, accessed October 15, 2003.

Chapter 8

Much of the thinking about forest resurgence was inspired by working in El Salvador with colleagues at Prisma, especially Susan Kandel, and Herman Rosa, under the auspices of a grant from the MacArthur Foundation. Other research and activist experiences occurred in Brazil over decades with the Kayapo, rubber tappers, and mostly recently with Kilombo inhabitants. This paper was written while the author held a Guggenheim Foundation Fellowship.

1. I use the rise of steam transport as the inaugural point of modern globalization and imperialism, but there is ample scholarship that describes longer and earlier engagement. The classic in this regard is (Grove 1995).
2. A critique of modernization theory and development is much too enormous a task to take on here.

Chapter 9

1. While the success of the Green Revolution (in productivity terms alone) is frequently attributed to the genetic changes in the crops themselves, the fundamental evolutionary principle that adaptation occurs within the framework of particular environmental circumstances holds here as much as in other cases of evolution. The environmental constraints included irrigation, chemical fertilizer (especially nitrogen-based), and pes-

ticides. That varieties evolved to be more efficient in this environmental background than in a background of low water, low nitrogen, and abundant pests, is not in any way surprising. One could state, as a general evolutionary principle, that the performance of a variety adapted to (for example) environment A is better than a variety not adapted to environment A. Suggesting that there is something remarkable about the fact that the adapted variety performs better in environment A than the unadapted variety betrays a naivety about the basic structure of biological evolution. To suggest that the Green Revolution was a great success in terms of production makes this an elementary error.

2. This section follows closely the interpretation of Patel (2008).
3. This statement will undoubtedly attract scorn. For the interested reader, the argument of Badgley et al. (2007) is unassailable.
4. Nelson (1994) has been the only one to my knowledge that has approached this problem experimentally, documenting a "farmer first" approach to the development of pest management in tomato production in Nicaragua. Morales (2002; Morales and Perfecto 2000; Morales, Perfecto, and Ferguson 2001) has been one of the most persistent advocates of an organic approach to understanding the dialectical relationship between farmer and researcher.

Chapter 11

1. See http://www.conservation.org/where/priority_areas/hotspots/Pages/hotspots_main .aspx. Accessed April 10, 2013.

Chapter 12

1. The Penan population may be divided into Eastern and Western Penan (Needham 1972); those living around the Kelabit Highlands are Eastern Penan.
2. Harrisson carried out surveys of megaliths in the Kelabit Highlands (e.g., see Harrisson 1958) and also excavated some of these, but this work was never written up. A survey of megalithic monuments for the broader Apad Uat highland area was made by Schneeberger (1979) before World War II.
3. Although it is possible that this was at least partly borrowed from the Bible story of the flood, after the Kelabit became Christian in the 1950s, there are also echoes of this story on the other side of the border among related peoples, who tell of a culture hero who broke through a barrier in a gorge and drained a huge lake (Schneeberger 1979, 51).
4. The term "anthropic" refers to changes that are brought about by human activities, but not necessarily deliberately; the term "anthropogenic" refers to deliberate changes brought about by humans.
5. All dates are reported as uncalibrated; see Barker et al. (2008) for a full list of calibrated ages.

Chapter 13

We wish to acknowledge the contributions of Lisa Haddad, Sotiria Cherpelis, H. M. Mary Joo, and Morgan Potter, who participated in the Botanical Field Methods course at New

York University and helped to collect the field data described in this chapter. Also, we wish to thank Fred Valdéz for generously welcoming us into his field camp, Jon Lohse for sharing archaeological data, Somayeh Tarighat for helping to edit the manuscript, and the Programme for Belize who provided encouragement and access to the Rio Bravo Reserve.

Chapter 14

This is a revised version of a paper presented at the The Social Life of Forests Conference, University of Chicago, May 29, 2008 to June 1, 2008. The author is grateful to the organizers for their support, and most of all to Professor Kathleen Morrison for continued comments and interaction. Ullas Karanth, Harini Nagendra, Nitin Sethi, Ghazala Shahabuddin, and MD Madhusudan have all helped in many ways. All the usual disclaimers apply.

1. Supportive of the legal route but more open to local livelihood issues was George B. Schaller. See also Rosencranz and Lele (2008).
2. Also see http://www.sanctuaryasia.com/kidsfortigers/index.php. And see http://www.vanashakti.in for arguments against the Forest Rights Act that combine biodiversity concerns and water security.
3. These groups have challenged the act in the Supreme Court (Gopalakrishnan, personal communication, 2007). The Wildlife Lobby, letters and press releases are available at http://forestrightsact.awardspace.com.
4. For example, see the issue of *Seminar* no. 566, September 2007. *Nature, Wildlife, People*: http://www.india-seminar.com.
5. Historians differ on their estimates of population in 1600 and on how much land was arable. Guha (2001) gives a figure of 114 million and Habib (1982) about 145 to 150 million.
6. This is in contrast to Han agriculture in China, with its antagonism to elephants. See Elvin (2004, 17).
7. The point is made for British knowledge systems in the early-to-mid-nineteenth century but will be equally valid in case of princely India. See Arnold (2006, 8).
8. See Nehru (1987, 562). The note is undated but is almost certainly from February 1948.
9. This issue has surprisingly never been explored in depth. But note the rate of food grain production was 4.7 percent from 1965 to 1981, up from 2.2 percent from 1961 to 1965. See Panagariya (2008, 73).
10. Krishnan's ideas were crystallized in an important note: see Krishnan (1972). Also see Guha (2000).
11. The incident has been fictionalized in Ghosh (2004).
12. See Chatterjee (2008). Far from being "unable to gain access to political society" (61), tribal, forest, and pastoral groups are doing so more systematically than in the past.
13. The alliances were against mining in Kudremukh, Karnataka. Wildlife First and the Kudremukh Foundation joined forces with the prosperous peasantry downstream. For ecological impacts see Krishnaswamy et al. (2003), Rangarajan (2006), and Praveen Bhargava, Wildlife First! (pers. comm.).

Chapter 17

1. *Parkia biglobosa* is also a multipurpose tree, used for human food (fermented seeds make a protein rich condiment), fodder, medicine, green manure, and fuelwood. The most significant use of gum Arabic is as a stabilizer in the food industry. The tree's gum is readily soluble in water, which makes it an ideal stabilizer in soft drinks and candies. In ancient trade, gum Arabic was used in poultices, as fodder, famine food, and importantly, as a binding medium to mix pigments in illuminated manuscripts. Gum Arabic, known to Medieval Europeans, was introduced across trade routes that connected Senegal and Mali with the Iberian Peninsula.

2. Over the past centuries, the tree has spread to the Gambia, the Democratic Republic of the Congo, Ghana, and Uganda due to human influence (Hall et al. 1996). The subspecies *paradoxa* dominates in the western Sudan while ssp. *nilotica* is found in the Democratic Republic of the Congo, Ethiopia, Uganda, and the Sudan.

3. "Recalcitrant seeds" refers to the fact that seed viability drops very rapidly. For shea trees, viability declines within a week of seed removal from the fruit and is completely lost within three to six weeks (Hall et al. 1996). Out-crossing complicates the selection for "superior," "true-to-type" individuals stemming from heterozygous parents.

4. *Vitellaria paradoxa* subsp. *nilotica* populations in Uganda are known for a lower proportion of stearic acid, which results in a less viscous butter that does not hold its form. It is usually sold in liquid form in plastic jugs (Maranz and Wiesman 2003).

5. The Mossi plain covers one-quarter of Burkina Faso's territory (70,668 km^2) and encompasses half of the country's population.

6. Male farmers cited tree health, yield, fruit characteristics, growth, competitive effect on other crops, and resistance to mistletoe as factors affecting tree selection (Lovett and Haq 2000a).

Chapter 18

1. NSF Grant # 0434043. HSD: "Understanding Dynamic Resource-Management Systems and Land-Cover Transitions in Montane Mainland Southeast Asia," http://www2.eastwestcenter.org/environment/MMSEA/.

   NASA Grant # NNG04GH59G (Project ID: 12 IDS/03–0365–0079): "The role of land-cover change in MMSEA in altering regional hydrological processes under a changing climate," http://research.eastwestcenter.org/mmsea.

2. http://www.peets.com/shop/tea_detail.asp?rdir=1&id=1335.

3. Jiaqing reign period, 1796 to 1820; Daoguang reign period, 1820 to 1850.

4. Interview with Zhang Yi 张毅, Jinghong, September 19, 2006.

5. Interview with Zhang Shixin, General Manager, Menghai Tea Factory, Menghai, September 15, 2006.

6. Interviews at Menghai Tea Research Institute, 2006.

7. Interview with Fan Jie, Yunnan Chayuan Company, Kunming, September 4, 2006.

8. Ibid.

9. Interview with Zhang Yi, 2006.

10. Interview with Quan Cun'an, September 10, 2006.
11. Interview with Xishuangbanna Commission for Reform and Development 西双版纳发展与改革委员会, Jinghong, September 19, 2006.
12. This is not the place for an investigation into whether or not the Ben were a distinct ethnic and cultural group as claimed. In this context, it is the new or recovered claim to a Ben identity that is significant.
13. Interview with Quan Cun'an, September 10, 2006.

Chapter 19

This chapter is based on the work of numerous collaborators. In the NSF-funded project, my collaborators in China include Dr. Xu Jianchu, Kunming Institute of Botany; Mr. Lu Xing and Mrs. Zhang Peifang, Yunnan University; Dr. Nicholas Menzies, UCLA; and Dr. Janet Sturgeon, University of British Columbia. In Thailand they include Dr. David Thomas, World Agroforestry Center, Chiang Mai; and Dr. Benchaphun Ekasingh, Chiang Mai University. In Laos they include Mr. Khamla Phanvilay, Dr. Sithong Thomanivong, and Dr. Yayoi Fujita, National University of Laos. The spatial information and modeling was done by Mr. John Vogler, East-West Center, Hawaii. Project collaborators on the NASA-funded project include Dr. Tom Giambelluca and Dr. Alan Ziegler, University of Hawaii; Dr. Peter Troch and Dr. Maite Guardiola, University of Arizona; and Dr. Omer Sen, Istanbul Technical University. The views and opinions expressed in the chapter are mine alone, and all mistakes and omissions are my responsibility.

1. This research was supported by NASA Grant # NNG04GH59G (Project ID: 12 IDS/03–0365–0079), "The role of land-cover change in MMSEA in altering regional hydrological processes under a changing climate," http://research.eastwestcenter.org/mmsea.
2. This research was supported by NSF Grant # 0434043. HSD: Understanding Dynamic Resource-Management Systems and Land-Cover Transitions in Montane Mainland Southeast Asia, http://www2.eastwestcenter.org/environment/MMSEA/.
3. This section is based on Sturgeon and Menzies (2008) and Sturgeon (2005). See also Menzies, chapter 18, this volume.
4. This section is based on Sturgeon and Menzies (2008).
5. This section is based on Thomas (2005) and Thomas et al. (2008).

Chapter 20

1. Interview with a farmer, Mon, 24 February 2005.
2. Interview with a rubber farmer, Oudomxin, 27 September 2005.
3. Ibid.
4. Interview with village leader, Oudomxin, 8 April 2005.
5. Interview with village leader, Donechai, 12 July 2005
6. Interview with village leader, Donechai, 10 April 2005.

7. Interview with the owner of the agricultural extension and trade company, Muang Sing, 10 July 2005.
8. Interview with a representative of the company, Muang Sing, 26 September 2005.
9. Interview with the head of DAFO, Muang Sing, 11 July 2005.
10. Ibid.
11. Ibid.
12. Interview with the head of PAFO, Luang Namtha, 8 July 2005.
13. Interview with the deputy head of PPIO, Luang Namtha, 8 July 2005.

Chapter 21

1. Ghanaians use the term "citizen" to refer both to national citizenship and to membership in an ancestral village or hometown. Interviewed in the United States, some expatriate Ghanaians told their interlocutor that they knew the names and approximate locations of their hometowns, but had never actually been there (Mindelyn Buford, personal communication). When ancestry is conflated with chiefly jurisdiction, personal status is oxymoronic: community members are both local citizens and subjects of the stool (chiefly office).
2. Land ownership in Asante is less a legal fact than a complex mosaic of overlapping claims to access, use, and/or allocate particular tracts of land, giving rise to frequent debates, disputes, and outright conflicts. Examples of the dynamics of land ownership and contestation, and their implications for power relations and socioeconomic difference, are discussed later in the chapter.
3. In the southern regions of Ghana, chiefly offices are referred to as "stools"; in the north, they are called "skins."
4. http://www.ghanaweb.com/GhanaHomePage/NewsArchive/artikel.php?ID=68311.
5. The term "forest rent" was coined by François Ruf to denote unearned value derived from forest resources, including the rich subsoils of virgin forest growth (Ruf 1995; Austin 2006). On the history of cocoa tribute see, inter alia, Arhin (1986); Boni (2005, 2006); Dunn and Robertson (1973).
6. State Concessions Act, 1962.
7. Under the constitution, chiefs exercise exclusive jurisdiction over "chieftaincy affairs"— that is, succession to and removal from chiefly office, as well as matters of ceremonial protocol (*Constitution of the Republic of Ghana, 1992*, Art. 270). On the rise of chiefly influence since the late 1960s, see especially Arhin (2001).
8. In addition to commercial tree crops such as cocoa, coffee, and oil palm, agroforestry schemes have encouraged farmers to grow fruit trees in addition to food and other annual crops, and plant various kinds of woody shrubs as live fences around their gardens and as a source of fodder for goats and sheep (Okali and Berry 1983).

Chapter 23

I gratefully acknowledge the support of Gray Tappan, a geographer with the US Geological Survey, EROS Center, South Dakota. His remote sensing studies on long-term trends in vegetation in Niger revealed the scale of re-greening by farmers. His work on land use and land cover studies is a valuable source of information and inspiration.

1. The impacts mentioned are based on an exploratory study by Larwanou, Abdoulaye, and Reij (2006). This report, as well as other information about the re-greening, can be downloaded from www.frameweb.org/nigerregeneration.

Chapter 24

1. This theory has been widely disseminated in popular media (e.g., American Broadcasting Corporation 2002; British Broadcasting Corporation 2002).
2. Initial Xinguano occupations date sometime between 200 BC and 800 AD, but the exact date is uncertain because the two earliest radiocarbon dates are from mixed deposits, and in one case (Beta-Analytic Laboratory number 176143) the dates are demonstrably out of sequence.
3. Manioc, fish, and piqui fruit constitute 90 percent or more of the diet (Carneiro 1957).
4. Materials and methods are available as supporting material at Science Online: http://www.sciencemag.org/cgi/content/full/301/5640/1710/DC1.
5. The Kuikuro study area is about the same as the Kuikuro (Carib Xinguano) traditional territory. It is about 1000 km^2 and is located in an area of traditional Xinguano occupations that is 25,000 to 35,000 km.^{2}
6. Several additional large sites are known in the study area, on the basis of indigenous knowledge of dark-earth locations, and numerous smaller occupation sites are also known, including small roadside hamlets along the major roads.
7. Here, "galactic" describes the regional site clusters organized around a center or hub (X11 and X13), with several major residential sites connected to it. See supporting material online.
8. See online supporting material: http://www.sciencemag.org/cgi/content/full/301/5640/1710/DC1.
9. Dark earth forms in compost areas, although widespread burning produces minor but extensive soil alterations.
10. See online supporting material text.

Chapter 25

1. All these data were collected through interviews with adult household members in Pucallpa and in the community of Las Mercedes in February 2010.
2. The history of extreme inflation and several changes in currency over the last three decades have up to now proven too challenging for us to make an exact income comparison.

Chapter 26

This article is based on a talk given at the Social Life of Forests conference at the University of Chicago, May 2008. The article is improved by comments from participants of the conference. The fieldwork was supported by grants from the Tinker Foundation to MAP and RRS and a Fulbright Scholarship to RRS. We are indebted to our field assistants in Peru for primary data collection.

Chapter 27

This chapter is based on work supported by the National Science Foundation Human Social Dynamics Program Grant No. BCS-0527578. We received indispensable support from many institutions and individuals in Brazil and the United States. We particularly would like to thank the Núcleo de Altos Estudos Amazônicos of the Universidade Federal do Pará, and especially Drs. Edna Castro and David McGrath, and the personnel of the Instituto de Pesquisas Ambientais na Amazônia (IPAM). Many of the field data presented were gathered by a skilled group of field researchers from Ponta de Pedras including Valois Delcastagne (coordinator), Elaine Gouvea, Cimara Evangelista, Leandro Gouvea, Socorro Tavares DelCastagne, Juciane Ribeiro, Edwilson Ribeiro, and Leidiane Lobato da Silva. Thanks to project collaborators Miguel Pinedo-Vasquez, Christine Padoch, Robin Sears, Peter Deadman, and Sandra Costa. We thank their dedication to this study and the support of numerous households in rural and urban areas. We appreciate the support of the Department of Anthropology and the Anthropological Center for Training and Research on Global Environmental Change (ACT) at Indiana University, particularly Vonnie Peischl, Linda Barchet, Jessica Chelekis, Scott Hetrick, and Rodrigo Pedrosa, and the support offered by the Laboratoire d'anthropologie sociale (LAS) at College de France during the academic year 2008–2009.

1. The "social life of forests" is defined here as a multidimensional concept including the history of forest landscape change, the consumption chains of forest resources, the forms of forest management and uses, the cultural value and symbolic meanings of forests and forest species, and the institutional arrangements of access and ownership of forest resources.
2. Obviously, other forces of forest change not considered for this particular analysis are also at play—for instance, environmental and climate change.
3. Households of family members with which one maintains social and economic connections and communication.

Chapter 28

We thank Kathleen Morrison, Christine Padoch, and Susanna Hecht for the invitation to participate in this symposium. We are also grateful to Chicago Wilderness members whose work has provided the basis for this chapter. We are grateful to Jon Markel for providing figures 28.1 and 28.3, and to Carol Freeman for providing the photographs for figure 28.2.

REFERENCES

Abizaid, C. 2005. "An Anthropogenic Meander Cutoff Along the Ucayali River, Peruvian Amazon." Geographical Review 95 (1): 122–35.

Achard, F., H. Eva, H. Stibig, P. Mayaux, J. Gallego, T. Richards, and J. Malingreau. 2002. "Determination of Deforestation Rates of the World's Humid Tropical Forests." Science 297 (5583): 999.

Acosta, P., C. Calderon, P. Fajnzylber, and H. Lopez. 2008. "What Is the Impact of International Remittances on Poverty and Inequality in Latin America?" World Development 36 (1): 89–114.

Adam, J. G. 1948. "Les Reliques Boisées et les Essences Des Savanes Dans la Zone Préforestière en Guinée Française" [Relics and Woody Species in the Savannah Area Preforest in French Guinea]. Bulletin de la société Botanique Française 98:22–26.

Adams, R. E. W. 1995. "A Regional Perspective on the Lowland Maya of Northeast Petén and Northwestern Belize." Paper presented at 60th Annual Meeting of the Society for American Archaeology, Minneapolis, MN.

Adeney, J. M., N. L. Christensen, and S. L. Pimm. 2009. "Reserves Protect against Deforestation Fires in the Amazon." Plos One 4 (4): e5014.

Agrawal, A., and K. Sivaramakrishnan. 2000. Agrarian Environments: Resources, Representations, and Rule in India. Durham, NC: Duke University Press.

———. 2001. "Introduction: Agrarian Environments." In Agrarian Environments: Resources, Representations, and Rule in India, edited by A. Agrawal and K. Sivaramakrishnan, 1–22. Durham, NC: Duke University Press.

Aide, T. M., and H. R. Grau. 2004. "Globalization, Migration, and Latin American Ecosystems." Science 305 (5692): 1915–16.

Aide, T. M., J. K. Zimmerman, L. Herrera, M. Rosario, and M. Serrano. 1995. "Forest Recovery in Abandoned Tropical Pastures in Puerto Rico." Forest Ecology and Management 77 (1–3): 77–86.

Aide, T. M., J. K. Zimmerman, M. Rosario, and H. Marcano. 1996. "Forest Recovery in Abandoned Cattle Pastures Along an Elevational Gradient in Northeastern Puerto Rico." Biotropica 28 (4): 537–48.

Aiton, W. 1811. A Treatise on the Origin, Qualities and Cultivation of Moss-Earth, with Directions for Converting It into Manure. AIR: Printed by Wilson and Paul and sold for the author by them; Brash and Reid, Glasgow; A. Constable and Co Edinburgh;

Longman, Hurst, Rees, Orme, and Co. London; and by other Booksellers in Town and Country.

Aiyar, M. S. 2003. "Can the Congress Find a Future?" Seminar 526:14–22.

Albion, R. G. (1926) 2000. Forests and Sea Power: The Timber Problem of the Royal Navy, 1652–1862. Annapolis: Naval Institute Press.

Albritton Jonsson, F. 2013. Enlightenment's Frontier: The Scottish Highlands and the Origins of Environmentalism. New Haven, CT: Yale University Press.

Alchian, A. A., and H. Demsetz. 1973. "The Property Rights Paradigm." The Journal of Economic History 33 (1): 16–27.

Alexiades, M., ed. 2009. Mobility and Migration in Indigenous Amazonia: Contemporary Ethnoecological Perspectives. Oxford: Berghahn.

Ali, I. 1966. "A Critical Review of Malayan Silviculture in the Light of Changing Demand and Form of Timber Utilization." Malayan Forester 29 (4): 228–33.

Allman, J. M. 1993. The Quills of the Porcupine: Asante Nationalism in an Emergent Ghana. Madison: University of Wisconsin Press.

———. 1997. "Fathering, Mothering and Making Sense of Ntamoba: Reflections on the Economy of Child-Rearing in Colonial Asante." Africa 67:296–321.

Alvard, M., J. B. Alcorn, R. E. Bodmer, R. Hames, K. Hill, J. Hudson, R. L. Lyman, R. K. Puri, E. A. Smith, and A. M. Stearman. 1995. "Intraspecific Prey Choice by Amazonian Hunters." Current Anthropology 36:789–818.

Amanor, K. 1999. Global Restructuring and Land Rights in Ghana: Forest Food Chains, Timber, and Rural Livelihoods. Research Report 108, vol. no 108. Uppsala: Nordiska Afrikainstitutet.

———. 2007. "Conflicts and the Reinterpretation of Customary Tenure in Ghana." In Conflicts over Land and Water in Africa, edited by B. Derman and R. Odgaard, 33–59. Oxford: James Currey.

American Broadcasting Corporation. 2002. "Fertile Secret: How Did Ancient Amazonians Turn Sand into Rich Soil? Garbage." September 19, 2002, available online at http://abcnews.go.com/sections/scitech/DyeHard/dyehard020919.html.

Amster, M. 2003. "New Sacred Lands: The Making of a Christian Prayer Mountain in Highland Borneo." In Sacred Places and Modern Landscapes, edited by R. Lukens-Bull, 131–60. Tempe: Arizona State University.

Anderson, A. B. 1990. "Extraction and Forest Management by Rural Inhabitants in the Amazon Estuary." In Alternatives for Deforestation, edited by A. B. Anderson, 65–85. New York: Columbia University Press.

Anderson, A. B., I. Mousasticoshvily, and D. S. Macedo. 1993. Impactos Ecológicos e Sócio-Econômicos da Exploração Seletiva de Virola no Estuário Amazônico [Ecological Impacts and Socio-Economic Exploitation Selective Virola in the Amazon Estuary]. Brazil: WWF.

Anderson, B. 1870. Narrative of a Journey to Musardu, the Capital of the Western Mandingoes. New York: S.W. Green.

Anderson, B. R. 1984. Imagined Communities: Reflections on the Origin and Spread of Nationalism. Brooklyn, NY: Verso.

———. 1991. Imagined Communities : Reflections on the Origin and Spread of Nationalism (2nd ed.). New York: Verso.

Anderson, D., and R. Grove. 1989. Conservation in Africa : People, Policies, and Practice. New York: Cambridge University Press.

Anderson, J. 1777. Miscellaneous Observations on Planting and Training Timber-Trees; Particularly Calculated for the Climate of Scotland, by Agricola. Edinburgh: Printed for C. Elliot.

Anderson, M. L., and C. J. Taylor. 1967. A History of Scottish Forestry. London,: Nelson.

Andrade, M. C. D. O. 1980. The Land and People of Northeast Brazil. Albuquerque: University of New Mexico Press.

Andrews, E. 1930. The Bombay Burmah Trading Corporation Limited in Burmah, Siam and Java: Teak the Cutting and Marketing. London: s.n., WorldCat record.

Andrews, E. W. 1943. The Archaeology of Southwestern Campeche. Washington, DC: Carnegie Institution of Washington.

Angelson, A. 2007. Forest Cover Change in Space and Time: Combining the von Thunen and Forest Transition Theories. Washington, DC: The World Bank.

Ankersen, T., and G. Barnes. 2005. "Inside the Polygon: Emerging Community Tenure Systems and Forest Resource Extraction." In Working Forests in the Neotropics: Conservation through Sustainable Management, edited by D. J. Zarin, J. R. R. Alavalapatti, F. E. Putz, and M. Schmink, 156–77. New York: Columbia University Press.

Anshari, G., A. Peter Kershaw, and S. Van Der Kaars. 2001. "A Late Pleistocene and Holocene Pollen and Charcoal Record from Peat Swamp Forest, Lake Sentarum Wildlife Reserve, West Kalimantan, Indonesia." Palaeogeography, Palaeoclimatology, Palaeoecology 171 (3–4): 213–28.

Anshari, G., A. P. Kershaw, S. V. D. Kaars, and G. Jacobsen. 2004. "Environmental Change and Peatland Forest Dynamics in the Lake Sentarum Area, West Kalimantan, Indonesia." Journal of Quaternary Science 19 (7): 637–55.

Antlitz, D. 2005. "The Ecological Condition and Management Needs of Natural Areas in the Forest Preserve District of Cook County." CW Journal 3:23–28.

Arce-Nazario, J. A. 2007. "Landscape Images in Amazonian Narrative: The Role of Oral History in Environmental Research." Conservation and Society 5 (1): 115–33.

Arhin, K. 1986. The Expansion of Cocoa Production: The Working Condition of Migrant Cocoa Farmers in the Western and the Central Regions. Legon: Institute of Africa Studies, University of Ghana.

———. 1994. "The Economic Implications of Transformations in Akan Funeral Rites." Africa 64:307–22.

———. 2001. Transformations in Traditional Rule in Ghana (1951–1996). Accra: Sedco Enterprise.

Arima, E. Y., R. T. Walker, S. G. Perz, and M. Caldas. 2005. "Loggers and Forest Fragmentation: Behavioral Models of Road Building in the Amazon Basin." Annals of the Association of American Geographers 95 (3): 525–41.

Arnold, C. 2002. "The Reconstitution of Property: Property as a Web of Interests." Harvard Environmental Law Review 26 (2): 281.

Arnold, D. 2006. The Tropics and the Traveling Gaze: India, Landscape, and Science, 1800–1856. Seattle: University of Washington Press.

Ashmore, W., and A. B. Knapp. 1999. Archaeologies of Landscape: Contemporary Perspectives. Malden: Blackwell Publishers.

Asner, G. P., M. Keller, R. Pereira, and J. C. Zweede. 2002. "Remote Sensing of Selective Logging in Amazonia—Assessing Limitations Based on Detailed Field Observations, Landsat Etm+, and Textural Analysis." Remote Sensing of Environment 80 (3): 483–96.

Associated Press. 1999. "Biological Bounty at Former Nuclear Bomb Factory." June 24.

———. 2001. "Getting Rid of Radioactive Weeds." May 4.

Atchara, R. 2009. "Constructing the Meanings of Land Resource and a Community in the Context of Globalization." PhD diss., Chiang Mai University, Chiang Mai.

Athas, W. F. 1996. Investigation of Excess Thyroid Cancer Incidence in Los Alamos County. Santa Fe: New Mexico Department of Health.

Atran, S. 1993. "Itza Maya Tropical Agro-Forestry." Current Anthropology 34 (5): 633–700.

Atran, S., A. F. Chase, S. L. Fedick, G. Knapp, H. McKillop, J. Marcus, N. B. Schwartz, and M. C. Webb. 1993. "Itza Maya Tropical Agro-Forestry." Current Anthropology 34 (5): 633–700.

Austin, G. 1996. "'No Elders Were Present': Commoners and Private Ownership in Asante, 1807–96." The Journal of African History 37 (1): 1–30.

———. 2005. "The Political Economy of the Natural Environment in West African History: Asante and Its Savanna Neighbors in the 19th and 20th Centuries." In Land Rights and the Politics of Belonging in West Africa, edited by R. Kuba and C. Lentz, 189–207. Leiden: E. J. Brill.

Azam, J. P., and F. Gubert. 2005. "Migrant Remittances and Economic Development in Africa: A Review of Evidence." Paper presented at the African Economic Research Consortium (AERC) Plenary Session, Nairobi.

Aziz, S. A., W. F. Laurance, and R. Clements. 2010. "Forests Reserved for Rubber?" Frontiers in Ecology and the Environment 8 (4): 178.

Baden-Powell, B. H. 1873. Memorandum on the Supply of Teak and Other Timbers in the Burma Market. Calcutta: Government Printing.

———. 1874a. The Forest System of British Burma. Calcutta: Government Printing.

———. 1874b. "On the Defects of the Existing Forest Law." In Report of the Proceedings of the Forest Conference 1873–74, edited by B. H. Baden-Powell and J. S. Gamble. 3–29. Calcutta: Government Printing.

Badgley, C., J. Moghtader, E. Quintero, E. Zakem, M. J. Chappell, K. Aviles-Vazquez, A. Samulon, and I. Perfecto. 2007. "Organic Agriculture and the Global Food Supply." Renewable Agriculture and Food Systems 22 (02): 86–108.

Bak, H., and W. Hölbling. 2003. "Natures Nation" Revisited : American Concepts of Nature from Wonder to Ecological Crisis. European Contributions to American Studies, vol. 49. Amsterdam: VU University Press.

Baker, D. 2003. "The Environmental Kuznets Curve." Journal of Economic Perspectives 17 (1): 226–27.

Baker, J. 1995. "Migration in Ethiopia and the Role of the State." In The Migration Experience in Africa, edited by J. Baker and T. Akin Aina, 234–56. Sweden: The Nordic Africa Institute.

Balée, W. L. 1989. "The Culture of Amazonian Forests." Advances in Economic Botany 7:1–21.

———. 1994. Footprints of the Forest: Ka'apor Ethnobotany— the Historical Ecology of Plant Utilization by an Amazonian People. Biology and Resource Management in the Tropics Series. New York: Columbia University Press.

———. 1998. Advances in Historical Ecology. New York: Columbia University Press.

———. 2006. "The Research Program of Historical Ecology." Annual Review of Anthropology 35:1–24.

Balée, W. L., and C. L. Erickson. 2006a. Time and Complexity in Historical Ecology: Studies in the Neotropical Lowlands. New York: Columbia University Press.

———. 2006b. "Time, Complexity, and Historical Ecology." In Time and Complexity in Historical Ecology : Studies in the Neotropical Lowlands, edited by W. L. Balée and C. L. Erickson, 1–20. New York: Columbia University Press.

Balick, M. J., and R. Mendelsohn. 1992. "Assessing the Economic Value of Traditional Medicines from Tropical Rain Forests." Conservation Biology 6 (1): 128–30.

Balick, M. J., M. Nee, and D. E. Atha. 2001. Checklist of the Vascular Plants of Belize, with Common Names and Uses. Memoirs of the New York Botanical Garden V. 85. Bronx: New York Botanical Garden Press.

Ball, J. W., and R. G. Kelsay. 1992. "Prehistoric Intrasettlement Land Use and Residual Soil Phosphate Levels in the Upper Belize Valley, Central America." In Gardens of Prehistory: The Archaeology of Settlement Agriculture in Greater Mesoamerica, edited by T. W. Killion, 234–62. Tuscaloosa: University of Alabama Press.

Baptista, S. R. 2008. "Metropolitanization and Forest Recovery in Southern Brazil: A Multiscale Analysis of the Florianópolis City-Region, Santa Catarina State, 1970 to 2005." Ecology and Society 13 (2):5.

Baptista, S. R., and T. K. Rudel. 2006. "A Re-Emerging Atlantic Forest? Urbanization, Industrialization and the Forest Transition in Santa Catarina, Southern Brazil." Environmental Conservation 33 (3): 195–202.

Barber, C. V., and K. Talbot. 2003. "The Chainsaw and the Gun: The Role of Military in Deforesting Indonesia." Journal of Sustainable Forestry 16 (3/4): 131–60.

Barbier, E. B. 2008. "In the Wake of Tsunami: Lessons Learned from the Household Decision to Replant Mangroves in Thailand." Resource and Energy Economics 30 (2): 229–49.

Barbier, E. B., J. C. Burgess, and A. Grainger. 2010. "The Forest Transition: Towards a More Comprehensive Theoretical Framework." Land Use Policy 27 (2): 98–107.

Barbieri, A. F., and D. L. Carr. 2005. "Gender-Specific out-Migration, Deforestation and Urbanization in the Ecuadorian Amazon." Global and Planetary Change 47 (2–4): 99–110.

Barker, G. 2005. "The Archaeology of Foraging and Farming at Niah Cave, Sarawak." Asian Perspectives 44 (1): 90–106.

Barker, G., H. Barton, I. Datan, B. Davenport, J. M. S. Jones, J. Langub, L. Lloyd-Smith, B. Nyíri, and B. Upex. 2008. "The Cultured Rainforest Project: The First (2007) Field Season." Sarawak Museum Journal 65 (86): 121–90.

Barkin, D. 2002. "The Reconstruction of a Modern Mexican Peasantry." Journal of Peasant Studies 30 (1): 73–90.

———. 2004. "Who Are the Peasants?" Latin American Research Review 39 (3): 270–81.

Barlow, J., C. A. Peres, L. M. P. Henriques, P. C. Stouffer, and J. M. Wunderle. 2006. "The Responses of Understorey Birds to Forest Fragmentation, Logging and Wildfires: An Amazonian Synthesis." Biological Conservation 128 (2): 182–92.

Barr, C., D. Brown, and A. Casson. 1999. Corporate Debt and the Indonesian Forestry Sector. Washington, DC: CIFOR.

Barros, A. C., and C. Uhl. 1995. "Logging along the Amazon River and Estuary: Patterns, Problems, and Potential." Forest Ecology and Management 77:87–105.

Barry, D., and I. Monterrosso. 2007. "Community-Based Forestry and the Changes in Tenure and Access Rights in the Mayan Biosphere Reserve, Guatemala." Paper presented at RECOFT, TC International Conference, Poverty Reduction and Forests: Tenure, Market, and Policy Reforms. Bangkok.

Barton, H. 2005. "The Case for Rainforest Foragers: The Starch Record at Niah Cave, Sarawak." Asian Perspectives 44 (1): 56–72.

Bassett, T. J., C. Blanc-Pamard, and J. Boutrais. 2007. "Constructing Locality: The Terroir Approach in West Africa." Africa 77: 104–29.

Bassett, T. J., and D. Crummey. 2003. African Savannas: Global Narratives and Local Knowledge of Environmental Change. Oxford, Portsmouth: James Currey; Heinemann.

Bauer, A. M. 2010. Socializing Environments and Ecologizing Politics: The Production of Nature and Social Differentiation in Iron Age Northern Karnataka. PhD diss., University of Chicago, Dept. of Anthropology.

Bauer, A. M., and Morrison, K. D. 2013. "Assessing Anthropogenic Soil Erosion with Multi-Spectral Satellite Imagery: An Archaeological Case Study of Long Term Land Use." In Koppal District, Karnataka, South Asian Archaeology 2007, edited by D. Frenez and M. Tosi, 67–76. Oxford: British Archaeological Reports.

Baum, G. A., and H. J. Weimer. 1992. Participation et Développement Socio-Économique Comme Conditions Préalables Indispensables D'une Implication Active Des Populations Riveraines Dans la Conservation de la Forêt Classée de Ziama [Participation and Socio-Economic Development as Essential Prerequisites Active Involvement in the Riparian Populations in Conservation Forest Reserve Ziama]. Conakry: République de Guinée; Deutsche Forst-Consult/Neu-Isenburg/RFA/KfW.

Bayly, C. A. 1990. Indian Society and the Making of the British Empire. New York: Cambridge University Press.

Beach, T., S. Luzzadder-Beach, N. Dunning, J. Jones, J. Lohse, T. Guderjan, S. Bozarth, S. Millspaugh, and T. Bhattacharya. 2009. "A Review of Human and Natural

Changes in Maya Lowland Wetlands over the Holocene." Quaternary Science Reviews 28:1710–24.

Bebbington, A. 1999. "Capitals and Capabilities: A Framework for Analyzing Peasant Viability, Rural Livelihoods and Poverty." World Development 27 (12): 2021–44.

Beck, H. 2005. "Seed Predation and Dispersal by Peccaries Throughout the Neotropics and Its Consequences: A Review and Synthesis." In Seed Fate, edited by P. M. Forget, J. E. Lambert, P. E. Hulme, and S. B. V. Wall, 77–115. Cambridge: CABI Publishing.

Becker, B. K. 2005. "Geopolítica da Amazônia [Amazon Geopolitics]." Estudos Avança-dos 19 (53): 71–86.

Beckerman, S. 1979. "The Abundance of Protein in Amazonia: A Reply to Gross." American Anthropologist 81 (3): 533–60.

Beinart, W. 2000. "African History and Environmental History." African Affairs 99 (395): 269–302.

Beinart, W., and L. Hughes. 2007. Environment and Empire. The Oxford History of the British Empire Companion Series. New York: Oxford University Press.

Bellemare, J., G. Motzkin, and D. R. Foster. 2002. "Legacies of the Agricultural Past in the Forested Present: An Assessment of Historical Land-Use Effects on Rich Mesic Forests." Journal of Biogeography 29 (10–11): 1401–20.

Bellingham, P. J., and A. D. Sparrow. 2000. "Resprouting as a Life History Strategy in Woody Plant Communities." Oikos 89 (2): 409–16.

Bellwood, P. S. 2005. First Farmers : The Origins of Agricultural Societies. Malden: Blackwell.

Benchaphun, E., K. T. Ngam, T. Promburom, S. Sinchaikul, and N. Thinrach. 2005. "Production Systems and Land Use Productivity of Farmers in Chiang Mai, Chiang Rai and Lamphun." Journal of Agricultural Economics 24 (2): 49–69.

Bengtsson, J., S. G. Nilsson, A. Franc, and P. Menozzi. 2000. "Biodiversity, Distur-bances, Ecosystem Function and Management of European Forests." Forest Ecol-ogy and Management 132 (1): 39–50.

Benya, E. S. J. 1979. "Forestry in Belize, Part II: Modern Times and Transition." Beliz-ean Studies 7:13–28.

Berry, S. 2001. Chiefs Know Their Boundaries: Essays on Property, Power, and the Past in Asante, 1896–1996. Portsmouth, Oxford: Heinemann; James Currey.

Bhargava, M. 1999. State, Society, and Ecology: Gorakhpur in Transition, 1750–1830. Delhi: Manohar.

Bhat, D. M., K. S. Murali, and N. H. Ravindranath. 2001. "Formation and Recovery of Secondary Forests in India: A Particular Reference to Western Ghats in South India." Journal of Tropical Forest Science 13 (4): 601–20.

Bhattarai, M., and M. Hammig. 2004. "Governance, Economic Policy, and the Envi-ronmental Kuznets Curve for Natural Tropical Forests." Environment and Develop-ment Economics 9:367–82.

Bindra, P. 2008. Tigers Forever. Ranthambhore Foundation. Available from http://www.sanctuaryasia.com/kidsfortigers/index.php.

Blackett, T. 2003. "What Is a Brand?" In Brands and Branding, edited by R. Clifton, J. Simmons and S. Ahmad, 13–25. London: Profile.

Blumler, M. A. 1998. "Biogeography of Land-Use Impacts in the Near East." In Nature's Geography: New Lessons for Conservation in Developing Countries, edited by K. S. Zimmerer and K. Young, 215–36. Madison: University of Wisconsin Press.

Boerner, J., A. Mendoza, and S. A. Vosti. 2007. "Ecosystem Services, Agriculture, and Rural Poverty in the Eastern Brazilian Amazon: Interrelationships and Policy Prescriptions." Ecological Economics 64 (2): 356–73.

Boffa, J. 1999. Agroforestry Parklands in Sub-Saharan Africa. FAO Conservation Guide, No. 34. Rome: Food and Agriculture Organization of the UN.

Boffa, J. M. 1991. "Land and Tree Management and Tenure Policies in Burkina Faso's Agroforestry System." MS thesis, Purdue University.

———. 1995. "Productivity and Management of Agroforestry Parklands in the Sudan Zone of Burkina Faso, West Africa." PhD diss., Purdue University.

Boffa, J. M., D. M. Knudson, G. Yameogo, and P. Nikiema. 1996. "Shea Nut (*Vitellaria paradoxa*) Production and Collection in Agroforestry Parklands of Burkina Faso." In Domestication and Commercialization of Non-Timber Forest Products in Agroforestry Systems, Non-Wood Forest Products, edited by R. R. B Leakey, A. B. Temu, M. Melnyk, and P. Vantomme, 9:110–22. Rome: Food and Agriculture Organization of the UN.

Bolland, O. N. 1977. The Formation of a Colonial Society: Belize, from Conquest to Crown Colony. Baltimore: Johns Hopkins University Press.

Bommel, P., T. Bonaudo, T. Barbosa, J. B. da Veiga, M. V. Pak, and J. F. Tourrand. 2010. "The Complex Relationship between Cattle Ranching and the Forest in Brazilian Amazonia. A Multiagent Modelling Approach." Cahiers Agricultures 19 (2): 104–11.

Boni, S. 2005. Clearing the Ghanaian Forest : Theories and Practices of Acquisition, Transfer and Utilisation of Farming Titles in the Sefwi-Akan Area. Legon: Institute of African Studies.

———. 2006. "Indigenous Blood and Foreign Labor: The Ancestralization of Land Rights in Sefwi (Ghana)." In Land and the Politics of Belonging in West Africa, edited by R. Kuba and C. Lentz, 161–86. Leiden: E. J. Brill.

Booth, R. 2008. "'Black Gold' Coffee Leaves a Bitter Taste for Some." The Guardian, March 22.

Botkin, D. B. 1990. Discordant Harmonies: A New Ecology for the Twenty-First Century. New York: Oxford University Press.

Botshon, A. 2007. Saving Sterling Forest : The Epic Struggle to Preserve New York's Highlands. Albany: State University of New York Press.

Bouahom, B., L. Douangsavanh, and J. Rigg. 2004. "Building Sustainable Livelihoods in Laos: Untangling Farm from Non-Farm, Progress from Distress." Geoforum 35 (5): 607–19.

Bourdieu, P. 1984. Distinction : A Social Critique of the Judgement of Taste. Cambridge, MA: Harvard University Press.

———. 1994. Distinction. New York: Routledge and Kegan Paul.

Boutcher, W. 1775. A Treatise on Forest-Trees; Containing Not Only the Best Methods of Their Culture Hitherto Practised, But a Variety of New and Useful Discoveries, the Result of Many Repeated Experiments: As Also, Plain Directions for Removing Most of the Valuable Kinds of Forest Trees. Edinburgh: Printed by R. Fleming.

Bowie, K. A. 1992. "Unraveling the Myth of the Subsistence Economy: Textile Production in Nineteenth-Century Northern Thailand." The Journal of Asian Studies 51 (4): 797–823.

Bowker, G. C. 2000. "Biodiversity Datadiversity." Social Studies of Science 30 (5): 643–83.

Brandis, D. 1888. "Teak." In Encyclopaedia Brittanica, 9th ed., 103–05. Edinburgh: Adam and Charles Black.

Brawn, J. D., and D. F. Stotz. 2001. "The Importance of the Chicago Region and the Chicago Wilderness Initiative to Avian Conservation." In Urbanization and Ornithology, edited by J. Marzluff and R. Donnelly, 509–22. Washington, DC: Island Press.

Bray, D. B., L. Merino-Perez, P. Negreros-Castillo, G. Segura-Warnholtz, J. M. Torres-Rojo, and H. F. M. Vester. 2003. "Mexico's Community-Managed Forests as a Global Model for Sustainable Landscapes." Conservation Biology 17 (3): 672–77.

Breman, H., and J. J. Kessler. 1995. Woody Plants in Agro-Ecosystems of Semi-Arid Regions: With an Emphasis on the Sahelian Countries, vol. 23. New York: Springer-Verlag.

British Broadcasting Corporation. 1992. "The Secret of El Dorado." December 19, 2002. Available online at http://www.bbc.co.uk/science/horizon/2002/eldorado.shtml.

Brncic, T. M., K. J. Willis, D. J. Harris, and R. Washington. 2007. "Culture or Climate? The Relative Influences of Past Processes on the Composition of the Lowland Congo Rainforest." Philosophical Transactions of the Royal Society B: Biological Sciences 362 (1478): 229–42.

Broadbent, E. N., G. P. Asner, M. Keller, D. E. Knapp, P. J. C. Oliveira, and J. N. Silva. 2008. "Forest Fragmentation and Edge Effects from Deforestation and Selective Logging in the Brazilian Amazon." Biological Conservation 141 (7): 1745–57.

Brockington, D. 2002. Fortress Conservation: The Preservation of the Mkomazi Game Reserve, Tanzania. Bloomington: Indiana University Press.

———. 2003. "Injustice and Conservation: Is 'Local Support' Necessary for Sustainable Protected Areas?" Policy Matters 12:22–30.

Brockington, D., R. Duffy, and J. Igoe. 2008. Nature Unbound: Conservation, Capitalism and the Future of Protected Areas. Sterling, VA: Earthscan.

Brokaw, N. V. L., and E. P. Mallory. 1990. Trees of Rio Bravo: A Guide to Trees of the Rio Bravo Conservation and Management Area, Belize. Manomet: Manomet Bird Observatory.

Bromley, D. W. 1992. "The Commons, Property, and Common-Property Regimes." In Making the Commons Work: Theory, Practice, and Policy, edited by D. W. Bromley, 3–16. San Francisco: ICS Press.

Brondizio, E. S. 2004. "From Staple to Fashion Food: Shifting Cycles, Shifting Opportunities in the Development of the Açaí Fruit (*Euterpe oleracea* Mart.) Economy in the Amazon Estuary." In Working Forests in the American Tropics: Conservation through Sustainable Management?, edited by D. Zarin et al., 348–61. New York: Columbia University Press.

———. 2008. The Amazonian Caboclo and the Açaí Palm: Forest Farmers in the Global Market. New York: New York Botanical Garden Press.

———. 2011. "Forest Resources, Family Networks and the Municipal Disconnect: Examining Recurrent Underdevelopment in the Amazon Estuary." In The Amazonian Várzea: The Decade Past and the Decade Ahead, edited by M. Pinedo-Vasquez, M. Ruffino, C. Padoch, and E. S. Brondizio, 297–32. Dordrecht, The Netherlands: Springer Publishers copublication with the New York Botanical Garden Press.

Brondízio, E.S., and A.D. Siqueira. 1997. "From Extractivists to Forest Farmers: Changing Concepts of Agricultural Intensification and Peasantry in the Amazon Estuary." Research in Economic Anthropology 18:233–79.

Brookfield, H. C. 2001. Exploring Agrodiversity. New York: Columbia University Press.

Brosius, J. 1991. "Foraging in Tropical Rain Forests: The Case of the Penan of Sarawak, East Malaysia (Borneo)." Human Ecology 19 (2): 123–50.

———. 1997. "Endangered Forest, Endangered People: Environmentalist Representations of Indigenous Knowledge." Human Ecology 25 (1): 47–69.

Browder, J. D., and B. J. Godfrey. 1997. Rainforest Cities: Urbanization, Development and Globalization of the Brazilian Amazon. New York: Columbia University Press.

Brown, D. W. 1999. Addicted to Rent: Corporate and Spatial Distribution of Forest Resources in Indonesia: Implications for Forest Sustainability and Government Policy. Jakarta: Indonesia–UK Tropical Forest Management Programme, Provincial Forest Management Programme.

Brown, K. 2003. "'Trees, Forests and Communities': Some Historiographical Approaches to Environmental History on Africa." Area 35 (4): 343–56.

Browne, F. G. 1952. "Kerangas Lands of Sarawak." Malay Forester 15:61–73.

Bruner, A. G., R. E. Gullison, R. E. Rice, and G. A. B. Da Fonseca. 2001. "Effectiveness of Parks in Protecting Tropical Biodiversity." Science 291 (5501): 125.

Bryan, R. 1994. Soil Erosion, Land Degradation and Social Transition: Geoecological Analysis of a Semi-Arid Tropical Region, Kenya. Advances in Geoecology, 27. Cremlingen, Germany: Catena.

Bryant, R. 1996. "Asserting Sovereignty through Natural Resource Use: Karen Forest Management on the Thai-Burmese Border." In Resources, Nations and Indigenous Peoples: Case Studies from Australasia, Melanesia and Southeast Asia, edited by R. Howitt, 32–41. Melbourne: Oxford University Press.

———. 2006. "Burma and the Politics of Teak: Dissecting a Resource Curse." In A History of Natural Resources in Asia : The Wealth of Nature, edited by P. Boomgard and G. Bankoff, 143–62. London: Palgrave Macmillan.

Bryant, R. L. 1997. The Political Ecology of Forestry in Burma, 1824–1994. London: Hurst.

Bryceson, D. 1996. "Deagrarianization and Rural Employment in Sub-Saharan Africa: A Sectoral Perspective." World Development 24 (1): 97–111.

———. 1999. "African Rural Labour, Income Diversification and Livelihood Approaches: A Long-Term Development Perspective." Review of African Political Economy 26 (80): 171–89.

Bryceson, D. F., J. E. Mooij, and C. Kay. 2000. Disappearing Peasantries?: Rural Labour in Africa, Asia and Latin America. Exeter: Intermediate Technology.

Buckingham, S., and R. Kulcur. 2009. "Gendered Geographies of Environmental Injustice." Antipode 41 (4): 659–83.

Buhaug, H., and P. Lujala. 2005. "Accounting for Scale: Measuring Geography in Quantitative Studies of Civil War." Political Geography 24 (4): 399–418.

Burkill, H. M. 1984. The Useful Plants of West Tropical Africa. Kew: Royal Botanical Gardens. 1:61 and 2:338.

Burkill, H. M., J. M. Dalziel, and J. Hutchinson. 1985. The Useful Plants of West Tropical Africa, 5 vols. Kew: Royal Botanic Gardens.

Burma Forest Department. 1899. Progress Report of Forest Administration in British Burma for 1898–9. Rangoon: Government Printing.

———. 1947. "Empire Forests and the War: Burma." In Fifth British Empire Forestry Conference Statements, 1–26. London: n.p.

Burman, R. J. J. 1992. "The Institution of Sacred Groves." Journal of the Indian Anthropological Society 27:219–38.

Bush, R. 2010. "Food Riots: Poverty, Power and Protest." Journal of Agrarian Change 10 (1): 119–29.

Butler, R. 2007. "Borneo's Rainforest Protected." Retrieved from http://news.mongabay.com/2007/0212-borneo.html.

Byron, N. 2001. "Keys to Smallholder Forestry." Forests, Trees and Livelihoods 11 (4): 279–94.

Caldas, M., R. Walker, E. Arima, S. Perz, S. Aldrich, and C. Simmons. 2007. "Theorizing Land Cover and Land Use Change: The Peasant Economy of Amazonian Deforestation." Annals of the Association of American Geographers 97 (1): 86–110.

Cameron, E., and J. Macpherson. 1776. The Fingal of Ossian, an Ancient Epic Poem in Six Books. Translated from the Original Galic Language by Mr. James Macpherson; and New Rendered into Heroic Verse by Ewen Cameron. Warrington: William Eyres.

Campaign for Survival and Dignity. 2007. "Second Open Letter to the Bombay Natural History Society, the Wildlife Protection Society of India and the Conservation Action Trust and Wildlife First." Available from http://forestrightsact.awardspace.com/. Direct URL: http://forestrightsact.awardspace.com/updates/bnhs_open_letter_2.rtf.

Campbell, D. G., A. Ford, K. S. Lowell, J. Walker, Jeffery K. Lake, C. Ocampo-Raeder, A. Townsmith, and M. Balick. 2006. "The Feral Forests of the Eastern Petén." In Time and Complexity in Historical Ecology, edited by W. Balee and C. L. Erickson, 21–56. New York: Columbia University Press.

Caner, L., D. L. Seen, Y. Gunnell, B. R. Ramesh, and G. Bourgeon. 2007. "Spatial Heterogeneity of Land Cover Response to Climatic Change in the Nilgiri Highlands (Southern India) since the Last Glacial Maximum." The Holocene 17 (2): 195–205.

Caner, L., F. Toutain, G. Bourgeon, and A.-J. Herbillon. 2003. "Occurrence of
Sombric-Like Subsurface A Horizons in Some Andic Soils of the Nilgiri Hills
(Southern India) and Their Palaeoecological Significance." Geoderma 117 (3–4):
251–65.

Capers, R. S., R. L. Chazdon, A. R. Brenes, and B. V. Alvarado. 2005. "Successional
Dynamics of Woody Seedling Communities in Wet Tropical Secondary Forests."
Journal of Ecology 93 (6): 1071–84.

Cardillo, M., G. M. Mace, J. L. Gittleman, K. E. Jones, J. Bielby, and A. Purvis. 2008.
"The Predictability of Extinction: Biological and External Correlates of Decline in
Mammals." Proceedings of the Royal Society B: Biological Sciences 275 (1641): 1441.

Caribbean Central American Research Council (CCARC). 2007. "Finzmos: Indigenous
Lands Demarcation of the Miskitu Territory in Honduras." Paper presented at
Seminar, University of Texas, Austin.

Carneiro, R. L. 1957. Subsistence and Social Structure; an Ecological Study of the
Kuikuru Indians. Ann Arbor: University of Michigan Press.

———. 1983. "The Cultivation of Manioc among the Kuikuru of the Upper Xingú." In
Adaptive Responses of Native Amazonians, edited by R. Hames and W. Vickers,
65–111. New York: Academic Press.

Carneiro, R. L., and A. F. C. Wallace. 1960. "Slash-and-Burn Agriculture: A Closer
Look at Its Implications for Settlement Patterns." In Men and Cultures: Selected
Papers, edited by A. Wallace, 229. Philadelphia: University of Pennsylvania Press.

Carney, J. A. 1988. "Struggles over Crop Rights and Labour within Contract Farming
Households in a Gambian Irrigated Rice Project." Journal of Peasant Studies 15 (3):
334–49.

Carr, D. 2009. "Population and Deforestation: Why Rural Migration Matters." Progress in Human Geography 33 (3): 355–78.

Carson, R. 1962. Silent Spirng. New York: Houghton Mifflin.

Casid, J. H. 2005. Sowing Empire: Landscape and Colonization. Minneapolis: University of Minnesota Press.

Castree, N., and B. Braun. 1998. "The Construction of Nature and the Nature of
Construction: Analytical and Political Tools for Building Survivable Futures." In
Remaking Reality: Nature at the Millennium, edited by N. Castree and B. Braun,
3–42. London: Routledge.

Caton, H. 1985. "The Preindustrial Economics of Adam Smith." The Journal of Economic History 45 (04): 833–53.

Caviglia-Harris, J. L., D. Chambers, and J. R. Kahn. 2009. "Taking The 'U' out of
Kuznets: A Comprehensive Analysis of the Ekc and Environmental Degradation."
Ecological Economics 68 (4): 1149–59.

CEAM (El Centro de Estudios Amazónicos). 2003. Moxos: Una Limnocultura [Moxos:
A Limnoculture]. Barcelona: Centre d'Estudis Amazònics.

Cederlöf, G., and K. Sivaramakrishnan. 2006. Ecological Nationalisms: Nature, Livelihoods, and Identities in South Asia. Seattle: University of Washington Press.

Chagnon, N. A., and R. B. Hames. 1979. "Protein Deficiency and Tribal Warfare in
Amazonia: New Data." Science 203 (4383): 910–13.

Chakrabarti, R. 2006. "The Sundarbans, Badamiyan, History, Nature and Landscape." Paper presented at Nature in the Raj, August 18–19, Nehru Memorial Museum and Library, New Delhi.

Chalfin, B. 2000. "Risky Business: Economic Uncertainty, Market Reforms and Female Livelihoods in Northeast Ghana." Development and Change 31 (5): 987–1008.

———. 2001. "Border Zone Trade and the Economic Boundaries of the State in North-East Ghana." Africa 71 (2): 202–24.

———. 2004. Shea Butter Republic: State Power, Global Markets, and the Making of an Indigenous Commodity. New York: Routledge.

Chambers, R., N. C. Saxena, and T. Shah. 1989. To the Hands of the Poor: Water and Trees. London: Intermed. Tech. Publishers.

Chape, S., J. Harrison, M. Spalding, and I. Lysenko. 2005. "Measuring the Extent and Effectiveness of Protected Areas as an Indicator for Meeting Global Biodiversity Targets." Philosophical Transactions of the Royal Society B: Biological Sciences 360 (1454): 443–55.

Chapin, M. 2004. "A Challenge to Conservationists." World Watch 17 (6): 17–31.

———. 2006. "Mapping Indigenous Lands: Issues and Considerations." Working Paper, Centre for the Support of Native Lands.

Chapin, M., Z. Lamb, and B. Threlkeld. 2005. "Mapping Indigenous Lands." Annual Review of Anthropology 34:619–38.

Chaplin, J. E. 2003. "Nature and Nation—Natural History in Context." Transactions of The American Philosophical Society 93:75–96.

Charras, M. 1982. De la Forêt Maléfique à L'herbe Divine : la Transmigration en Indonésie, les Balinais à Sulawesi [In the Evil Forest, Grass Divine: Transmigration in Indonesia, Balinese Sulawesi], vol. 5. Paris, Ann Arbor: Editions de la Maison des sciences de l'homme ; University Microfilms International.

Chatterjee, P. 2008. "Democracy and Economic Transformation in India." Economic and Political Weekly 19:53–62.

Chaturvedi, A. N. 1992. "Management of Secondary Forests." Wasteland News 7 (2): 40–44.

Chazdon, R. L. 2003. "Tropical Forest Recovery: Legacies of Human Impact and Natural Disturbances." Perspectives in Plant Ecology, Evolution and Systematics 6 (1–2): 51–71.

———. 2008a. "Beyond Deforestation: Restoring Forests and Ecosystem Services on Degraded Lands." Science 320 (5882): 1458–60.

———. 2008b. "Chance and Determinism in Tropical Forest Succession." In Tropical Forest Community Ecology, edited by W. Carson and S. A. Schnitzer, 384–408. New York: Blackwell Publishing.

Chazdon, R. L., S. Careaga, C. Webb, and O. Vargas. 2003. "Community and Phylogenetic Structure of Reproductive Traits of Woody Species in Wet Tropical Forests." Ecological Monographs 73 (3): 331–48.

Chazdon, R. L., S. G. Letcher, M. Van Breugel, M. Martínez-Ramos, F. Bongers, and B. Finegan. 2007. "Rates of Change in Tree Communities of Secondary Neotropical Forests Following Major Disturbances." Philosophical Transactions of the Royal Society B: Biological Sciences 362 (1478): 273–89.

Chazdon, R. L., A. Redondo Brenes, and B. Vilchez Alvarado. 2005. "Effects of Climate and Stand Age on Annual Tree Dynamics in Tropical Second-Growth Rain Forests." Ecology 86 (7): 1808–815.

Chernela, J. M. 1993. The Wanano Indians of the Brazilian Amazon: A Sense of Space, 1st ed. Austin: University of Texas Press.

Chevalier, A. 1909. Rapport Sur les Nouvelles Recherches Sur les Plantes a Caoutchouc de la Guinée Française [New Research Report on Rubber Plants of French Guinea]. Dakar: Archives du Senegal.

———. 1933. "Les Bois Sacrés Des Noirs de L'afrique Tropicale Comme Sanctuaries de la Nature [The Sacred Wood the Blacks of Africa as Tropical Nature Sanctuaries]." Revue de la Société de Biogéographie:37–42.

———. 1948. "Nouvelles Recherches Sur L'arbre a Beurre Du Soudan, Butyrospermum Parkii [New Research on the Shea Butter Tree from Sudan, Butyrospermum Parkii]." Revue de Botanique Appliquée 28:241–56.

Chhatre, A., and A. Agrawal. 2009. "Trade-Offs and Synergies between Carbon Storage and Livelihood Benefits from Forest Commons." Proceedings of the National Academy of Sciences of the United States of America 106 (42): 17667–70.

Chhatre, A., and V. K. Saberwal. 2006. Democratizing Nature: Politics, Conservation, and Development in India. New Delhi: Oxford University Press.

Chicago Historical Society. 2005. Planning, City and Regional. The Electronic Encyclopedia of Chicago. Retrieved from http://www.encyclopedia.chicagohistory.org/pages/973.html

Chicago Metropolis 2020. 2001. "2001 Metropolis Index: Measuring Progress toward Shared Regional Goals." Retrieved from http://www.chicagometropolis2020.org/indicators/cm-2020/about-report.htm.

Chicago Metropolitan Area for Planning. 2010. Go to 2040. Chicago: Author.

Chicago Wilderness. 1997. An Atlas of Biodiversity. Chicago: Chicago Region Biodiversity Council.

———. 1999. Biodiversity Recovery Plan. Chicago: Chicago Region Biodiversity Council.

———. 2006. The State of Our Chicago Wilderness: A Report Card on the Health of the Region's Ecosystems. Chicago: Chicago Region Biodiversity Council.

Choudhary, K. 2000. "Development Dilemma: Resettlement of Gir Maldharis." Economic and Political Weekly 35 (30): 2662–68.

Choudhury, S. R. 1970. "Let Us Count Our Tigers." Cheetal 14 (2): 41–51.

Clark, D. A. 2007. "Detecting Tropical Forests' Responses to Global Climatic and Atmospheric Change: Current Challenges and a Way Forward." Biotropica 39 (1): 4–19.

Clark, G. 1999. "Negotiating Asante Family Survival in Kumasi, Ghana." Africa 69 (1): 66–86.

Cleaver, K. 1992. "Deforestation in the Western and Central African Forest: The Agricultural and Demographic Causes, and Some Solutions." In Conservation of West and Central Africa's Rainforests, Environment Paper No. 1, edited by K. Cleaver, M. Munashighe, M. Dyson, N. Egli, A. Peuker, and F. Wencélius. 65–78. Washington, DC: World Bank.

Clement, C. R. 1988. "Domestication of the Pejibaye Palm (Bactris Gasipaes): Past and Present." Advances in Economic Botany 6: 155–74.

———. 1999. "1492 and the Loss of Amazonian Crop Genetic Resources. I. The Relation between Domestication and Human Population Decline." Economic Botany 53 (2): 188–202.

———. 2006. "Fruit Trees and the Transition to Food Production in Amazonia." In Time and Complexity in Historical Ecology, edited by W. Baleé and C. L. Erickson, 165–85. New York: Columbia University Press.

Clements, F. E. 1916. Plant Succession : An Analysis of the Development of Vegetation. Publication/Carnegie Institution of Washington vol. no 242. Washington, DC: Carnegie Institution of Washington.

Cochrane, M. A., and W. F. Laurance. 2008. "Synergisms among Fire, Land Use, and Climate Change in the Amazon." Ambio 37 (7–8): 522–27.

Cohen, D. 2006. Household Gods: The British and Their Possessions. New Haven, CT: Yale University Press.

Colchester, M. 2008. Beyond Tenure: Rights-Based Approaches to Peoples and Forest Areas: Some Lessons from the Forest Peoples Programme. Moreton-in-Marsh: FPP and RRI.

Cole, M. A. 2003. "Development, Trade, and the Environment: How Robust Is the Environmental Kuznets Curve?" Environment and Development Economics 8:557–80.

Collier, P., and A. Hoeffler. 2004. "Greed and Grievance in Civil War." Oxford Economic Papers 56 (4): 563.

Collins, N. M., Sayer, J. A., and Whitmore, T. C., eds. 1991. The Conservation Atlas of Tropical Forests: Asia and the Pacific. London and Basingstoke: Macmillan.

Collins, S. L. 1992. "Fire Frequency and Community Heterogeneity in Tallgrass Prairie Vegetation." Ecology 73 (6): 2001–006.

Colson, F., J. Bogaert, A. Carneiro, B. Nelson, E. R. Pinage, and R. Ceulemans. 2009. "The Influence of Forest Definition on Landscape Fragmentation Assessment in Rondonia, Brazil." Ecological Indicators 9 (6): 1163–68.

Comaroff, J., and J. L. Comaroff. 2000. "Millennial Capitalism: First Thoughts on a Second Coming." Public Culture 12 (2): 291–343.

———. 2008. Ethnicity Inc. Chicago: University of Chicago Press.

Comber, A., P. Fisher, and R. Wadsworth. 2004. "Integrating Land-Cover Data with Different Ontologies: Identifying Change from Inconsistency." International Journal of Geographical Information Science 18 (7): 691–708.

———. 2005a. "Comparing the Consistency of Expert Land Cover Knowledge." International Journal of Applied Earth Observation and Geoinformation 7 (3): 189–201.

———. 2005b. What Is Land Cover? Environment and Planning B—Planning and Design 32 (2): 199–209.

Commercial Club of Chicago. (1909) 1970. Plan of Chicago by D. H. Burnham and E. H. Bennett, edited by C. Moore. New York: Da Capo Press.

Conklin, B. A., and L. R. Graham. 1995. "The Shifting Middle Ground: Amazonian Indians and Eco-Politics." American Anthropologist 97 (4): 695–710.

Connell, J. H. 1978. "Diversity in Tropical Rain Forests and Coral Reefs." Science 199 (4335): 1302.

Connell, J. H., and R. O. Slatyer. 1977. "Mechanisms of Succession in Natural Communities and Their Role in Community Stability and Organization." American Naturalist 111 (982): 1119–44.

Coomes, O. T. 1992. "Making a Living in the Amazon Rain Forest: Peasants, Land, and Economy in the Tahuayo River Basin of Northeastern Peru." PhD diss., Department of Geography, University of Wisconsin, Madison.

———. 1996. "State Credit Programs and the Peasantry under Populist Regimes: Lessons from the Apra Experience in the Peruvian Amazon." World Development 24 (8): 1333–46.

Coppel, C. A. 1983. Indonesian Chinese in Crisis, vol. 8. Oxford: Oxford University Press.

Cormier-Salem, M. C., and T. J. Bassett. 2007. "Introduction: Nature as Local Heritage in Africa: Longstanding Concerns, New Challenges." Africa: The Journal of the International African Institute 77 (1): 1–17.

Cosgrove, D. E., and S. Daniels. 1988. The Iconography of Landscape: Essays on the Symbolic Representation, Design, and Use of Past Environments, vol. 9. New York: Cambridge University Press.

Costa, S. M., and E. S. Brondizio. 2011. "Floodplain Cities of the Brazilian Amazon: Characterization and Tendencies along the Solimões, Amazonas, and Estuarine Regions." In The Amazonian Várzea: The Decade Past and the Decade Ahead, M. Pinedo-Vasquez, M., M. Ruffino, C. Padoch,. E. S. Brondizio, 297–332. Dordrecht, The Netherlands: Springer Publishers, copublication with the New York Botanical Garden Press.

Coulibaly-Lingani, P., M. Tigabu, P. Savadogo, P. C. Oden, and J. M. Ouadba. 2009. "Determinants of Access to Forest Products in Southern Burkina Faso." Forest Policy and Economics 11 (7): 516–24.

Cousins, B. 2000. "Tenure and Common Property Resources in Africa." In Evolving Land Rights, Policy and Tenure in Africa, edited by C. Toulmin and J. Quan, 151–79. London: DFID / IIED / NRI.

———. 2007. "More Than Socially Embedded: The Distinctive Character of 'Communal Tenure' Regimes in South Africa and Its Implications for Land Policy." Journal of Agrarian Change 7 (3): 281–315.

Cousins, N. 1945. Modern Man Is Obsolete. New York: Viking Press.

Cowles, H. C. 1901. "The Physiographic Ecology of Chicago and Vicinity; a Study of the Origin, Development, and Classification of Plant Societies (Concluded)." Botanical Gazette 31 (3): 145–82.

———. (1912) 2007. "Conservation of Our Forests." In Henry Chandler Cowles: Pioneer Ecologist, edited by V. M. Cassidy, 257–61. Chicago: Kedzie Sigel Press.

Crawford, S., and E. Ostrom. 1995. "A Grammar of Institutions." American Political Science Review 89 (3): 582–600.

Cronon, W. 1991. Nature's Metropolis : Chicago and the Great West, 1st ed. New York: W.W. Norton.

————. 1995. "The Trouble with Wilderness, or Getting Back to the Wrong Nature." In Uncommon Ground: Toward Reinventing Nature, edited by W. Cronon, 69–90. New York: W.W. Norton & Co.

————. 1995. Uncommon Ground: Toward Reinventing Nature, 1st ed. New York: W.W. Norton & Co.

Crook, R. 1986. "Decolonization, the Colonial State, and Chieftaincy in the Gold Coast." African Affairs 85 (338): 75.

Crook, R., S. Affou, D. N. A. Hammond, A. F. Vanga, and M. O. Yeboah. 2005. The Law, Legal Institutions and the Protection of Land Rights in Ghana and Cote d'Ivoire. Developing a More Effective and Equitable System. Final Report SSRU Project R. 7993.

Crumley, C. L. 1994. Historical Ecology: Cultural Knowledge and Changing Landscapes, 1st ed. School of American Research Advanced Seminar Series. Seattle: School of American Research Press, distributed by the University of Washington Press.

Curran, L. M., S. N. Trigg, A. K. Mcdonald, D. Astiani, Y. M. Hardiono, P. Siregar, I. Caniago, and E. Kasischke. 2004. "Lowland Forest Loss in Protected Areas of Indonesian Borneo." Science 303 (5660): 1000–003.

Daly, H. E. 1996. Beyond Growth: The Economics of Sustainable Development. Boston: Beacon.

Dalziel, J. M. 1937. The Useful Crops of West Tropical Africa. London: Crown Agents.

Dávalos, L. M. 2001. "The San Lucas Mountain Range in Colombia: How Much Conservation Is Owed to the Violence?" Biodiversity and Conservation 10:69–78.

Davidson, J. 2002. Violence and Politics in West Kalimantan, Indonesia. PhD diss., University of Washington. Seattle.

Davidson, J. S., and D. Kammen. 2002. "Indonesia's Unknown War and the Lineages of Violence in West Kalimantan." Indonesia 73:53–87.

Davis, D. K. 2005. "Potential Forests: Degradation Narratives, Science, and Environmental Policy in Protectorate Morocco, 1912–1956." Environmental History 10 (2): 211–38.

De Janvry, A., and E. Sadoulet. 2001. "Income Strategies among Rural Households in Mexico: The Role of Off-Farm Activities." World Development 29 (3): 467–80.

De Jong, W. 2001. "Tree and Forest Management in the Floodplains of the Peruvian Amazon." Forest Ecology and Management 150:125–34.

————. 2010. "Forest Rehabilitation and Its Implication for Forest Transition Theory." Biotropica 42:3–9.

De Koninck, R., and S. Dery. 1997. "Agricultural Expansion as a Tool of Population Redistribution in Southeast Asia." Journal of Southeast Asian Studies 28 (1): 1–26.

De Sherbinin, A., L. K. Vanwey, K. Mcsweeney, R. Aggarwal, A. Barbieri, S. Henry, L. M. Hunter, W. Twine, and R. Walker. 2008. "Rural Household Demographics, Livelihoods and the Environment." Global Environmental Change—Human Policy Dimensions 18 (1): 38–53.

De Witte, M. 2001. Long Live the Dead!: Changing Funeral Celebrations in Asante, Ghana. Amsterdam: Aksant Academic Publishers.

Del Castillo, I. Y. 2003. "Debates about Lo Andino in Twentieth Century Peru." In Imaging the Andes: Shifting Margins of a Marginal World, edited by T. Salman and A. Zoomers, 40–63. Amsterdam, The Netherlands: Askant Academic Publishers. CEDLA Latin American Studies, no. 91.

Dean, M. 1999. Governmentality: Power and Rule in Modern Society. London: Sage Publications.

DeBoer, W. R., K. Kintigh, and A. G. Rostoker. 1996. "Ceramic Seriation and Site Reoccupation in Lowland South America." Latin American Antiquity 7 (3): 263–78.

Deere, C. D., and M. Leon. 2001. "Institutional Reform of Agriculture under Neoliberalism: The Impact of the Women's and Indigenous Movements." Latin American Research Review 36 (2): 31–64.

DeFries, R., T. Rudel, M. Uriarte, and M. Hansen. 2010. "Deforestation Driven by Urban Population Growth and Agricultural Trade in the Twenty-First Century." Nature Geoscience 3 (3): 178–81.

Deloche, J. 1993. Trade and Transportation Routes in India before Steam Navigation, Volume I: Land Routes. Delhi: Oxford University Press.

Denevan, W., and C. Padoch, eds. 1987. Swidden-Fallow Agroforestry in the Peruvian Amazon. Advances in Economic Botany vol. 5. New York: New York Botanical Garden.

Denevan, W. M. 1966. The Aboriginal Cultural Geography of the Llanos de Mojos of Bolivia, vol. 48. Berkeley: University of California Press.

———. 1984. "Ecological Heterogeneity and Horizontal Zonation of Agriculture in the Amazonian Floodplain." In Frontiers Expansion in Amazonia, edited by M. Schmink and C. H. Wood, 311–36. Gainesville: University of Florida Press.

———. 1991. "Prehistoric Roads and Causeways of Lowland Tropical America." In Ancient Road Networks and Settlement Hierarchies in the New World, edited by C. Trombold, 230–42. Cambridge: Cambridge University Press.

———. 1992a. "The Pristine Myth: The Landscape of the Americas in 1492." Annals of the Association of American Geographers 82 (3): 369–85.

———. 1992b. "Stone vs. Metal Axes: The Ambiguity of Shifting Cultivation in Prehistoric Amazonia." Journal of the Steward Anthropological Society 20 (1–2): 153–65.

———. 1996. "A Bluff Model of Riverine Settlement in Prehistoric Amazonia." Annals of the Association of American Geographers 86 (4): 654–81.

———. 2000. Cultivated Landscapes of Native Amazonia and the Andes. Oxford Geographical and Environmental Studies. New York: Oxford University Press.

Deng, X. 1993. Zhongguo Zhiqing Meng [The Dream of China's Educated Youth]. Beijing: People's Literature Publishing House.

Dennis, P., and J. Grey. 1996. Emergency and Confrontation: Australian Military Operations in Malaya & Borneo 1950–1966. The Official History of Australia's Involvement in Southeast Asian Conflicts, 1948–1975, vol. 5. St. Leonards, N. S. W., Australia: Allen & Unwin, in association with the Australian War Memorial.

Deshpande, G. P. 2002. Selected Writings of Jyotiba Phule. New Delhi: Leftword Books.

Desmarais, A. A. 2007. La Vía Campesina: Globalization and the Power of Peasants. Ann Arbor: Pluto Press.

Dharamkumarsinh, K. S. 1959. A Field Guide to the Big Game Censuses in India. Indian Board for Wildlife Pamphlet no. 2. Delhi: Indian Board for Wildlife.

———. 1978. "The Changing Wildlife of Kathiawar." Journal of the Bombay Natural History Society 75 (3): 632–50.

Di Gessa, S., P. Poole, and T. Bending. 2008. "Participatory Mapping as a Tool for Empowerment: Experiences and Lessons Learned from the Ilc Network." Rome: Knowledge for Change Series.

Diamond, J. 1988. "Express Train to Polynesia." Nature 336:307–08.

Dietz, S., and W. N. Adger. 2003. "Economic Growth, Biodiversity Loss and Conservation Effort." Journal of Environmental Management 68 (1): 23–35.

Digby, S. 1971. War-Horse and Elephant in the Dehli Sultanate: A Study of Military Supplies. Oxford: Orient Monographs.

Dirzo, R., and P. H. Raven. 2003. "Global State of Biodiversity and Loss." Annual Review of Environment and Resources 28:137–67.

Divyabhanusinh. 2005. The Story of Asia's Last Lions. Mumbai: MARG.

———. 2006. "Junagadh State and Its Lions: Conservation in Princely India, 1879–1947." Conservation and Society 4 (4): 522–40.

Dobson, N. 1973. A History of Belize. London: Longman Caribbean.

Doolittle, A. A. 2007. "Native Land Tenure, Conservation, and Development in a Pseudo-Democracy: Sabah, Malaysia." Journal of Peasant Studies 34 (3): 474–97.

Dove, M. R. 1985. "The Agroecological Mythology of the Javanese and the Political Economy of Indonesia." Indonesia 39:1–36.

———. 1992. "Foresters' Beliefs about Farmers: A Priority for Social Science Research in Social Forestry." Agroforestry Systems 17:13–41.

———. 1993. "Rubber Eating Rice, Rice Eating Rubber." Paper presented at Agrarian Studies Seminar, Connecticut.

———. 2006. "Indigenous People and Environmental Politics." Annual Review of Anthropology 35:191–208.

Dowie, M. 2009. Conservation Refugees: The Hundred Year Conflict between Global Conservation and Native People. Cambridge, MA: MIT Press.

Downie, J. 1959. An Economic Policy for British Honduras. Belize City: British Honduras Government House.

Dozon, J. P. 1985. La Société Bété, Côte D'ivoire [Bete Society, Ivory Coast]. Paris: Karthala-ORSTOM.

Drake, W. D. 1993. "Towards Building a Theory of Population-Environment Dynamics: A Family of Transitions." In Population-Environment Dynamics, edited by G. Ness, W. D. Drake, and S. R. Brechin, 305–55. Ann Arbor: University of Michigan Press.

Drayton, R. H. 2000. Nature's Government: Science, Imperial Britain, and the 'Improvement' of the World. New Haven, CT: Yale University Press.

Drechsel, P., and S. Dongus. 2010. "Dynamics and Sustainability of Urban Agriculture: Examples from Sub-Saharan Africa." Sustainability Science 5 (1): 69–78.

Drummond, M. A., and T. R. Loveland. 2010. "Land-Use Pressure and a Transition to Forest-Cover Loss in the Eastern United States." Bioscience 60 (4): 286–98.

Duncan, J. S., and N. Duncan. 2004. Landscapes of Privilege: The Politics of the Aesthetic in an American Suburb. New York: Routledge.

Dunn, J., and A. F. Robertson. 1973. Dependence and Opportunity: Political Change in Ahafo. African Studies Series vol. 9. Cambridge: Cambridge University Press.

Dunning, N., and T. Beach. 2000. "Stability and Instability in Prehispanic Maya Landscapes." In Imperfect Balance: Landscape Transformations in the Precolumbian Americas, ed. D. L. Lentz, 179–202. New York: Columbia University Press.

Dunning, N., V. Scarborough, F. Valdez Jr., S. Luzzadder-Beach, T. Beach, and J. G. Jones. 1999. "Temple Mountains, Sacred Lakes, and Fertile Fields: Ancient Maya Landscapes in Northwestern Belize." Antiquity 72:650–60.

Dupré, G. 1991. "Les Arbres, le Fourré et le Jardin: les Plantes Dans la Société de Aribinda, Burkina Faso [Trees, the Thicket and Garden: Plants in Aribinda Society Burkina Faso]." In Savoirs Paysans et Dévéloppement, edited by G. Dupré, 181–94. Paris: Karthala-ORSTOM.

Durán, A., and Bracco, R. 2000. Arqueologia de las Tierras Bajas [Archaeology of the Lowlands]. Montevideo: Ministerio de Educacion y Cultura, Comision Nacional de Arqueologia.

Durán Coirolo, A., and R. B. Boksar. 2000. La Arqueología de las Tierras Bajas. Montevideo: Comisión Nacional de Arqueología, Ministerio de Educación y Cultura.

Duryea, M. L. 2000. "Restoring the Urban Forest Ecosystem: An Introduction." In Restoring the Urban Forest Ecosystem, edited by M. L. Duryea, E. Kampf Binelli, and L. V. Korhnak, 1–17. Gainesville: School of Forest Resources and Conservation.

Dutt, B., R. Koleta, and V. Hoshing. 2007. "The Hunter and the Hunted: Conservation with Marginalized Communities." In Making Conservation Work: Securing Biodiversity in This New Century, edited by G. Shahabuddin and M. Rangarajan, 241–89. Delhi: Permanent Black.

Duvall, C. S. 2003. "Symbols, Not Data: Rare Trees and Vegetation History in Mali." Geographical Journal 169 (4): 295–312.

Dwyer, J., E. Mcpherson, H. Schroeder, and R. Rowntree. 1992. "Assessing the Benefits and Costs of the Urban Forest." Journal of Arboriculture 18:227.

Dwyer, J. F., D. J. Nowak, M. H. Noble, and S. M. Sisinni. 2000. Connecting People with Ecosystems in the 21st Century: An Assessment of Our Nation's Urban Forests. Portland: US Forest Service, Pacific Northwest Research Station.

Earthrights International. 2005. "The Price of Luxury: The Global Teak Trade and Forced Labour in Burma." Retrieved from http://www.earthrights.org/teak/indepth.shtml.

Eaton, R. M. 1993. The Rise of Islam and the Bengal Frontier, 1204–1760. Comparative Studies on Muslim Societies vol. 17. Berkeley: University of California Press.

Ebeling, J., and M. Yasué. 2008. "Generating Carbon Finance through Avoided Deforestation and Its Potential to Create Climatic, Conservation and Human Development Benefits." Philosophical Transactions of the Royal Society B 363:1917–24.

Economic and Political Weekly Correspondents. 1985. "Project Tiger and People: A Report on Similipal." Economic and Political Weekly 20 (33): 1380–83.

Edwards, P. N. 1996. The Closed World : Computers and the Politics of Discourse in Cold War America. Cambridge, MA: MIT Press.

Ekasingh, B., K. T. Ngam, T. Promburom, S. Sinchaikul, and N. Thinrach. 2005. "Production Systems and Land Use Productivity of Farmers in Chiang Mai, Chiang Rai and Lamphun." Journal of Agricultural Economics 24 (2): 49–69.

Elias, M., and J. Carney. 2005. "Shea Butter, Globalization, and Women of Burkina Faso." In A Companion to Feminist Geography, edited by L. Nelson and J. Seager, 93–108. Malden, MA: Blackwell.

Ellis, F. 1998. "Household Strategies and Rural Livelihood Diversification." Journal of Development Studies 35 (1): 1–38.

Elmhirst, R. 2004. "Labour Politics in Migrant Communities: Ethnicity and Women's Activism in Tangerang, Indonesia." In Labour in Southeast Asia: Local Processes in a Globalised World, edited by R. Elmhirst and R. Saptari, 387–406. New York: Routledge Curzon.

Elmqvist, T., M. Wall, A.-L. Berggren, L. Blix, A. Fritioff, and U. Rinman. 2001. "Tropical Forest Reorganization after Cyclone and Fire Disturbance in Samoa: Remnant Trees as Biological Legacies." Conservation Ecology 5 (2): 10.

Elvin, M. 2004. The Retreat of the Elephants: An Environmental History of China. New Haven, CT: Yale University Press.

Emery, K. F. 2007. "Assessing the Impact of Ancient Maya Animal Use." Journal for Nature Conservation 15 (3): 184–95.

Endicott, K. 1997. "Review: Violence and the Dream People." The Journal of Asian Studies 56 (1): 262–63.

Environnement et Développement du Tiers-Monde (ENDA). 1992. Avenir Des Terroirs: la Ressource Humaine [The Future of Terroirs: Human Resource]. Dakar: ENDA/GRAF.

Erickson, C. L. 1995. "Archaeological Methods for the Study of Ancient Landscapes of the Llanos de Mojos in the Bolivian Amazon." In Archaeology in the Lowland American Tropics: Current Analytical Methods and Applications, edited by P. W. Stahl, 66–95. Cambridge: Cambridge University Press.

———. 2000a. "An Artificial Landscape-Scale Fishery in the Bolivian Amazon." Nature 408 (6809): 190–93.

———. 2000b. "Lomas de Occupación en los Llanos de Moxos." In La Arqueología de las Tierras Bajas [The Archaeology of the Lowlands], edited by A. Durán Coirolo and R. B. Boksar, 207–26. Montevideo: Comisión Nacional de Arqueología, Ministerio de Educación y Cultura.

———. 2001. "Pre-Columbian Roads of the Amazon." Expedition 43 (2): 21–30.

———. 2002. "Large Moated Settlements: A Late Pre-Columbian Phenomenon in the Amazon." Paper presented at 2nd Annual Meeting of the Society for the Anthropology of Lowland South America, St. Johns College, Annapolis, Maryland.

———. 2003. "Historical Ecology and Future Explorations." In Amazonian Dark Earths: Origins, Properties, Management, edited by J. Lehmann, D. C. Kern, B. Glaser, and W. I. Woods, 455–500. Netherlands: Kluwer.

———. 2004. "Historical Ecology and Future Explorations." In Amazonian Dark Earths, edited by J. Lehmann, D. C. Kern, B. Glaser, and W. I. Woods, 455–500. Dordrecht: Kluwer.

———. 2006. "The Domesticated Landscapes of the Bolivian Amazon." In Time and Complexity in Historical Ecology: Studies in the Neotropical Lowlands, edited by W. L. Balée and C. L. Erickson, 235–78. New York: Columbia University Press.

———. 2009. "Agency, Roads, and the Landscapes of Everyday Life in the Bolivian Amazon." In Landscapes of Movement: The Anthropology of Roads, Paths, and Trails, edited by J. Snead, C. L. Erickson, and A. Darling, 204–31. Philadelphia: University of Pennsylvania Museum of Archaeology and Anthropology Press.

Erickson, C. L., and W. L. Balée. 2006. "The Historical Ecology of a Complex Landscape in Bolivia." In Time and Complexity in Historical Ecology: Studies in the Neotropical Lowlands, edited by W. L. Balée and C. L. Erickson, 187–233. New York: Columbia University Press.

Erickson, C. L., and J. Walker. 2009. "Pre-Columbian Roads as Landscape Capital." In Landscapes of Movement: The Anthropology of Roads, Paths, and Trails, edited by J. Snead, C. L. Erickson, and A. Darling, 232–52. Philadelphia: University of Pennsylvania Museum of Archaeology and Anthropology Press.

Escobar, A. 2008. Territories of Difference : Place, Movements, Life, Redes. Durham, NC: Duke University Press.

European Commission. 2003a. Interpretation Manual of European Union Habitats-Eur 25. Brussels: European Commission, DG Environment.

———. 2003b. Natura 2000 and Forests: "Challenges and Opportunities": Interpretation Guide. Brussels: European Commission, DG Environment.

European Environmental Agency. 2004. High Nature Value Farmland: Characteristics, Trends and Policy Challenges. Luxembourg: Publications of the European Communities.

———. 2005. Agriculture and Environment in Eu-15—the Irena Indicator Report. Luxembourg: Publications of the European Communities.

———. 2006. Progress Towards Halting the Loss of Biodiversity by 2010. Luxembourg: Publications of the European Communities.

———. 2008. European Forests—Ecosystem Conditions and Sustainable Use. Luxembourg: Publications of the European Communities.

Eva, H. D., E. E. de Miranda, C. M. Di Bella, V. Gond, O. Huber, M. Sgrenzaroli, S. Jones, A. Coutinho, A. Dorado, and M. Guimarães. 2002. A Vegetation Map of South America. Luxembourg: Joint Research Centre, European Commission.

Evans, J. 1982. Plantation Forestry in the Tropics. Oxford: Clarendon Press.

Eyre, S. R. 1968. Vegetation and Soil: A World Picture, 2nd ed. London: Edward Arnold.

Fairhead, J., T. Geysbeek, S. Holsoe, and M. Leach. 2003. African-American Exploration in West Africa: Four Nineteenth-Century Diaries. Bloomington: Indiana University Press.

Fairhead, J., and M. Leach. 1994a. "Contested Forests: Modern Conservation and Historical Land Use in Guinea's Ziama Reserve." African Affairs 93 (373): 481–512.

———. 1994b. "Reversing Landscape History: Power, Policy and Socialised Ecology in West Africa's Forest-Savanna Mosaic." Unpublished Manuscript.

———. 1996. Misreading the African Landscape: Society and Ecology in a Forest-Savanna Mosaic. African Studies Series vol. 90. New York: Cambridge University Press.

———. 2003. Science, Society and Power: Environmental Knowledge and Policy in West Africa and the Caribbean. New York: Cambridge University Press.

Falk, N. E. 1973. "Wilderness and Kingship in Ancient South Asia." History of Religions 13 (1): 1–15.

Farina, A., R. Santolini, G. Pagliaro, S. Scozzafava, and I. Schipani. 2005. "Eco-Semiotics: A New Field of Competence for Ecology to Overcome the Frontier between Environmental Complexity and Human Culture in the Mediterranean." Israel Journal of Plant Sciences 53 (3): 167–75.

Farley, K. A. 2007. "Grasslands to Tree Plantations: Forest Transition in the Andes of Ecuador." Annals of the Association of American Geographers 97 (4): 755–71.

Farmer, B. H. 1974. Agricultural Colonization in India since Independence. Delhi: Published for the Royal Institute of International Affairs by Oxford University Press.

Fay, C., and G. Michon. 2003. "Redressing Forestry Hegemony—Where a Forestry Regulatory Framework Is Best Replaced by an Agrarian One." Paper presented at Rural Livelihoods, Forests and Biodiversity, May 19–23, Bonn, Germany.

Fearnside, P. M. 2008. "The Roles and Movements of Actors in the Deforestation of Brazilian Amazonia." Ecology and Society 13 (1): 23.

Feder, G., and D. Feeny. 1993. "The Theory of Land Tenure and Property Rights." In The Economics of Rural Organization: Theory, Practice, and Policy, edited by J. E. Stiglitz, K. Hoff, and A. Braverman. 240–58. New York: Oxford University Press.

Fedick, S. L. 1996. The Managed Mosaic: Ancient Maya Agriculture and Resource Use. Salt Lake City: University of Utah Press.

Feeley, K. J., and M. R. Silman. 2008. "Unrealistic Assumptions Invalidate Extinction Estimates." Proceedings of the National Academy of Sciences 105 (51): E121.

Ferrars, M., and B. Ferrars. 1900. Burma. New York: S. Low, Marston and Co. ; Dutton.

Field Museum. n.d. Journey through Calumet. Available from http://www.fieldmuseum.org/calumet/.

Figueroa, F., V. Sanchez-Cordero, J. A. Meave, and I. Trejo. 2009. "Socioeconomic Context of Land Use and Land Cover Change in Mexican Biosphere Reserves." Environmental Conservation 36 (3):180–91.

Finan, F., E. Sadoulet, and A. de Janvry. 2005. "Measuring the Poverty Reduction Potential of Land in Rural Mexico." Journal of Development Economics 77 (1): 27–51.

Fitzpatrick, D. 2005. "'Best Practice' Options for the Legal Recognition of Customary Tenure." Development and Change 36 (3): 449–75.

———. 2006. "Evolution and Chaos in Property Rights Systems: The Third World Tragedy of Contested Access." The Yale Law Journal 115 (5): 996–1048.

Flint, C. 2005. The Geography of War and Peace: From Death Camps to Diplomats. New York: Oxford University Press.

Fofana, S., Y. Camara, M. Barry, and A. Sylla. 1993. Etude Relative au Feu Auprés Des Populations Des Bassins Versants Types Du Haut Niger [Fire Study Relative to the

Populations of Watershed Types of the High Niger]. Conakry: Republique de Guinée; Programme d'Aménagement des Bassins Versants Haut-Niger.

Folan, W. J., L. A. Fletcher, and E. R. Kintz. 1979. "Fruit, Fiber, Bark, and Resin: Social Organization of a Maya Urban Center." Science 204 (4394): 697–701.

Food and Agriculture Organization (FAO). 1978. Forestry for Local Community Development. FAO Forestry Paper 7. Rome: UN Food and Agriculture Organization.

———. 1993. Forest Resources Assessment, 1990 : Tropical Countries. FAO Forestry Paper no. 112, vol. 124. Rome: UN Food and Agriculture Organization.

———. 2001. Global Forest Resources Assessment 2000. FAO Forestry Paper no. 140. Rome: UN Food and Agriculture Organization.

———. 2002. Land Tenure and Rural Development. FAO Land Tenure Studies No. 3. Rome: UN Food and Agriculture Organization.

———. 2006a. Global Forest Resources Assessment 2005. FAO Forestry Paper No. 147. Rome: UN Food and Agriculture Organization.

———. 2006b. Global Forest Resources Assessment 2005: National Report No. 001— India. Rome: UN Food and Agriculture Organization.

———. 2007. Faostat Online Statistical Service. United Nations Food and Agriculture Organization.

Food and Agriculture Organization (FAO) and J.-P. Lanly. 1982. Tropical Forest Resources. FAO Forestry Paper no. 30, vol. 30. Rome: UN Food and Agriculture Organization.

Ford, C. 2007. "Nature's Fortunes: New Directions in the Writing of European Environmental History." The Journal of Modern History 79 (1): 112–33.

Ford, A., and R. Nigh. 2009. "Origins of the Maya Forest Garden: Maya Resource Management." Journal of Ethnobiology 29 (2): 213–36.

Foresta, R. A. 1991. Amazon Conservation in the Age of Development: The Limits of Providence. Gainesville: University of Florida Press.

Forsyth, T., and A. Walker. 2008. Forest Guardians, Forest Destroyers: The Politics of Environmental Knowledge in Northern Thailand. Seattle: University of Washington Press.

Fortmann, L., and J. W. Bruce, eds. 1988. Whose Trees?: Proprietary Dimensions of Forestry. Boulder: Westview Press.

Foster, D. R. 2002. "Insights from Historical Geography to Ecology and Conservation: Lessons from the New England Landscape." Journal of Biogeography 29:1269–75.

———. 2008. "The Natural and Cultural History of a Temperate Forest: Local Stories, Global Connections, and Environmental Opportunity and Challenge." Paper presented at the conference, Social Life of Forests: New Frameworks for Studying Change, May 30–31, University of Chicago.

Foster, D. R., B. Hall, S. Barry, S. Clayden, and T. Parshall. 2002. "Cultural, Environmental and Historical Controls of Vegetation Patterns and the Modern Conservation Setting on the Island of Martha's Vineyard, USA." Journal of Biogeography 29 (10–11): 1381–400.

Foster, D. R., G. Motzkin, D. Bernardos, and J. Cardoza. 2002. "Wildlife Dynamics in the Changing New England Landscape." Journal of Biogeography 29:1337–57.

Foucar, E. C. V. 1956. I Lived in Burma. London: D. Dobson.

Foucault, M. 1991. "Governmentality." In The Foucault Effect: Studies in Governmentality, edited by G. Burchell, C. Gordon, and P. Miller, 87–104. Chicago: University of Chicago Press.

Fox, J., D. Mcmahon, M. Poffenberger, and J. Vogler. 2008. Land for My Grandchildren: Land Use and Tenure Change in Ratanakiri: 1989–2007. Phnom Penh: Community Forestry International (CFI) and the East West Center.

Fox, J., K. Suryanata, P. Hershock, and A. H. Pramono. 2008. "Mapping Boundaries, Shifting Power: The Socio-Ethical Dimensions of Participatory Mapping." In Contentious Geographies: Environmental Knowledge, Meaning, Scale, edited by M. Goodman, M. Boykoff, and K. Evered. 203–17. Oxford: Ashgate Publishing Co.

Fox, J., and J. B. Vogler. 2005. "Land-Use and Land-Cover Change in Montane Mainland Southeast Asia." Environmental Management 36 (3): 394–403.

Franke, R. W., and B. H. Chasin. 1980. Seeds of Famine: Ecological Destruction and the Development Dilemma in the West African Sahel. Montclair, NJ: Allanheld, Osmun.

Fraser, C. 2010. Rewilding the World. New York: Metropolitan.

Freedman, P. 2008. Out of the East: Spices and the Medieval Imagination. New Haven, CT: Yale University Press.

Fresquez, P. R., D. R. Armstrong, and L. H. Pratt. 1997. "Radionuclides in Bees and Honey within and around Los Alamos National Laboratory." Journal of Environmental Science Health A32:1309–23.

Fresquez, P. R., D. R. Armstrong, and J. G. Salazar. 1994. Tritium Concentrations in Bees and Honey at Los Alamos National Laboratory. Medium: ED; Size: 11 p.

Fresquez, P. R., W. R. Velasquez, and L. Naranjo Jr. 2000. Effects of the Cerro Grande Fire (Smoke and Fallout Ash) on Soil Chemical Properties within and around Los Alamos National Laboratory. Los Alamos: Los Alamos National Laboratory.

Freudenberger, M. S., J. A. Carney, and A. R. Lebbie. 1997. "Resiliency and Change in Common Property Regimes in West Africa: The Case of the Tongo in the Gambia, Guinea, and Sierra Leone." Society and Natural Resources 10 (4): 383–402.

Fujita, Y., and K. Phanvilay. 2008. "Land and Forest Allocation in Lao People's Democratic Republic: Comparison of Case Studies from Community-Based Natural Resource Management Research." Society and Natural Resources 21 (2): 120–33.

Fujita, Y., S. Thongmanivong, T. Vongvisouk, K. Phanvilay, and H. Chanthavong. 2007. "Dynamic Land Use Change in Sing District, Luang Namtha Province, Lao PDR." Unpublished Report for International Program for Research on the Interactions between Population, Development, and the Environment (Pripode). Vientiane: Faculty of Forestry, National University of Laos.

Gadgil, M., and R. Guha. 1992. This Fissured Land: An Ecological History of India. Berkeley, CA: University of California Press.

———. 2000. The Use and Abuse of Nature. Omnibus ed., 2 vols. New York: Oxford University Press.

Galindo-González, J., S. Guevara, and V. J. Sosa. 2000. "Bat- and Bird-Generated Seed Rains at Isolated Trees in Pastures in a Tropical Rainforest." Conservation Biology 14 (6): 1693–703.

Garcia, R., B. Soares-Filho, and D. Sawyer. 2007. "Socioeconomic Dimensions, Migration, and Deforestation: An Integrated Model of Territorial Organization for the Brazilian Amazon." Ecological Indicators 7 (3): 719–30.

Garcia-Barrios, L., Y. M. Galvan-Miyoshi, I. A. Valdivieso-Perez, O. R. Masera, G. Bocco, and J. Vandermeer. 2009. "Neotropical Forest Conservation, Agricultural Intensification, and Rural Out-Migration: The Mexican Experience." Bioscience 59 (10): 863–73.

Gardiner, J. 1942. "The Teak Industry of Burma." Australian Timber Journal 8:736–37.

Gascoigne, J. 1998. Science in the Service of Empire: Joseph Banks, the British State and the Uses of Science in the Age of Revolution. Cambridge: Cambridge University Press.

Geary, G. 1886. Burma, after the Conquest, Viewed in Its Political, Social, and Commercial Aspects, from Mandalay. London: S. Low, Marston, Searle, & Rivington.

GeoAmazonia. 2009. GeoAmazonia: Environment Outlook in Amazonia. Panama City: United Nations Environment Programme; Amazon Cooperation Treaty Organization.

Ghana. 1992. Constitution of the Republic of Ghana. Accra: Ghana Publishing Corp.

———. 1999. National Land Policy. Accra: Ministry of Lands and Forestry.

Ghate, R., and K. Beazley. 2007. "Aversion to Relocation: A Myth?" Conservation and Society 5 (3): 331.

Ghosh, A. 2004. The Hungry Tide. Delhi: Ravi Dayal.

Gilbert, M. 1988. "The Sudden Death of a Millionaire: Conversion and Consensus in a Ghanaian Kingdom." Africa 58 (3): 291–314.

Gillespie, T. W. 2001. "Application of Extinction and Conservation Theories for Forest Birds in Nicaragua." Conservation Biology 15 (3): 699–709.

Gilliland, H. B., ed. 1971. Flora of Malaysia. Vol III, Grasses. Singapore: University of Singapore Press.

Gillson, L., and K. J. Willis. 2004. "'As Earth's Testimonies Tell': Wilderness Conservation in a Changing World." Ecology Letters 7 (10): 990–98.

Glacken, C. J. 1967. Traces on the Rhodian Shore; Nature and Culture in Western Thought from Ancient Times to the End of the Eighteenth Century. Berkeley: University of California Press.

Glantz, M. H. 1976. The Politics of Natural Disaster : The Case of the Sahel Drought. Praeger Special Studies in International Economics and Development. New York: Praeger.

Glaser, B. 2007. "Prehistorically Modified Soils of Central Amazonia: A Model for Sustainable Agriculture in the Twenty-First Century." Philosophical Transactions of the Royal Society B-Biological Sciences 362 (1478): 187–96.

Glaser, B., and W. I. Woods, eds. 2004. Amazonian Dark Earths: Explorations in Space and Time. Berlin: Springer-Verlag.

Glassman, J. 2006. "Primitive Accumulation, Accumulation by Dispossession, Accumulation by 'Extra-Economic' Means." Progress in Human Geography 30 (5): 608.

Gleditsch, N. P., P. Wallensteen, M. Eriksson, M. Sollenberg, and H. Strand. 2002. "Armed Conflict 1946–2001: A New Dataset." Journal of Peace Research 39:615–37.

Glennemeier, K. 2004. "The State of Our Wooded Lands : Results from the Chicago Wilderness Woods Audit." CW Journal 2:16–22.

Global Witness. 2003. A Conflict of Interests: The Uncertain Future of Burmaís Forests. London: Global Witness.

"Gloster: Made for Life." Promotional leaflet distributed at London Garden Centres, 2005–2006.

Goddard, M. A., A. J. Dougill, and T. G. Benton. 2010. "Scaling up from Gardens: Biodiversity Conservation in Urban Environments." Trends in Ecology and Evolution 25 (2): 90–98.

Godwin, G. 1940. Our Woods in War: A Survey of Their Vital Role in Defense. London: Acorn.

Gold, A. G., and B. R. Gujar. 2003. In the Time of Trees and Sorrows: Nature, Power, and Memory in Rajasthan. Delhi: Oxford University Press.

Golinski, J. 2007. British Weather and the Climate of Enlightenment. Chicago: University of Chicago Press.

Gómez-Pompa, A., J. S. Flores, and M. A. Fernández. 1990. "The Sacred Cacao Groves of the Maya." Latin American Antiquity 1 (3): 247–57.

Gómez-Pompa, A., J. S. Flores, and V. Sosa. 1987. The "Pet Kot": A Man-Made Tropical Forest of the Maya." Interciencia 12 (1): 10–15.

Goulding, M., and R. Barthem. 2003. The Smithsonian Atlas of the Amazon. Washington DC: Smithsonian Institution.

Government of India. 1868. "Seasoning of Timber by Girdling Previous to Felling." London: Her Majesty's Stationary Office.

Graf, W. L. 1994. Plutonium and the Rio Grande: Environmental Change and Contamination in the Nuclear Age. New York: Oxford University Press.

Gragson, T. L. 1992. "The Use of Palms by the Pume Indians of Southwestern Venezuela." Principes 36 (3): 133–42.

Graham, P. 1812. General View of the Agriculture of Stirlingshire : With Observations on the Means of Its Improvement. Edinburgh: Printed for G. & W. Nicol.

Grainger, A. 1984. "Quantifying Changes in Forest Cover in the Humid Tropics: Overcoming Current Limitations." Journal of World Forest Resource Management 1:3–63.

———. 1993. "Population as Concept and Parameter in the Modelling of Tropical Land Use Change." In Population-Environment Dynamics, edited by G. Ness, W. D. Drake, and S. R. Brechin, 71–101. Ann Arbor: University of Michigan Press.

———. 1995. "The Forest Transition: An Alternative Approach." Area 27:242–51.

———. 1996a. "The Degradation of Tropical Rain Forest in Southeast Asia: Taxonomy and Appraisal." In Land Degradation in the Tropics, edited by M. E. Eden and J. T. Parry, 61–75. London: Mansell.

———. 1996b. "Integrating the Socio-Economic and Physical Dimensions of Degraded Tropical Lands in Global Climate Change Mitigation Assessments." In Forest Ecosystems, Forest Management and the Global Climate Cycle, edited by M. Apps and D. T. Price. 335–48. Berlin: Springer Verlag.

———. 2008. "Difficulties in Tracking the Long Term Global Trand in Tropical Forest Area." Proceedings of the National Academy of Sciences of the United States of America 105:818–23.

———. 2009. "Measuring the Planet to Fill Terrestrial Data Gaps." Proceedings of the National Academy of Sciences of the United States of America 106 (49): 20557–58.

———. 2010. "The Bigger Picture—Tropical Forest Change in Context, Concept and Practice." In Reforesting Landscapes: Linking Pattern and Process, edited by H. Nagendra and J. Southworth, 15–43. Berlin: Springer.

Grainger, A., D. H. Boucher, P. C. Frumhoff, W. F. Laurance, T. Lovejoy, J. Mcneely, M. Niekisch, P. Raven, N. S. Sodhi, and O. Venter. 2009. "Biodiversity and Redd at Copenhagen." Current Biology 19 (21): R974–76.

Grainger, A., and B. S. Malayang III. 2006. "A Model of Policy Changes to Secure Sustainable Forest Management and Control of Deforestation in the Philippines." Forest Policy and Economics 8 (1): 67–80.

Grau, H., and M. Aide. 2008. "Globalization and Land-Use Transitions in Latin America." Ecology and Society 13 (2):16.

Grau, H. R., T. M. Aide, J. K. Zimmerman, and J. R. Thomlinson. 2004. "Trends and Scenarios of the Carbon Budget in Postagricultural Puerto Rico (1936–2060)." Global Change Biology 10 (7): 1163–79.

Grau, H. R., T. M. Aide, J. K. Zimmerman, J. R. Thomlinson, E. Helmer, and X. M. Zou. 2003. The Ecological Consequences of Socioeconomic and Land-Use Changes in Postagriculture Puerto Rico. Bioscience 53 (12): 1159–68.

Grau, H. R., M. E. Hernandez, J. Gutierrez, N. I. Gasparri, M. C. Casavecchia, E. E. Flores-Ivaldi, and L. Paolini. 2008. "A Peri-Urban Neotropical Forest Transition and its Consequences for Environmental Services." Ecology and Society 13:35.

Green, W. 1991. Lutte Contre les Feux de Brousse [Fighting the Brush Fires]. Report for the Project DERIK, Développement Rural Intégré de Kissidougou.

Greenberg, J. 2002. A Natural History of the Chicago Region. Chicago: University of Chicago Press.

Greene, S. 2006. "Getting over the Andes: The Geo-Eco-Politics of Indigenous Movements in Peru's Twenty-First Century Inca Empire." Journal of Latin American Studies 38:327–54.

Greenough, P. R. 2003. "Pathogens, Pugmarks, and Political Emergency: The 1970s South Asian Debate on Nature." In Nature in the Global South: Environmental Projections in South and Southeast Asia, edited by P. R. Greenough and A. L. Tsing. 201–30. Durham, NC: Duke University Press.

Greenough, P. R., and A. L. Tsing. 2003. Nature in the Global South: Environmental Projects in South and Southeast Asia. Durham, NC: Duke University Press.

Greller, A. M. 2000. "Vegetation in the Floristic Regions of North and Central America." In Imperfect Balance: Landscape Transformations in the Precolumbian Americas, ed. D. L. Lentz. New York: Columbia University Press.

Grieg-Gran, M., I. Porras, and S. Wunder. 2005. "How Can Market Mechanisms for Forest Environmental Services Help the Poor? Preliminary Lessons from Latin America." World Development 33 (9): 1511–27.

Griffin, R. 1996. The Nature of Fascism. Reprinted ed. London [u.a.]: Routledge.

Groninger, J. W., and C. M. Ruffner. 2010. "Hearts, Minds, and Trees: Forestry's Role in Operation Enduring Freedom." Journal of Forestry 108 (3): 141–47.

Gross, D. R. 1975. "Protein Capture and Cultural Development in the Amazon Basin." American Anthropologist 77 (3): 526–49.

Grove, R. 1995. Green Imperialism : Colonial Expansion, Tropical Island Edens, and the Origins of Environmentalism, 1600–1860. New York: Cambridge University Press.

Guardiola-Claramonte, M., P. A. Troch, A. D. Ziegler, T. W. Giambelluca, J. B. Vogler, and M. A. Nullet. 2008. "Local Hydrologic Effects of Introducing Non-Native Vegetation in a Tropical Catchment." Ecohydrology 1 (1): 13–22.

Guariguata, M., R. Chazdon, J. Denslow, J. Dupuy, and L. Anderson. 1997. "Structure and Floristics of Secondary and Old-Growth Forest Stands in Lowland Costa Rica." Plant Ecology 132 (1): 107–20.

Guariguata, M. R., J. J. R. Adame, and B. Finegan. 2000. "Seed Removal and Fate in Two Selectively Logged Lowland Forests with Constrasting Protection Levels." Conservation Biology 14 (4): 1046–54.

Guevara, S., J. Laborde, and G. Sanchez-Rios. 2004. "Rain Forest Regeneration beneath the Canopy of Fig Trees Isolated in Pastures of Los Tuxtlas, Mexico." Biotropica 36 (1): 99–108.

———. 2005. "Los Árboles Que la Selva Dejó Átras [Forest Trees Left Behind]." INCI 30 (10): 595–601.

Guevara, S., S. E. Purata, and E. Maarel. 1986. "The Role of Remnant Forest Trees in Tropical Secondary Succession." Plant Ecology 66 (2): 77–84.

Guha, R. 1989. The Unquiet Woods: Ecological Change and Peasant Resistance in the Himalaya. New York: Oxford University Press.

———. 2000. Nature's Spokesman: M. Krishnan and Indian Wildlife. New York: Oxford University Press.

———. 2006. How Much Should a Person Consume?: Environmentalism in India and the United States. Berkeley: University of California Press.

Guha, S. 1999. Environment and Ethnicity in India, 1200–1991. Cambridge Studies in Indian History and Society vol. 4. New York: Cambridge University Press.

———. 2001. Health and Population in South Asia: From Earliest Times to the Present. Delhi: Permanent Black.

———. 2002. "Claims on the Commons: Political Power and Natural Resources in Pre-Colonial India." Indian Economic and Social History Review 39 (2–3): 181.

Gunnell, Y., K. Anupama, and B. Sultan. 2007. "Response of the South Indian Runoff-Harvesting Civilization to Northeast Monsoon Rainfall Variability during the Last 2000 Years: Instrumental Records and Indirect Evidence." The Holocene 17 (2): 207–15.

Günter, S., M. Weber, R. Erreis, and N. Aguirre. 2007. "Influence of Distance to Forest Edges on Natural Regeneration of Abandoned Pastures: A Case Study in the Tropical Mountain Rain Forest of Southern Ecuador." European Journal of Forest Research 126 (1): 67–75.

Guo, H., C. Padoch, Y. Fu, Z. Dao, and K. Coffey. 2002. "Household Level Agrobiodi-
versity Assessment." In Cultivating Biodiversity—Understanding, Analysing and
Using Agricultural Diversity, edited by H. Brookfield, C. Padoch, H. Parsons, and
M. Stocking, 70–77. London: ITDG Publishing.

Gutierrez, M. L., and D. Juhé-Beaulaton. 2002. "Histoire Du Parc à Néré (Parkia
Biglobosa Jaċqu. Benth.) Sur le Plateau D'abomey (Bénin): de Sa Conservation Pour
la Production et la Commercialisation D'un Condiment, L'afitin [History of Néré
Park on Plateau D'Abomey (Benin) (Parkia biglobosa Benth Jacqu.): Conservation
for Production and Marketing on Condiment, the afiti]." Cahiers d'Outre-mer
220:453–74.

Habib, I. 1982a. An Atlas of the Mughal Empire: Political and Economic Maps with
Detailed Notes, Bibliography and Index. New York: Oxford University Press.

———. 1982b. "The Geographical Background." In The Cambridge Economic History
of India, Vol. I, C. 1200-1750, edited by T. Raychaudhuri and I. Habib, 1–13. Cambridge: Cambridge University Press.

———. 1982c. "Population." In The Cambridge Economic History of India Vol. I, C.
1200-1750, edited by T. Raychaudhuri and I. Habib, 163–170. Cambridge: Cambridge University Press.

Hack, K. 2001. Defence and Decolonisation in Southeast Asia: Britain, Malaya and
Singapore, 1941–68. Richmond, Surrey: Curzon.

Hair, P. E. H. 1962. "An Account of the Liberian Hinterland C. 1780." Sierra Leone
Studies 16:218–26.

Hale, C. 2002. "Does Multiculturalism Menace? Governance, Cultural Rights, and the
Politics of Identity in Guatemala." Journal of Latin American Studies 34 (3): 485–524.

———. 2005. "Neoliberal Multiculturalism: The Remaking of Cultural Rights and
Racial Dominance in Central America." PoLAR: Political and Legal Anthropology
Review 28 (1): 10–19.

Hales, P. B. 1997. Atomic Spaces: Living on the Manhattan Project. Urbana: University
of Illinois Press.

Hall, J. B. 1996. Vitellaria paradoxa: A Monograph. School of Agricultural and Forest
Sciences Publication vol. 8. Bangor: University of Wales.

Hamilton, T. 1761. A Treatise on the Manner of Raising Forest Trees. Edinburgh:
Printed for G. Hamilton and J. Balfour.

Hansen, M. C., S. V. Stehman, P. V. Potapov, T. R. Loveland, J. R. G. Townshend,
R. S. Defries, K. W. Pittman, B. Arunarwati, F. Stolle, and M. K. Steininger. 2008.
"Humid Tropical Forest Clearing from 2000 to 2005 Quantified by Using Multi-
temporal and Multiresolution Remotely Sensed Data." Proceedings of the National
Academy of Sciences 105 (27): 9439.

Hanson, T., S. Brunsfeld, and B. Finegan. 2006. "Variation in Seedling Density and
Seed Predation Indicators for the Emergent Tree Dipteryx Panamensis in Continuous and Fragmented Rain Forest." Biotropica 38 (6):770–74.

Haraway, D. 1991. "Situated Knowledges: The Science Question in Feminism and the
Privilege of Partial Perspective." In Simians, Cyborgs, and Women: The Reinvention of Nature, 183–202. New York: Routledge.

Harper, T. N. 1997. "The Politics of the Forest in Colonial Malaya." Modern Asian Studies 31 (01): 1–29.

Harris, M., and S. Nugent. 2004. Some Other Amazonians: Perspectives on Modern Amazonia. London: Institute for the Study of the Americas.

Harrisson, T. 1958. "A Living Megalithic in Upland Borneo." Sarawak Museum Journal VIII:694–702.

———. 1959. World Within: A Borneo Story. London: Cresset Press.

Harvey, C., and W. Haber. 1998. "Remnant Trees and the Conservation of Biodiversity in Costa Rican Pastures." Agroforestry Systems 44 (1): 37–68.

Hayashida, F. M. 2005. "Archaeology, Ecological History, and Conservation." Annual Review of Anthropology 34:43–65.

Hayden, B. 2003. "Were Luxury Foods the First Domesticates? Ethnoarchaeological Perspectives from Southeast Asia." World Archaeology 34: 458–69.

Hecht, S. B. 1985. "Environment, Development and Politics—Capital Accumulation and the Livestock Sector in Eastern Amazonia." World Development 13 (6): 663–84.

———. 1993. "The Logic of Livestock and Deforestation in Amazonia." Bioscience 43 (10): 687–95.

———. 2003. "Indigenous Soil Management and the Creation of Amazonian Dark Earths: Implications of Kayapó Practice." In Amazonian Dark Earths: Origin, Properties, Management, edited by J. Lehmann, D. C. Kern, B. Glaser, and W. I. Woods, 355–72. Netherlands: Kluwer Academic Publishers.

———. 2009. "The New Rurality: Forest Recovery, Peasantries and the Paradoxes of Landscape." Land Use Policy 17:141–60.

———. 2011. The Scramble for the Amazon: Imperial Contests and the Tropical Odessey of Euclides da Cunha. Chicago: University of Chicago Press.

Hecht, S. B., and A. Cockburn. 1989. The Fate of the Forest: Developers, Destroyers, and Defenders of the Amazon. New York: Verso.

Hecht, S. B., S. Kandel, I. Gomes, N. Cuellar, and H. Rosa. 2006. "Globalization, Forest Resurgence, and Environmental Politics in El Salvador." World Development 34 (2): 308–23.

Hecht, S. B., and S. S. Saatchi. 2007. "Globalization and Forest Resurgence: Changes in Forest Cover in El Salvador." Bioscience 57 (8): 663–72.

Heckenberger, M. J. 1998. "Manioc Agriculture and Sedentism in Amazonia: The Upper Xingu Example." Antiquity 72:633–47.

———. 2001. "Epidemias, Índios Bravos e Brancos: Contato Cultural e Etnogénese do Alto Xingú [Epidemics, Indian Braves and White: Cultural CONTACT Ethnogenesis and the Upper Xingu]." In Os Povos Indígenas do Alto Xingú: História e Cultura, edited by B. Franchetto and M. J. Heckenberger, 77–110. Rio de Janeiro: Editora UFRJ.

———. 2002. "Rethinking the Arawakan Diaspora: Hierarchy, Regionality, and the Amazonian Formative." In Comparative Arawakan Histories: Rethinking Language Family and Culture Area in Amazonia, edited by J. Hill and F. Santos-Granero, 99–122. Urbana: University of Illinois Press.

———. 2005. The Ecology of Power : Culture, Place, and Personhood in the Southern Amazon, A.D. 1000–2000. New York: Routledge.

Heckenberger, M., and E. G. Neves. 2009. "Amazonian Archaeology." Annual Review of Anthropology 38:251–66.

Heckenberger, M. J., J. Christian Russell, J. R. Toney, and M. J. Schmidt. 2007. "The Legacy of Cultural Landscapes in the Brazilian Amazon: Implications for Biodiversity." Philosophical Transactions of the Royal Society B: Biological Sciences 362 (1478): 197–208.

Heckenberger, M. J., A. Kuikuro, U. T. Kuikuro, J. C. Russell, M. Schmidt, C. Fausto, and B. Franchetto. 2003. "Amazonia 1492: Pristine Forest or Cultural Parkland?" Science 301 (5640): 1710.

Heckenberger, M. J., J. B. Petersen, and E. Goes Neves. 2001. "Of Lost Civilizations and Primitive Tribes, Amazonia: Reply to Meggers." Latin American Antiquity 12:328–33.

Heckenberger, M. J., J. B. Petersen, and E. G. Neves. 1999. "Village Size and Permanence in Amazonia: Two Archaeological Examples from Brazil." Latin American Antiquity 10 (4): 353–76.

Heidhues, M. S. 2003. Goldiggers, Farmers, and Traders in the "Chinese Districts" of West Kalimantan, Indonesia. Ithaca: Cornell University, Southeast Asian Program Publications.

Henderson, D. M., and J. H. Dickson. 1994. A Naturalist in the Highlands: James Robertson, His Life and Travels in Scotland, 1767–1771. Edinburgh: Scottish Academic Press.

Henderson, A., and G. Galeano. 1996. "*Euterpe, Prestoea,* and *Neonicholsonia* (Palmae)." Flora Neotropica, Monograph 72:1–89.

Heneghan, L., C. Mulvaney, K. Ross, L. Umek, C. Watkins, L. M. Westphal, and D. H. Wise. 2012. "Lessons Learned from Chicago Wilderness—Implementing and Sustaining Conservation Management in an Urban Setting." Diversity 4 (1): 74–93.

Heneghan, L., L. Umek, B. Bernau, K. Grady, J. Iatropulos, D. Jabon, and M. Workman. 2009. "Ecological Research Can Augment Restoration Practice in Urban Areas Degraded by Invasive Species—Examples from Chicago Wilderness." Urban Ecosystems 12 (1): 63–77.

Herlihy, P. H., and G. Knapp. 2003. "Maps of, by, and for the Peoples of Latin America." Human Organization 62 (4): 303–14.

Hertsgaard, M. 2009. "Regreening Africa." The Nation, November 19. Available at http://www.thenation.com/article/regreening-africa.

Hiernaux, P., A. Ayantunde, A. Kalilou, E. Mougin, B. Gerard, F. Baup, M. Grippa, and B. Djaby. 2009. "Trends in Productivity of Crops, Fallow and Rangelands in Southwest Niger: Impact of Land Use, Management and Variable Rainfall." Journal of Hydrology 375 (1–2): 65–77.

Hiernaux, P., L. Diarra, V. Trichon, E. Mougin, N. Soumaguel, and F. Baup. 2009. "Woody Plant Population Dynamics in Response to Climate Changes from 1984 to 2006 in Sahel (Gourma, Mali)." Journal of Hydrology 375 (1–2): 103–13.

Hightower, J. 1973. Hard Tomatoes, Hard Times: A Report of the Agribusiness Accountability Project on the Failure of America's Land Grant College Complex. Cambridge, MA: Schenkman.

Hill, A. M. 1989. "Chinese Dominance of the Xishuangbanna Tea Trade: An Interregional Perspective." Modern China 15 (3): 321–45.

Hiraoka, M. 1985. "Floodplain Farming in the Peruvian Amazon." Geographical Review of Japan 58 (1): 1–23.

———. 1989. "Agricultural Systems on the Floodplains of the Peruvian Amazon." In Fragile Lands in Latin America: Stategies for Sustainable Development, edited by J. O. Browder, 75–101. Boulder: Westview Press.

———.1994. "Mudanças nos padrões econômicos de uma população ribeirinha do estuário do Amazonas [Changes in Economic Patterns of a Riverine Population of the Amazon Estuary]." In Povos das Águas: Realidade e Perspectivas na Amazônia. MPEG/Universidade Federal do Para, edited by L. Furtado, A. F. Mello, and W. Leitão, 133–57. Belém, Para, Brazil.

Hirsch, P. 1990. Development Dilemmas in Rural Thailand. Oxford: Oxford University Press.

———. 1993. Political Economy of Environment in Thailand. Manila: Journal of Contemporary Asia Publishers.

Hobbs, R. J., S. Arico, J. Aronson, J. S. Baron, P. Bridgewater, V. A. Cramer, P. R. Epstein, J. J. Ewel, C. A. Klink, and A. E. Lugo. 2006. "Novel Ecosystems: Theoretical and Management Aspects of the New Ecological World Order." Global Ecology and Biogeography 15 (1): 1–7.

Hobbs, R. J., and L. F. Huenneke. 1992. "Disturbance, Diversity, and Invasion: Implications for Conservation." Conservation Biology 6 (3): 324–37.

Hockings, P. 1980. Ancient Hindu Refugees: Badaga Social History 1550–1975. Delhi, India: Vikas Publishing House.

Hodgson, S. 2004. "Land and Water—the Rights Interface." FAO Legal Papers Online 36. Retrieved from http://www.fao.org/legal/prs-ol/lp036.pdf

Holling, C. S. 1973. "Resilience and Stability of Ecological Systems." Annual Review of Ecology and Systematics 4 (1):1–23.

Hollis, D. 1997. Performance Assessment and Composite Analysis of Los Alamos National Laboratory Material Disposal Area G. Los Alamos: Los Alamos National Laboratory.

Homewood, K., and W. A. Rodgers. 1991. Maasailand Ecology : Pastoralist Development and Wildlife Conservation in Ngorongoro, Tanzania. Cambridge Studies in Applied Ecology and Resource Management. New York: Cambridge University Press.

Hope, G., A. P. Kershaw, S. V. D. Kaars, S. Xiangjun, P.-M. Liew, L. E. Heusser, H. Takahara, M. Mcglone, N. Miyoshi, and P. T. Moss. 2004. "History of Vegetation and Habitat Change in the Austral-Asian Region." Quaternary International 118–119:103–26.

Hopkins, J. F. P., and N. Levtzion. 2000. Corpus of Early Arabic Sources for West African History. Princeton, NJ: Markus Wiener Publishers.

Hopwood, S. F. 1935. "The Influence of the Growing Use of Substitutes for Timber Upon Forest Policy with Special Reference to Burma." Indian Forester 61:558–72.

House, S., and C. Dingwall. 2003. "A Nation of Planters: Introducing the New Trees, 1650–1900." In People and Woods in Scotland; a History, edited by T. C. Smout, 128–57. Edinburgh: Edinburgh University Press.

Howard, A. L. 1923. "Commercial Prospects of Burma Woods." Asiatic Review 19:396–401.

Howe, H. F. 1994. "Managing Species Diversity in Tallgrass Prairie: Assumptions and Implications." Conservation Biology 8 (3): 691–704.

———. 1995. "Succession and Fire Season in Experimental Prairie Plantings." Ecology 76 (6): 1917–25.

Hudson, A. B. 1977. "Linguistic Relations among Borneo Peoples with Special Reference to Sarawak: An Interim Report." In Studies in Third World Societies Vol 3. Sarawak: Linguistics and Development Problems, edited by M. D. Zamora, V. Sutlive, and N. Altshuler, 1–44. Williamsburg: Borneo Research Council.

Hughes, J. 2008. "A Land without Tigers: How Maharajah Ganga Singh of Bikaner Compensated for Disadvantages of His State and Status." Workshop on Political Ecology. March 6–7, Dept. of Sociology, Delhi School of Economics, University of Delhi.

Hughes, R. F., J. B. Kauffman, and V. C. J. Jaramillo. 1999. "Biomass, Carbon, and Nutrient Dynamics of Secondary Forests in a Humid Tropical Region of México." Ecology 80 (6): 1892–907.

Huijun, G., C. Padoch, K. Coffey, C. Aiguo, and F. Yongneng. 2002. "Economic Development, Land Use and Biodiversity Change in the Tropical Mountains of Xishuangbanna, Yunnan, Southwest China." Environmental Science and Policy 5 (6): 471–79.

Hutton, J. 1788. "The Theory of Rain." Transactions of the Royal Society of Edinburgh: 41–86.

———. 1797. Elements of Agriculture. Edinburgh: National Library of Scotland.

Hyman, E. L. 1991. "A Comparison of Labor-Saving Technologies for Processing Shea Nut Butter in Mali." World Development 19 (9): 1247–68.

Illinois Natural History Survey. 1994. The Changing Illinois Environment: Critical Trends Volume 3. Ecological Resources. Springfield: Illinois Department of Energy and Natural Resources.

Inda, J. X. 2005. Anthropologies of Modernity : Foucault, Governmentality, and Life Politics. Malden, MA: Blackwell.

Inouye, R. S., N. J. Huntly, D. Tilman, J. R. Tester, M. Stillwell, and K. C. Zinnel. 1987. "Old-Field Succession on a Minnesota Sand Plain." Ecology 68 (1): 12–26.

Instituto Brasileiro de Geografia e Estatística (IBGE). 1950–2006. "Produção da Extração Vegetal e da Silvicultura." Departamento Agropecuário, Diretoria de Pesquisas, Instituto Brasileiro de Geografia e Estatística, Rio de Janeiro, Brazil. Data from 1950–1990 available at the archives of IBGE-Belém. Data from 1990–2006 available online at URL www.ibge.gov.br.

Instituto Nacional de Estadística e Informática (INEI). 2007. Censos Nacionales 2005: X de Población y V de Vivienda [Extraction Plant Production and Forestry]. Available at http://www.inei.gob.pe/.

Instituto National de Pesquisas Espaciais, Projeto Desmatamento (INPE-PRODES). 2008. "Instituto National de Pesquisas Espaciais, Projeto Desmatamento 1997–2007 INPE, Coordenação-Geral de Observação da Terra [National Institute for Space Research, INPE 1997–2007 Deforestation Project, General Coordination of

Earth Observation]." São José dos Campos, Brazil. Available at http://www.obt .inpe.br/prodes/ (in Portuguese).

Integrated Regional Information Networks. 2002. Angola: One of the Worst Places in the World for Children. Retrieved from http://reliefweb.int/node/100965.

Inter-American Development Bank. 2007. Remittances 2007: A Bend in the Road or a New Direction? Washington, DC: Author.

International Conference on the Protection of Fauna and Flora of Africa. 1933. "Letter from Society for the Preservation of Wild Fauna in the Empire to the Under Secretary of State, 18 December."

International Land Coalition. (2008). Participatory Mapping as a Tool for Empowerment: Experiences and Lessons Learned from the ILC Network. Rome: ILC.

Irawan, S., and L. Tacconi. 2009. "Reducing Emissions from Deforestation and Forest Degradation (Redd) and Decentralized Forest Management." International Forestry Review 11 (4): 427–38.

Iriarte, J., I. Holst, O. Marozzi, C. Listopad, E. Alonso, A. Rinderknecht, and J. Montaña. 2004. "Evidence for Cultivar Adoption and Emerging Complexity during the Mid-Holocene in the La Plata Basin." Nature 432 (7017): 614–17.

Iverson, L. R. 1991. "Forest Resources of Illinois: What Do We Have and What Are They Doing for Us? Symposium Proceedings: Our Living Heritage." Illinois Natural History Survey Bulletin 34:361–74.

Izikowitch, K. G. (1944) 2004. Over the Misty Mountain : A Journey from Tonkin to the Lamet in Laos. Bangkok: White Lotus.

Jacoby, K. 2001. Crimes against Nature: Squatters, Poachers, Thieves, and the Hidden History of American Conservation. Berkeley: University of California Press.

Jalais, A. 2005. "Dwelling on Morichjhanpi: When Tigers Became 'Citizens,' Refugees 'Tiger-Food.'" Economic and Political Weekly:1757–62.

Janowski, M. 1988. "The Motivating Forces Behind Changes in the Wet Rice Agricultural System in the Kelabit Highlands." Sarawak Gazette 1504:9–20.

———. 1995. "The Hearth-Group, the Conjugal Couple and the Symbolism of the Rice Meal among the Kelabit of Sarawak." In About the House: Levi-Strauss and Beyond, edited by J. Carsten and S. Hugh-Jones. Cambridge: Cambridge University Press.

———. 1996. "The Kelabit Attitude to the Penan: Forever Children." La Ricerca Folklorica (34):55–58.

———. 2001. "Rice, Women, Men and the Natural Environment among the Kelabit of Sarawak." In Sacred Custodians of the Earth? Women, Spirituality and the Environment, edited by S. Tremayne and A. Low, 107–18. Oxford, New York: Berghahn.

———. 2004. "The Wet and the Dry: The Development of Rice Growing in the Kelabit Highlands, Sarawak." In Smallholders and Stockbreeders. Histories of Foodcrop and Livestock Farming in Southeast Asia, edited by P. Boomgard and D. Henley, 139–62. Leiden: KITLV Press.

———. 2005. "Rice as a Bridge between Two Symbolic Economies: Migration within and out of the Kelabit Highlands, Sarawak." In Histories of the Borneo Environment. Economic, Political and Social Dimensions of Change and Continuity, edited by R. Wadley, 245–70. Leiden: KITLV Press.

———. 2007. "Being 'Big,' Being 'Good': Feeding, Kinship, Potency and Status among the Kelabit of Sarawak." In Kinship and Food in Southeast Asia, edited by M. Janowski and F. Kerlogue, 93–120. Copenhagen: NIAS Press.

Janowski, M. 2003. The Forest, Source of Life: The Kelabit of Sarawak vol. 143. London: British Museum.

Janowski, M., and J. Langub. 2011. "Marks and Footprints in the Forest: The Kelabit and the Penan of Sarawak." In Why Cultivate? Understandings of Past and Present Adoption, Abandonment, and Commitment to Agriculture in Southeast Asia, edited by G. Barker and M. Janowski, 143–63. Leiden: KITLV Press.

Jansen, K. 2000. "Structural Adjustment, Peasant Differentiation and the Environment in Central America." In Disappearing Peasantries?: Rural Labour in Africa, Asia and Latin America, edited by D. F. Bryceson, J. E. Mooij, and C. Kay, 192–212. Exeter: Intermediate Technology.

Janzen, D. H. 1988. "Management of Habitat Fragments in a Tropical Dry Forest: Growth." Annals of the Missouri Botanical Garden 75 (1): 105–16.

———. 1997. Wildland Biodiversity Management in the Tropics. In Biodiversity 2: Understanding and Protecting Our Biological Resources, edited by M. Reaka-Kudla, D. Wilson, and E. O. Wilson, 411–31. Washington, DC: National Academy Press.

Jarvie, J., R. Kanaan, M. Malley, T. Roule, and J. Thomson. n.d. Conflict Timber: Dimensions of the Problem in Asia and Africa, Volume II—Asian Cases. Washington, DC: USAID; ARD.

Jianchu, X. 2006. "The Political, Social, and Ecological Transformation of a Landscape." Mountain Research and Development 26 (3): 254–62.

Jianchu, X., J. Fox, J. Vogler, Z. Yongshou, Y. Lixin, Q. Jie, and S. Leisz. 2005. "Land-Use and Land-Cover Change and Farmer Vulnerability in Xishuangbanna Prefecture in Southwestern China." Environmental Management 36 (3): 404–13.

Johns, A. 1999. Dreadful Visitations: Confronting Natural Catastrophe in the Age of Enlightenment. London: Routledge.

Johnsingh, A. J. T. 2006. Field Days: A Naturalist's Journey through South and Southeast Asia. Hyderabad, A.P., India: Universities Press: Distributed by Orient Longman.

Johnsingh, A. J. T., K. Sankar, and S. Mukherjee. 1997. "Saving Prime Tiger Habitat in Sariska Tiger Reserve." Cat News 27:3.

Johnson, E. W. 1999. Chicago Metropolis 2020: Preparing Metropolitan Chicago for the 21st Century. Chicago: Commerical Club of Chicago in Association with the American Academy of Arts and Sciences.

Johnson, S., and J. Boswell. 1984. A Journey to the Western Islands of Scotland and the Journal of a Tour to the Hebrides. New York: Penguin.

Jon, C. L., and P. N. Findlay. 2000. "A Classic Maya House-Lot Drainage System in Northwestern Belize." Latin American Antiquity 11'(2): 175–85.

Jonsson, H. 2005. Mien Relations : Mountain People and State Control in Thailand. Ithaca: Cornell University Press.

Juenong, W. 1990. 中国地方志茶叶历史资料. Beijing: 农业出版社. Beijing: Agricultural Publishing House of the Chinese Local Tea Historical Data.

Junqueira, A. B., G. H. Shepard, and C. R. Clement. 2010. "Secondary Forests on Anthropogenic Soils in Brazilian Amazonia Conserve Agrobiodiversity." Biodiversity and Conservation 19 (7): 1933–61.

Kabra, A. 2003. "Displacement and Rehabilitation of an Adivasi Settlement: Case of Kuno Wildlife Sanctuary, Madhya Pradesh." Economic and Political Weekly 38 (29): 3073–78.

———. 2006. "Wildlife Protection: Reintroduction and Relocation." Economic and Political Weekly 41 (14):1309.

Kahlheber, S. 1999. "Indications for Agroforestry. Archaeobotanical Remains of Crops and Woody Plants from Medieval Saouga, Burkina Faso." In The Exploitation of Plant Resources in Ancient Africa, edited by M. Van Der Veen, 89–100. New York: Kluwer Academic Pub.

Kaimowitz, D. 2003. Forests and War, Forests and Peace. Bogor, Indonesia: Center for International Forestry Research.

———. 2005. Vital Forest Graphics. United Nations Environment Program (UNEP).

Kalm, P. 1772. Travels into North America; Containing Its Natural History, and a Circumstantial Account of Its Plantations and Agriculture in General. Translated by J. R. Forster, 2d ed. London: T. Lowndes.

Kamete, A. Y. 1998. "Interlocking Livelihoods: Farm and Small Town in Zimbabwe." Environment and Urbanization 10:23–34.

Kammesheidt, L. 1999. "Forest Recovery by Root Suckers and Above-Ground Sprouts after Slash-and- Burn Agriculture, Fire and Logging in Paraguay and Venezuela." Journal of Tropical Ecology 15 (2): 143–57.

Kandeh, H. B. S., and P. Richards. 1996. "Rural People as Conservationists: Querying Neo-Malthusian Assumptions About Biodiversity in Sierra Leone." Africa: Journal of the International African Institute 66 (1): 90–103.

Kant, S. 2009. "Recent Global Trends in Forest Tenures." Forestry Chronicle 85 (6): 849–58.

Karanth, K. K. 2007. "Making Resettlement Work: The Case of India's Bhadra Wildlife Sanctuary." Biological Conservation 139 (3–4): 315–24.

Karanth, K. U. 2001. The Way of the Tiger : The Natural History and Conservation of the Endangered Big Cat. Hyderabad: Orient Longman India.

Karanth, K. U., J. D. Nichols, N. Kumar, W. A. Link, and J. E. Hines. 2004. "Tigers and Their Prey: Predicting Carnivore Densities from Prey Abundance." Proceedings of the National Academy of Sciences of the United States of America 101 (14): 4854.

Kareiva, P., S. Watts, R. Mcdonald, and T. Boucher. 2007. "Domesticated Nature: Shaping Landscapes and Ecosystems for Human Welfare." Science 316 (5833): 1866–69.

Kasanga, K., and G. R. Woodman. 2004. "Ghana: Local Law Making and Land Conversion in Kumasi, Asante." In Local Land Law and Globalization: a Comparative Study of Peri-Urban Areas in Benin, Ghana and Tanzania, edited by G. R. Woodman, U. Wanitzek and H. Sippel, 153. Münster: Lit Verlag.

Kathirithamby-Wells, J. 2005. Nature and Nation: Forests and Development in Peninsular Malaysia. Honolulu: University of Hawaii Press.

Kaufman, H. 1960. The Forest Ranger: A Study in Administrative Behavior. Baltimore: Published for Resources for the Future by Johns Hopkins Press.

Kauppi, P. E., J. H. Ausubel, J. Fang, A. S. Mather, R. A. Sedjo, and P. E. Waggoner. 2006. "Returning Forests Analyzed with the Forest Identity." Proceedings of the National Academy of Sciences 103 (46): 17574.

Kaushik, H. 2008. "MP Seeks Lions from Zoos." The Times of India, 11 May.

Kelly, R. T. 1912. Burma Painted and Described. London: Adam and Charles Black.

Kessler, J. J., and C. Geerling. 1994. Profil Environnemental Du Burkina Faso [*Environmental Profile from Burkina Faso*]. Wageningen: Université agronomique, Département de l'Aménagement de la nature.

Killion, T. W. 1990. "Cultivation Intensity and Residential Site Structure: An Ethnoarchaeological Examination of Peasant Agriculture in the Sierra de los Tuxtlas, Veracruz, Mexico." Latin American Antiquity 1 (3): 191–215.

Kim, J. S., G. Sparovek, R. M. Longo, W. J. De Melo, and D. Crowley. 2007. "Bacterial Diversity of Terra Preta and Pristine Forest Soil from the Western Amazon." Soil Biology and Biochemistry 39 (2): 684–90.

Klare, M. T. 2002. Resource Wars: The New Landscape of Global Conflict. New York: Henry Holt.

Klein, N. 2000. No Logo: Taking Aim at the Brand Bullies. London: Flamingo.

Klooster, D. 2003. "Forest Transitions in Mexico: Institutions and Forests in a Globalized Countryside." Professional Geographer 55 (2): 227–37.

Knox, A., R. S. Meinzen-Dick, and M. Di Gregorio, eds. 2000. Collective Action, Property Rights, and Devolution of Natural Resource Management: Exchange of Knowledge and Implications for Policy. CAPRi Working Papers. Feldafing, Germany: International Food Policy Research Institute (IFPRI).

Koch, E. 1998. Dara-Shikoh Shooting Nilgais: Hunt and Landscape in Mughal Painting. Freer Gallery of Art, Arthur M. Sackler Gallery, Smithsonian Institution.

Koop, G., and Tole, L. 2001. "Deforestation, Distribution and Development." Global Environmental Change 11:193–202.

Kosek, J. 2006. Understories: The Political Life of Forests in Northern New Mexico. Durham, NC: Duke University Press.

Kotey, N. A., J. Mayers, IIED Forestry and Land Use Programme, and Ghana Ministry of Lands and Forestry. 1998. Falling into Place : Policy That Works for Forests and People vol. no 4. London: International Institute for Environment and Development.

Kothari, A., N. Pathak, F. Vania, and A. Islam. 2000. Where Communities Care. Community Based Wildlife and Ecosystem Management in South Asia. London: International Institue of Environment and Development.

Kotraiah, C. T. M., and A. L. Dallapiccola. 2003 King, Court, and Capital: An Anthology of Kannada Literary Sources from the Vijayanagara Period. Delhi: Manohar Press.

Krishnan, M. 1959. The Mudumalai Wildlife Sanctuary. Madras State Forest Department.

———. 1972. Delimit Sanctum Sanctorum in National Parks and Wildlife Sanctuaries and Exclude All Human Disturbance Including Forestry Operations. Delhi: Indian Board for Wildlife, IX Session, 20–23.

Krishnaswamy, J., V. Mehta, M. Bunyan, N. Patil, S. Naveenkumar, K. Karanth, N. Jain, P. Bhargav, and S. Gubbi. 2003. Impact of Iron Ore Mining in Kudremukh on Bhadra River Ecosystem. New Delhi: Tata McGraw-Hill.

Krüger, F. 1998. "Taking Advantage of Rural Assets as a Coping Strategy for the Urban Poor: The Case of Rural–Urban Interrelations in Botswana." Environment and Urbanization 10 (1): 119–34.

Kuletz, V. 1998. Tainted Desert: Environmental Ruin in the American West. London: Routledge.

Kull, C. A., C. K. Ibrahim, and T. C. Meredith. 2007. "Tropical Forest Transitions and Globalization: Neo-Liberalism, Migration, Tourism, and International Conservation Agendas." Society and Natural Resources 20 (8): 723–37.

Kumar, M. 2005. "Claims on Natural Resources: Exploring the Role of Political Power in Pre-Colonial Rajasthan, India." Conservation and Society 3 (1): 134.

Kunen, J. L. 2004. Ancient Maya Life in the Far West Bajo: Social and Environmental Change in the Wetlands of Belize. Anthropological Papers of the University of Arizona no. 69. Tucson: University of Arizona Press.

Kunstadter, P., and E. C. Chapman. 1978. "Problems of Shifting Cultivation and Economic Development in Northern Thailand." In Farmers in the Forest, edited by P. Kunstadter, E. C. Chapman, and S. Sabhasri, 3–23. Honolulu: East-West Center; University of Hawaii Press.

Kuo, F. E. 2003. "The Role of Arboriculture in a Healthy Social Ecology." Journal of Arboriculture 29 (3): 148–55.

Kuo, F. E., and A. Faber Taylor. 2004. "A Potential Natural Treatment for Attention-Deficit/Hyperactivity Disorder: Evidence from a National Study." American Journal of Public Health 94 (9): 1580.

Kuo, F. E., and W. C. Sullivan. 2001. "Environment and Crime in the Inner City: Does Vegetation Reduce Crime?" Environment and Behavior 33 (3): 343–67.

Kupperman, K. O. 1982. "The Puzzle of the American Climate in the Early Colonial Period." The American Historical Review 87 (5): 1262–89.

Laborde, J., S. Guevara, and G. Sánchez-Ríos. 2008. Tree and Shrub Seed Dispersal in Pastures: The Importance of Rainforest Trees Outside Forest Fragments. Quebec: Centre d'etudes nordique, Universite Laval.

Lambert, A. D. 1996. "Empire and Seapower: Shipbuilding by the East India Company at Bombay for the Royal Navy, 1805–1850." In Les Flottes Des Compagnies Des Indes, 1600-1857, edited by P. Haudrere, 149–71. Vincennes: Service Historique de la Marine.

Lambin, E. F. 1997. "Modelling and Monitoring Land-Cover Change Processes in Tropical Regions." Progress in Physical Geography 21 (3): 375.

———. 1999. "Monitoring Forest Degradation in Tropical Regions by Remote Sensing: Some Methodological Issues." Global Ecology and Biogeography 8 (3–4): 191–98.

Lambin, E. F., and P. Meyfroidt. 2010. "Land Use Transitions: Socio-Ecological Feedback versus Socio-Economic Change." Land Use Policy 27 (2): 108–18.

Lambin, E. F., B. L. Turner, H. J. Geist, S. B. Agbola, A. Angelsen, J. W. Bruce, O. T. Coomes, R. Dirzo, G. Fischer, and C. Folke. 2001. "The Causes of Land-Use and Land-Cover Change: Moving Beyond the Myths." Global Environmental Change 11 (4): 261–69.

Langstroth, R. P. 1996. "Forest Islands in an Amazonian Savanna of Northeastern Bolivia." PhD diss., University of Wisconsin, Madison.

Langub, J. 1989. "Some Aspects of the Life of the Penan." Sarawak Museum Journal 40 (3): 163–84.

———. 1993. "Hunting and Gathering: A View from within." In Change and Development in Borneo. Selected Papers from the First Extraordinary Session of the Borneo Research Council, August 4–9, 1990, edited by V. H. Sutlive. Williamsburg: Borneo Research Council.

Lanly, J. P., ed. 1981. Tropical Forest Resources Assessment Project (Gems). Rome: UN Food and Agriculture Organization; UN Environment Programme.

Lao Extensions for Agriculture Project. 2007. Contract Farming in Laos: Cases and Questions. Vientiane: National Agriculture and Forestry Extension Service.

Larwanou, M., M. Abdoulaye, and C. Reij. 2006. Etude de la Régénération Naturelle Assistée dans la Région de Zinder (Niger): une première exploration d'un phénomène spectaculaire. Ms [Study of Natural Regeneration in the Zinder Region (Niger): A First Exploration of a Spectacular Phenomenon]. Washington, DC: International Resources Group for the US Agency for International Development.

Larson, A. M., D. Barry, and G. R. Dahal. 2010. "New Rights for Forest-Based Communities? Understanding Processes of Forest Tenure Reform." International Forestry Review 12 (1): 78–96.

Lathrap, D. W. 1970. The Upper Amazon. London: Praeger.

———. 1973. "The Antiquity and Importance of Long-Distance Trade Relationships in the Moist Tropics of Pre-Columbian South America." World Archaeology 5 (2): 170–86.

———. 1974. "The Moist Tropics, the Arid Lands, and the Appearance of Great Art Styles in the New World." Art and the Environment in Native America 7:115–58.

———. 1987. "The Introduction of Maize in Prehistoric Eastern North America: The View from Amazonia and the Santa Elena Peninsula." In Emergent Horticultural Economies of the Eastern Woodlands, edited by W. F. Keegan, 345–71. Carbondale: Southern Illinois University Center for Archaeological Investigations.

Lathrap, D. W., A. Gebhart-Sayer, and A. Mester. 1985. "The Roots of the Shipibo Art Style: Three Waves on Imiriacocha or There Were Incas before the Incas." Journal of Latin American Lore 11 (1): 31–119.

Laungaramsri, P. 2001. Redefining Nature: Karen Ecological Knowledge and the Challenge to the Modern Conservation Paradigm. Chennai, India: Earthworm Books.

Laurance, W. F. 2008. "Theory Meets Reality: How Habitat Fragmentation Research Has Transcended Island Biogeographic Theory." Biological Conservation 141 (7): 1731–44.

Lavelle, P. 2000. Facing Reality at Hanford. Seattle, WA: Government Accountability Project.

Le Billon, P. 2000. "The Political Ecology of Transition in Cambodia 1989–1999: War, Peace and Forest Exploitation." Development and Change 31 (4): 785–805.

———. 2001. "The Political Ecology of War: Natural Resources and Armed Conflicts." Political Geography 20 (5): 561–84.

———. 2007. "Geographies of War: Perspectives on 'Resource Wars.'" Geography Compass 1 (2): 163–82.

Leach, M., and J. Fairhead. 1994. The Forest Islands of Kissidougou: Social Dynamics of Environmental Change in West Africa's Forest-Savanna Mosaic. Report to ESCOR. Overseas Development Administration, London.

———. 1995. "Ruined Settlements and New Gardens: Gender and Soil-Ripening among Kuranko Farmers in the Forest-Savanna Transition Zone." IDS bulletin 26 (1): 24–32.

———. 2000. "Fashioned Forest Pasts, Occluded Histories? International Environmental Analysis in West Africa Locales." In Forests: Nature, People, Power, edited by M. Doornbos, A. Saith, and B. White, 35–59. Oxford: Blackwell.

Leach, M., and R. Mearns. 1996. The Lie of the Land: Challenging Received Wisdom on the African Environment. Oxford: International African Institute in association with James Currey.

Leary, J. 1995. Violence and the Dream People: The Orang Asli in the Malayan Emergency, 1948–1960 vol. 95. Athens: Ohio University Center for International Studies.

Lee, K. 1995. "Apuntes Sobre las Obras Hidráulicas Prehispánicas de las Llanuras de Moxos: Una Opción Ecológica Inédita [Notes about the Prehispanic Water Works Moxos Plains: An Ecological Option Unpublished]." Unpublished Manuscript.

Lefevbre, H. 1991. The Production of Space. Malden, MA: Wiley-Blackwell.

Lehmann, J. 2003. Amazonian Dark Earths: Origin, Properties, Management. Boston: Kluwer Acadamic.

Leigh, M. 1998. "The Political Economy of Logging in Sarawak." In The Politics of Environment in Southeast Asia, Resources and Resistance, edited by P. Hirsch and C. Warren, 93–106. London: Routledge.

Lele, S. 2007. "A Defining Moment for Forests?" Economic and Political Weekly 42:2379–83.

Lele, S. M., and R. Jayakumar, eds. 2006. Hydrology and Watershed Services in the Western Ghats of India: Effects of Land Use and Land Cover Change. New Delhi: Tata McGraw-Hill Pub. Co.

Lentz, D. L. 1999. "Plant Resources of the Ancient Maya: The Paleoethnobotanical Perspective." In Reconstructing Ancient Maya Diet, edited by C. D. White, 3–18. Salt Lake City: University of Utah Press.

———. 2000. Imperfect Balance: Landscape Transformations in the Precolumbian Americas. New York: Columbia University Press.

Lentz, D. L., L. Haddad, S. Cherpelis, H. J. M. Joo, and M. Potter. 2002. "Long-Term Influences of Ancient Maya Agroforestry Practices on Tropical Forest Biodiversity in Northwestern Belize." In Ethnobiology and Biocultural Diversity: Proceedings of the Seventh International Congress of Ethnobiology, edited by J. R. Stepp, F. S. Wyndham, and R. Zarger, 431–41. Athens: University of Georgia Press.

Lentz, D. L., and B. Hockaday. 2009. "Tikal Timbers and Temples: Ancient Maya Agroforestry and the End of Time." Journal of Archaeological Science 36 (7): 1342–53.

Lentz, D. L., and C. R. Ramírez-Sosa. 2002. "Cerén Plant Resources: Abundance and Diversity." In Before the Volcano Erupted: The Cerén Village in Central America, edited by P. D. Sheets, 33–42. Austin: University of Texas Press.

Letcher, S. G., and R. L. Chazdon. 2009. "Rapid Recovery of Woody Biomass, Species Richness, and Species Composition in a Forest Chronosequence in Northeastern Costa Rica." Biotropica 41:608–17.

Levins, R. 1979. "Coexistence in a Variable Environment." American Naturalist 114 (6): 765–83.

Levtzion, N., and J. F. P. Hopkins. 1981. Corpus of Early Arabic Sources for West African History. Cambridge: Cambridge University Press.

Lewicki, T., and M. Johnson. 1974. West African Food in the Middle Ages: According to Arabic Sources. New York: Cambridge University Press.

Lewis, M. L. 2003. Inventing Global Ecology: Tracking the Biodiversity Ideal in India, 1945–1997. New Perspectives in South Asian History vol. 5. Hyderabad: Orient Longman.

Lewis, T. L. 2000. "Transnational Conservation Movement Organizations: Shaping the Protected Area Systems of Less Developed Countries." Mobilization: An International Quarterly 5 (1): 103–21.

Li, P. 1981. 近代西藏茶叶市场之争与云南茶的地位 [The Battle of the Tea Market in Modern Tibet. Status with the Yunnan Tea]. Yunnan Shehui Kexue 4:71–78.

Li, T. 1999. Transforming the Indonesian Uplands: Marginality, Power and Production. Amsterdam: Harwood.

———. 2002. "Local Histories, Global Markets: Cocoa and Class in Upland Sulawesi." Development and Change 33 (3): 415–37.

Linares, O. F. 1976. "'Garden Hunting' in the American Tropics." Human Ecology 4 (4): 331–49.

Lindsay, J. M. 1977. "Forestry and Agriculture in the Scottish Highlands 1700–1850: A Problem in Estate Management." Agricultural History Review 25:23–36.

Linhares, L. F. D. 2004. "Kilombos of Brazil—Identity and Land Entitlement." Journal of Black Studies 34 (6): 817–37.

Little, P. D. 1996. "Pastoralism, Biodiversity, and the Shaping of Savanna Landscapes in East Africa." Africa: Journal of the International African Institute 66 (1): 37–51.

———. 2001. Amazonia: Territorial Struggles on Perennial Frontiers. Baltimore: Johns Hopkins University Press.

Liu, N. 1990. An Investigation Report on Land Use and Economy of Xishuangbanna. Kumming: Yunnan People's Press.

Lohse, J. C., and P. N. Findlay. 2000. "A Classic Maya House-Lot Drainage System in Northwestern Belize." Latin American Antiquity 11 (2): 175–85.

Long, A. J., and P. K. R. Nair. 1999. "Trees Outside Forests: Agro-, Community, and Urban Forestry." New Forests 17 (1–3): 145–74.

López Mazz, J., 2001. "Las estructuras tumulares (cerritos) del litoral Atlántico uruguayo [The Burial Mounds (Cerritos) of the Uruguayan Atlantic Coast]." Latín American Antiquity 12 (3): 231–55.

Lovett, P. N., and N. Haq. 2000a. "Diversity of the Sheanut Tree (*Vitellaria paradoxa* cf. Gaertn.) in Ghana." Genetic Resources and Crop Evolution 47 (3): 293–304.

———. 2000b. "Evidence for Anthropic Selection of the Sheanut Tree (Vitellaria Paradoxa)." Agroforestry Systems 48 (3): 273–88.

Lowenthal, D. 1965. "Introduction." In Man and Nature: Or, Physical Geography as Modified by Human Action, edited by G. P. Marsh, 3–7. Cambridge: Belknap Press.

Lowie, R. 1948. "The Tropical Forest: An Introduction." In Handbook of South American Indians Vol. 3: The Tropical Forest Tribes, edited by J. Steward, 1–56. Washington, DC: Smithsonian Institution Bureau of American Ethnology.

Lugo, A. E., and E. Helmer. 2004. "Emerging Forests on Abandoned Land: Puerto Rico's New Forests." Forest Ecology and Management 190 (2–3): 145–61.

Lund, C. 2002. "Negotiating Property Institutions: On the Symbiosis of Property and Authority in Africa." In Negotiating Property in Africa, edited by K. Juul and C. Lund, 11–43. Portsmouth: Heinemann.

———. 2008. Local Politics and the Dynamics of Property in Africa. New York: Cambridge University Press.

Lundell, C. L. 1934. Preliminary Sketch of the Phytogeography of the Yuacatan Peninsula. Washington, DC: Carnegie Institution.

———. 1945. "Vegetation and Natural Resources of British Honduras." In Plants and Plant Science in Latin America, edited by F. Verdoorn, 270–73. Waltham: Chronica Botanica Company.

Lye, T.-P., W. de Jong, and K.-I. Abe. 2003. The Political Ecology of the Tropical Forests in Southeast Asia: Historical Perspectives vol. 6. Melbourne: Trans Pacific Press.

Lyttleton, C., P. Cohen, H. Rattanavong, B. Thongkhamhane, and S. Sisaemgrat. 2004. Watermelons, Bars, and Trucks: Dangerous Intersections in Northwest Lao PDR. Vientiane: Institute for Cultural Research of Laos ; Macquarie University.

Macarthur, R. H., and E. O. Wilson. 1967. The Theory of Island Biogeography. Monographs in Population Biology vol. 1. Princeton, NJ: Princeton University Press.

Macinnes, A. I. 1996. Clanship, Commerce and the House of Stuart, 1603–1788. East Linton, Scotland: Tuckwell Press.

Mackey, B., K. Kintzele, D. Aftandilian, R. Engel, and P. Heltne. 2008. Keeping Nature Alive: Toward a Code of Ethics for Biodiversity Conservation. Gland: International Union for Conservation of Nature (IUCN).

Mackie, J. A. C. 1974. Konfrontasi: The Indonesia-Malaysia Dispute, 1963–1966. London: Oxford University Press.

Macpherson, J. 1763. Fingal, an Ancient Epic Poem. In Six Books: Together with Several Other Poems, Composed by Ossian the Son of Fingal. Translated from the Gallic Language by James Macpherson. Dublin: Printed for Peter Wilson.

Madhusudan, M. 2005. "Of Rights and Wrongs: Wildlife Conservation and the Tribal Bill." Economic and Political Weekly 40 (47): 4893.

Madhusudan, M., and R. Arthur. 2007. "Succeeding Poorly or Failing Better?" Seminar 577:74.

Madhusudan, M., and T. S. Raman. 2003. "Conservation as If Biological Diversity Matters: Preservation versus Sustainable Use in India." Conservation and Society 1 (1): 49.

Madhusudan, M., K. Shanker, A. Kumar, C. Mishra, A. Sinha, R. Arthur, A. Datta, M. Rangarajan, R. Chellam, and G. Shahabuddin. 2006. "Science in the Wilderness: The Predicament of Scientific Research in India's Wildlife Reserves." Current Science 91 (8): 1015–19.

Madhusudan, M. D., and C. Mishra. 2003. "Why Big, Fierce Animals Are Threatened: Conserving Large Mammals in Densely Populated Landscapes." In Battles over Nature: Science and the Politics of Wildlife Conservation, edited by V. K. Saberwal and M. Rangarajan, 31–55. New Delhi: Permanent Black.

Maffi, L. 2005. "Linguistic, Cultural, and Biological Diversity." Annual Review of Anthropology 29:599–617.

Mahmud, Z. B. H. 1979. "The Evolution of Population and Settlement in the State of Kedah." In Essays on Linguistic, Cultural and Socio-Economic Aspects of the Malaysian State of Kedah, edited by D. Asmah Haji Omar, 120–53. Kuala Lumpur: University of Malaya.

Makhijani, A., and S. I. Schwartz. 1998. "Victims of the Bomb." In Atomic Audit: The Costs and Consequences of US Nuclear Weapons since 1940, edited by S. I. Schwartz, 408–11. Washington, DC: Brookings Institution Press.

Malhi, Y., J. T. Roberts, R. A. Betts, T. J. Killeen, W. Li, and C. A. Nobre. 2008. "Climate Change, Deforestation, and the Fate of the Amazon." Science 319 (5860): 169–72.

Mallick, R. 1999. "Refugee Resettlement in Forest Reserves: West Bengal Policy Reversal and the Marichjhapi Massacre." Journal of Asian Studies 58 (1):104–25.

Malthus, T. R. 1826. An Essay on the Principle of Population, 6th ed. London: John Murray.

Mann, C. 2005. 1491: New Revelations About the Americas before Columbus. New York: Knopf.

Mann, C. C. 2000. "Earthmovers of the Amazon." Science 287 (54): 786–89.

———. 2002. "1491." Atlantic Monthly 289 (3): 41.

———. 2008. "Ancient Earthmovers of the Amazon." Science 321:1148–52.

Maranz, S. 2009. "Tree Mortality in the African Sahel Indicates an Anthropogenic Ecosystem Displaced by Climate Change." Journal of Biogeography 36 (6): 1181–93.

Maranz, S., W. Kpikpi, Z. Wiesman, A. De Saint Sauveur, and B. Chapagain. 2004. "Nutritional Values and Indigenous Preferences for Shea Fruits (Vitellaria paradoxa cf. Gaertn. F.) in Ghana." Genetics and Crop Evolution 47:293–304.

Maranz, S., and Z. Wiesman. 2003. "Evidence for Indigenous Selection and Distribution of the Shea Tree, Vitellaria paradoxa, and Its Potential Significance to Prevailing Parkland Savanna Tree Patterns in Sub Saharan Africa North of the Equator." Journal of Biogeography 30 (10): 1505–16.

Mariotti, A., and E. Peterschmitt. 1994. "Forest Savanna Ecotone Dynamics in India as Revealed by Carbon Isotope Ratios of Soil Organic Matter." Oecologia 97 (4): 475–80.

Marris, E. 2006. "Putting the Carbon Back: Black Is the New Green." Nature 442 (7103): 624–26.

Marsh, G. P. (1864) 1965. Man and Nature, or, Physical Geography as Modified by Human Action. Cambridge: Belknap Press.

Martine, G., and A. Marshall. 2007. State of World Population 2007: Unleashing the Potential of Urban Growth. New York: United Nations Population Fund (UNFPA).

Masco, J. 2007. The Nuclear Borderlands. Princeton, NJ: Princeton University Press.

Massey, D. B. 2005. For Space. Thousand Oaks, CA: SAGE.

Mather, A. S. 1992. "The Forest Transition." Area 24 (4): 367–79.

———. 1993. Afforestation: Policies, Planning, and Progress. London: Belhaven Press.

———. 2005. "Assessing the World's Forests." Global Environmental Change 15 (3): 267–80.

———. 2007. "Recent Asian Forest Transitions in Relation to Forest-Transition Theory." International Forestry Review 9 (1): 491–502.

Mather, A. S., and C. L. Needle. 1998. "The Forest Transition: A Theoretical Basis." Area 30 (2): 117–24.

———. 1999. "Development, Democracy and Forest Trends." Global Environmental Change 9 (2): 105–18.

Mathews, A. S. 2008. "State Making, Knowledge, and Ignorance: Translation and Concealment in Mexican Forestry Institutions." American Anthropologist 110 (4): 484–94.

———. 2009. "Unlikely Alliances Encounters between State Science, Nature Spirits, and Indigenous Industrial Forestry in Mexico, 1926–2008." Current Anthropology 50 (1): 75–101.

Mathieu, P., M. Zongo, and L. Paré. 2003. "Monetary Land Transactions in Western Burkina Faso: Commoditization, Papers and Ambiguities." In Securing Land Rights in Africa, edited by T. A. Benjaminsen and C. Lund, 109–28. London: Frank Cass.

Matthew, H. C. G. 2000. The Nineteenth Century : The British Isles: 1815–1901. Short Oxford History of the British Isles. New York: Oxford University Press.

Maung, M. 1976. "Nationalist Movements in Burma 1920–1940: Changing Patterns of Leadership from Sangha to Laity." MA thesis, Australian National University.

Mayaux, P., F. Achard, and J. P. Malingreau. 1998. "Global Tropical Forest Area Measurements Derived from Coarse Resolution Maps at a Global Level: A Comparison with Other Approaches." Environmental Conservation 25 (1): 37–52.

Mayaux, P., E. Bartholomé, M. Massart, C. Van Cutsem, A. Cabral, A. Nonguierma, O. Diallo, C. Pretorius, M. Thompson, and M. Cherlet. 2003. A Land Cover Map of Africa. Luxembourg: Joint Research Centre, European Commission.

Mazzucato, V., M. Kabki, and L. Smith. 2006. "Transnational Migration and the Economy of Funerals: Changing Practices in Ghana." Development and Change 37 (5): 1047–72.

McAfee, K., and E. N. Shapiro. 2010. "Payments for Ecosystem Services in Mexico: Nature, Neoliberalism, Social Movements, and the State." Annals of the Association of American Geographers 100 (3): 579–99.

McAnany, P. 1992a. "Agricultural Tasks and Tools: Patterns of Stone Tool Discard Near Prehistoric Maya Residences Bordering Pulltrouser Swamp, Belize." In Gardens of

Prehistory: The Archaeology of Settlement Agriculture in Greater Mesoamerica, edited by T. W. Killion. Tuscaloosa: University of Alabama Press.

———. 1992b. "Obscured by the Forest: Property and Ancestors in Maya Lowland Society." In Property in Economic Context, edited by R. C. Hunt and A. Gilman, 184–213. Lanham: University Press of America.

———. 1995. Living with the Ancestors: Kinship and Kingship in Ancient Maya Society, 1st ed. Austin: University of Texas Press.

———.1998. "Obscured by the Forest: Property and Ancestors in Maya Lowland Society." In Property in Economic Context, edited by R.C. Hunt and A. Gilman, 73–87. Lanham, MD: University Press of America.

McCaskie, T. C. 2000. Asante Identities: History and Modernity in an African Village, 1850–1950. Edinburgh: Edinburgh University Press.

McColl, R. 1967. "A Political Geography of Revolution: China, Vietnam, and Thailand." Journal of Conflict Resolution 11 (2): 153–67.

McCoy, A. W., C. B. Read, and L. P. Adams. 1973. The Politics of Heroin in Southeast Asia. New York: Harper Colophon Books.

McDade, L., and G. S. Hartshorn. 1994. "La Selva Biological Station." In La Selva: Ecology and Natural History of a Neotropical Rain Forest, edited by L. Mcdade, K. S. Bawa, H. Hespenheide, and G. S. Hartshorn, 6–14. Chicago: University of Chicago Press.

McElwee, P. 2004. "You Say Illegal, I Say Legal: The Relationship between 'Illegal' Logging and Poverty, Land Tenure, and Forest Use Rights in Vietnam." Journal of Sustainable Forestry 19 (1/2/3): 97–135.

McKey, D., L. Emperaire, M. Elias, F. Pinton, T. Robert, S. Desmouiliere, and L. Rival. 2001. "Local Management and Regional Dynamics of Varietal Diversity of Cassava in Amazonia." Genetics Selection Evolution 33:S465–90.

McKey, D., S. Rostain, J. Iriarte, B. Glaser, J. J. Birk, I. Holst, and D. Renard. 2010. "Pre-Columbian Agricultural Landscapes, Ecosystem Engineers, and Self-Organized Patchiness in Amazonia." Proceedings of the National Academy of Sciences of the United States of America 107 (17): 7823–28.

McKillop, H. 1994. "Ancient Maya Tree Cropping." Ancient Mesoamerica 5 (01): 129–40.

McMichael, P. 1990. "Incorporating Comparison within a World-Historical Perspective: An Alternative Comparative Method." American Sociological Review 55 (3): 385–97.

———. 2006. "Reframing Development: Global Peasant Movements and the New Agrarian Question." Canadian Journal of Development Studies-Revue Canadienne D Etudes Du Developpement 27 (4): 471–83.

McNeely, J. 1994. "Lessons from the Past: Forests and Biodiversity." Biodiversity and Conservation 3 (1): 3–20.

Meggers, B. J. 1954. "Environmental Limitation on the Development of Culture." American Anthropologist 56 (5): 801–24.

———. 1971. Amazonia: Man and Culture in a Counterfeit Paradise. Chicago: Aldine.

———. 1979. "Climatic Oscillation as a Factor in the Prehistory of Amazonia." American Antiquity 44:252–66.

———. 1991. "Cultural Evolution in Amazonia." In Profiles in Cultural Evolution, edited by A. T. Rambo and K. Gillogly, 191–216. Anthropological Papers 85. Museum of Anthropology, University of Michigan, Ann Arbor.

———. 1995. "Amazonia on the Eve of European Contact: Ethnohistorical, Ecological, and Anthropological Perspectives." Revista de Arqueología Americana 8:91–115.

———. 2001. "The Continuing Quest for El Dorado: Round Two." Latin American Antiquity 12 (3): 304–25.

Meinzen-Dick, R. S., A. Knox, and M. Di Gregorio, eds. 2001. "Collective Action, Property Rights, and Devolution of Natural Resource Management: Exchange of Knowledge and Implications for Policy." Feldafing, Germany: Zentralstelle für Ernährung und Landwirtschaft. Available online at http://www.capri.cgiar.org/workshop_devolution.asp.

Meinzen-Dick, R., and E. Mwangi. 2009. "Cutting the Web of Interests: Pitfalls of Formalizing Property Rights." Land Use Policy 26 (1): 36–43.

Meinzen-Dick, R. S., and R. Pradhan. 2002. Legal Pluralism and Dynamic Property Rights. CAPRi Working Paper no. 22. Washington, DC: International Food Policy Research Institute (IFPRI).

Menzies, N. 1992. "Strategic Space: Exclusion and Inclusion in Wildland Policies in Late Imperial China." Modern Asian Studies 26 (04): 719–33.

———. 1994. Forest and Land Management in Imperial China. New York: St. Martins Press.

———. 2007. Our Forest, Your Ecosystem, Their Timber: Communities, Conservation, and the State in Community-Based Forest Management. New York: Columbia University Press,

Mesquita, R. C. G., K. Ickes, G. Ganade, and G. B. Williamson. 2001. "Alternative Successional Pathways in the Amazon Basin." Journal of Ecology 89 (4): 528–37.

Métraux, A. 1948. "The Tribes of Mato Grosso and Eastern Bolivia: The Paressi." Handbook of South American Indians 3:349–60.

Meyer, J.-Y., C. Lavergne, and D. R. Hodel. 2008. "Time Bombs in Gardens: Invasive Ornamental Palms in Tropical Islands, with Emphasis on French Polynesia (Pacific Ocean) and the Mascarenes (Indian Ocean)." Palms 52:23–35.

Meyer, S. M. 2006. The End of the Wild. Cambridge, MA: MIT Press.

Meyfroidt, P., and E. F. Lambin. 2008. "Forest Transition in Vietnam and Its Environmental Impacts." Global Change Biology 14 (6): 1319–36.

Michon, G., H. de Foresta, P. Levang, and F. Verdeaux. 2007. "Domestic Forests: A New Paradigm for Integrating Local Communities' Forestry into Tropical Forest Science." Ecology and Society 12 (2):1 [online].

Miles, L., and V. Kapos. 2008. "Reducing Greenhouse Gas Emissions from Deforestation and Forest Degradation: Global Land-Use Implications." Science 320 (5882): 1454.

Mill, J. S. 1848. Principles of Political Economy. London: J. W. Parker.

Mills, J. H., and T. A. Waite. 2009. "Economic Prosperity, Biodiversity Conservation, and the Environmental Kuznets Curve." Ecological Economics 68 (7): 2087–95.

Millspaugh, C. F. 1895. Contribution to the Flora of Yucatan. Chicago: [s.n.]. Publication (Field Columbian Museum) Botanical series 1 (1): 3–6.

————. 1903. Plantae Yucatanae. (Regionis Antillanae): Plants of the Insular, Coastal and Plain Regions of the Peninsula of Yucatan, Mexico. 2 vols, vol. v 3, no 1–2. Chicago: Field Museum of Natural History.

Milroy, A. J. W. 1934. "The Preservation of Wildlife in India. No. 3, Assam." Journal of the Bombay Natural History Society 37:97–104.

Ministry of Environment and Forests. 2005. Joining the Dots: The Report of the Tiger Task Force. New Delhi: Government of India.

Mishra, H. R., and J. Ottaway. 2008. The Soul of a Rhino: A Nepali Adventure with Kings and Elephant Drivers, Billionaires, and Bureaucrats, Shamans and Scientists, and the Indian Rhinoceros. Guilford, CT: Lyons Press.

Mittermeier, R. A., C. G. Mittermeier, N. Myers, and P. Robles Gil. 1999. Hotspots: Earth's Biologically Richest and Most Endangered Terrestrial Ecoregions. 1st English ed. Mexico City: CEMEX: Conservation International.

Mizuta, H. 2000. Adam Smith's Library: A Catalogue. Rev. ed. New York: Oxford University Press.

Moe, S., P. Wegge, and E. Kapela. 1990. "The Influence of Man-Made Fires on Large Herbivores in Lake Burungi Area in Northern Tanzania." The African Journal of Ecology 28:35–45.

Mohandass, D., and P. Davidar. 2009. "Floristic Structure and Diversity of a Tropical Montane Evergreen Forest (Shola) of the Nilgiri Mountains, Southern India." Tropical Ecology 50 (2): 219–29.

Molnar, A., S. Scherr, and A. Khare. 2004. Who Conserves the World's Forest Areas? Community-Driven Strategies to Protect Forest Areas and Respect Rights. Washington, DC: Forest Trends.

Montagnini, F., and R. O. Mendelsohn. 1997. "Managing Forest Fallows: Improving the Economics of Swidden Agriculture." Ambio 26 (2): 118–23.

Monterroso, I., and D. Barry. 2007. "Community-Based Forestry and the Changes in Tenure and Access Rights in the Mayan Biosphere Reserve, Guatemala." Paper presented at RECOFT, TC International Conference, Poverty Reduction and Forests: Tenure, Market and Policy Reforms, Bangkok.

Mora Camargo, S. 2003. Early Inhabitants of the Amazonian Tropical Rain Forest: A Study of Humans and Environmental Dynamics. University of Pittsburgh Latin American Archaeology Reports vol. 3. Pittsburgh: University of Pittsburgh, Department of Anthropology.

Morales, H. 2002. "Pest Management in Tropical Agroecosystems: Lessons for Pest Prevention Research and Extension." Integrated Pest Management Review 7:145–63.

Morales, H., and I. Perfecto. 2000. "Traditional Knowledge and Pest Management in the Guatemalan Highlands." Agriculture and Human Values 17:49–63.

Morales, H., I. Perfecto, and B. Ferguson. 2001. "Traditional Fertilization and Its Effect on Corn Insect Populations in the Guatemalan Highlands." Agriculture Ecosystems and Environment 84:145–55.

Moran, E. F. 1982. Human Adaptability: An Introduction to Ecological Anthropology. Boulder, CO: Westview Press.

———. 1993. Through Amazonian Eyes : The Human Ecology of Amazonian Populations. Iowa City: University of Iowa Press.

Moran, E. F., E. S. Brondizio, J. M. Tucker, M. C. da Silva-Forsberg, S. Mccracken,
and I. Falesi. 2000. "Effects of Soil Fertility and Land-Use on Forest Succession in
Amazônia." Forest Ecology and Management 139 (1–3): 93–108.

Morehead, F. T. 1944. The Forests of Burma, vol. no 5. London: Longmans, Green & Co.

Morgan, G. 1935. "How Grant Park Was Saved for People: Ward's Great Fight Waged."
Chicago Daily News, June 8.

Morris, R. C. 1937. "Close Seasons for Big Game: Are They Beneficial?" Journal of the
Bombay Natural History Society 39:621–22.

Morrison, K. D. 1995. Fields of Victory: Vijayanagara and the Course of Intensification, Contributions of the University of California Archaeological Research Facility, no. 53, Berkeley, CA.

———.2001. "Coercion, Resistance, and Hierarchy: Local Processes and Imperial Strat-
egies in the Vijayanagara Empire." In Empires: Perspectives from Archaeology and
History, edited by S. Alcock, T. D'Altroy, K. Morrison, and C. Sinopoli, 253–78. New
York: Cambridge University Press.

———. 2002a. "General Introduction: Historicizing Adaptation, Adapting to History." In Forager-Traders in South and Southeast Asia: Long-Term Histories,
edited by K. D. Morrison and L. L. Junker. Cambridge: Cambridge University
Press.

———. 2002b. "Introduction: South Asia." In Forager-Traders in South and Southeast
Asia: Long-Term Histories, edited by K. D. Morrison and L. L. Junker, 21–40. Cambridge: Cambridge University Press.

———. 2002c. "Pepper in the Hills: Upland-Lowland Exchange and the Intensification
of the Spice Trade." In Forager-Traders in South and Southeast Asia: Long-Term
Histories , edited by K. D. Morrison and L. L. Junker, 122–58. Cambridge: Cambridge University Press.

———. 2006. "Historicizing Foraging in Asia: Power, History, and Ecology of Holocene
Hunting and Gathering." In An Archaeology of Asia, edited by M. Stark and B.
Blackwell, 279–302. New York: Wiley.

———. 2009. Daroji Valley: Landscape History, Place, and the Making of a Dryland
Reservoir System. Vijayanagara Research Project Monograph Series 18. Delhi:
Manohar Press.

———. in press. "Opening up the Pre-Colonial: Primeval Forests, Baseline Thinking,
and Other Archives in Environmental History." In Shifting Ground: People, Animals, and Mobility in India's Environmental History, edited by M. Rangarajan and
K. Sivaramakrishnan. Delhi: Oxford University Press.

Morrison, K. D., and Lycett, M.T. 2013. "Forest Products in a Wider World: Early
Historic Connections across Southern India." In Connections and Complexity:
Papers in Honor of Gregory Possehl, edited by S. Abraham, T. Raczek, and U. Rizvi,
127–42. Walnut Creek, CA: Left Coast Press.

Mortimer, R. 1974. Indonesian Communism under Sukarno: Ideology and Politics,
1959–1965. London: Cornell University Press.

Mourão, L., M. A. G. Jardim, and M. Grossmann, eds. 2004. Açaí (Euterpe Oleracea Mart.): Possibilidades E Limites Para O Desenvolvimento Sustentavel No Estuario Amazônico [*Acai (Euterpe oleracea Mart.): Possibilities and Limits for Sustainable Development in the Amazon Estuary*]. Belém: Editora do Museu Paraense Emilio Goeldi.

Mulcahy, M. 2006. Hurricanes and Society in the British Greater Caribbean, 1624–1783. Baltimore: Johns Hopkins University Press.

Muñoz, I., M. Paredes, and R. Thorp. 2007. "Group Inequalities and the Nature and Power of Collective Action: Cases from Peru." World Development 35 (11): 1929–46.

Murphy, A. B. 2005. "Territorial Ideology and Interstate Conflict: Comparative Considerations." In The Geography of War and Peace: From Death Camps to Diplomats, edited by C. R. Flint, 280–96. New York: Oxford University Press.

Murthy, M. L. K. 1994. "Forest Peoples and Historical Traditions in the Eastern Ghats, South India." In Living Traditions: Studies in the Ethnoarchaeology of South Asia, edited by B. Allchin, 205–18. New Delhi, India: Oxford and IBH.

Muscat, R. J. 1990. Thailand and the United States: Development, Security, and Foreign Aid. New York: Columbia University Press.

Mwangi, E., and S. Dohrn. 2008. "Securing Access to Drylands Resources for Multiple Users in Africa: A Review of Recent Research." Land Use Policy 25 (2): 240–48.

Myat Tun, M. 1938. "Some Asiatic Displays at the Glasgow Exhibition." Asiatic Review 34:806–09.

Myers, N. 1980. Conversion of Tropical Moist Forests. Washington, DC: National Academy of Sciences; US National Research Council.

———. 1989. Deforestation Rates in Tropical Forests and Their Climatic Implications. London: Friends of the Earth.

Myers, N., R. A. Mittermeier, C. G. Mittermeier, G. A. B. da Fonseca, and J. Kent. 2000. "Biodiversity Hotspots for Conservation Priorities." Nature 403 (6772): 853–58.

Nagendra, H. 2007. "Drivers of Reforestation in Human-Dominated Forests." Proceedings of the National Academy of Sciences 104 (39): 15218.

Nagendra, H., S. Pareeth, and R. Ghate. 2006. "People within Parks—Forest Villages, Land-Cover Change and Landscape Fragmentation in the Tadoba Andhari Tiger Reserve, India." Applied Geography 26 (2): 96–112.

Nair, P. K. R. 1989. Agroforestry Systems in the Tropics. Forestry Sciences vol. 31. Boston: Kluwer Academic Publishers.

Nassauer, J. I., and P. Opdam. 2008. "Design in Science: Extending the Landscape Ecology Paradigm." Landscape Ecology 23:633–44.

Nasution, A. H. 1953. Fundamentals of Guerilla Warfare. Jakarta: Indonesian Armed Forces.

National Agriculture and Forestry Research Institute (NAFRI), National University of Lao (NUoL), and National Association of Food Equipment Suppliers (NAFES). 2006. "Summary and Recommendations." Paper presented at Workshop on Rubber Development in Lao PDR: Exploring Improved Systems for Smallholder Production, Vientiane.

National Archives of the Gambia. 1937–1942. Annual Report of the Department of Agriculture. Bathurst: Government Printing.

National Archives of India. 1933. "Mysore Residency." File 161 of 1934, Serial 44618.

National Archives of Scotland. 1770–1771. "John Williams Memorial of 1770–71." File E 727/46/22.

Naughton-Treves, L. 2004. "Deforestation and Carbon Emissions at Tropical Frontiers: A Case Study from the Peruvian Amazon." World Development 32 (1): 173–90.

Nasution, A. H. 1965. Fundamentals of Guerrilla Warfare. New York: Praeger.

Needham, R. 1972. "Penan." In Ethnic Groups of Insular Southeast Asia Vol. 1, edited by F. Lebar, 176–80. New Haven, CT: Human Relations Area Files Press.

Neeff, T., and F. Ascui. 2009. "Lessons from Carbon Markets for Designing an Effective Redd Architecture." Climate Policy 9 (3): 306–15.

Neginhal, S. G. 1974. Project Tiger, a Management Plan. Bangalore: Government of Karnataka, 34–35.

Nehru, J. 1987. "Protection of Lions." In Jawaharlal Nehru, Selected Works, Vol. V, edited by S. Gopal. New Delhi: Nehru Memorial Trust.

Nelson, K. C. 1994. "Participation, Empowerment, and Farmer Evaluations: A Comparative Analysis of Ipm Technology Generation in Nicaragua." Agriculture and Human Values 11 (2): 109–25.

Nepstad, D., P. Moutinho, and B. Soares-Filho. 2006. "The Amazon in a Changing Climate: Large-Scale Reductions of Carbon Emissions from Deforestation and Forest Impoverishment." Retrieved from http://www.whrc.org/resources/published _literature/pdf/Amazon-and-Climate-2006.pdf.

Netting, R. M. 1993. Smallholders, Householders: Farm Families and the Ecology of Intensive, Sustainable Agriculture. Stanford: Stanford University Press.

Neumann, K., S. Kahlheber, and D. Uebel. 1998. "Remains of Woody Plants from Saouga, a Medieval West African Village." Vegetation History and Archaeobotany 7 (2): 57–77.

Neumann, R. P. 1997. "Forest Rights, Privileges and Prohibitions: Contextualising State Forestry Policy in Colonial Tanganyika." Environment and History 3:45–68.

———. 1998. Imposing Wilderness: Struggles over Livelihood and Nature Preservation in Africa. California Studies in Critical Human Geography vol. 4. Berkeley: University of California Press.

———. 2001. "Africa's 'Last Wilderness': Reordering Space for Political and Economic Control in Colonial Tanzania." Africa 71 (4): 641–65.

———. 2004. "Nature-State-Territory: Toward a Critical Theorization of Conservation Enclosures." In Liberation Ecologies: Environment, Development, Social Movements, edited by R. Peet and M. Watts, 195–217. New York: Routledge.

Neves, E. G., and J. B. Petersen. 2006. "Political Economy and Pre-Columbian Landscape Transformation in Central Amazonia." In Time and Complexity in Historical Ecology: Studies in the Neotropical Lowlands, edited by W. L. Balée and C. L. Erickson. 279–310. New York: Columbia University Press.

Nevins, J., and N. L. Peluso. 2008. Taking Southeast Asia to Market: Commodities, Nature, and People in the Neoliberal Age. Ithaca: Cornell University Press.

Nevle, R. J., and D. K. Bird. 2008. "Effects of Syn-Pandemic Fire Reduction and Reforestation in the Tropical Americas on Atmospheric CO_2 during European Conquest." Palaeogeography Palaeoclimatology Palaeoecology 264 (1–2): 25–38.

Newman, J. L. 1975. Drought, Famine, and Population Movements in Africa. Foreign and Comparative Studies: Eastern Africa vol. 17. Syracuse, NY: Maxwell School of Citizenship and Public Affairs, Syracuse University.

Newton, M. S. 2000. A Handbook of the Scottish Gaelic World. Portland: Four Courts Press.

Nicholson, S. E. 1979. "The Methodology of Historical Climate Reconstruction and Its Application to Africa." The Journal of African History 20 (1): 31–49.

Nicholson, S. E., C. J. Tucker, and M. B. Ba. 1998. "Desertification, Drought, and Surface Vegetation: An Example from the West African Sahel." Bulletin of the American Meteorological Society 79 (5): 815–29.

Nielsen, J. O., and A. Reenberg. 2009. "Cultural Barriers to Climate Change Adaptation: A Case Study from Northern Burkina Faso." Global Environmental Change—Human and Policy Dimensions 20 (1): 142–52.

Nimuendajú, C. 1952. "The Tapajó." Kroeber Anthropological Society Papers 6:1–25.

Nisbet, J. 1901. Burma under British Rule and Before. 2 vols. Westminster: A. Constable.

Noordwijk declaration on climate change: Atmospheric pollution and climatic change. Ministerial Conference held at Noordwijk, the Netherlands, November 6–7, 1989. Leidschendam: Climate Conference Secretariat.

Nordenskiöld, E. 1913. "Urnengraber Und Mounds Im Bolivianischen Flachlande [Umengraber and Mounds in the Bolivian Lowlands]." Baessler Archiv 3:205–55.

———. 1916. "Die Anpassung Der Indianer an Die Verhältnisse in Den Überschwemmungsgebieten in Südamerika [The Adaptation of the Indians on the Floodplains in South America]." Ymer 36:138–55.

Norem, R. H., R. Yoder, and Y. Martin. 1989. "Indigenous Agricultural Knowledge and Gender Issues in Third World Agricultural Development." In Studies in Technologies and Social Change, No. 11, edited by D. M. Warren, L. J. Slikkerveen, and S. O. Titilola, 91–100. Ames: Technology and Social Change Program, Iowa State University.

Northeastern Illinois Planning Commission. 2004. Chicago Wilderness Green Infrastructure Vision Final Report. Published online by Chicago Wilderness, http://www.chicagowilderness.org/files/8513/3020/2252/Green_Infrastructure_Vision_Final_Report.pdf.

Nowak, D. J., D. E. Crane, and J. C. Stevens. 2006. "Air Pollution Removal by Urban Trees and Shrubs in the United States." Urban Forestry and Urban Greening 4 (3–4): 115–23.

Nugent, S. 1993. Amazonian Caboclo Society: An Essay on Invisibility and Peasant Economy. Providence, Oxford: Berg.

Nyanteng, V. K. 1995. "Prospects for Ghana's Cocoa Industry in the 21st Century." In Cocoa Cycles: The Economics of Cocoa Supply, edited by F. Ruf and P. S. Siswoputranto, 179–208. Cambridge: Woodhead Publishers Ltd.

O'Loughlin, J., and F. Witmer. 2010. "The Localized Geographies of Violence in the North Caucasus of Russia, 1999–2007." Annals of the Association of American Geographers 100 (4): 178–201.

O'Neill, K. 1998. "Building the Bomb." In Atomic Audit: The Costs and Consequences of US Nuclear Weapons since 1940, edited by S. I. Schwartz. Washington, DC: Brookings Institution Press.

Ocampo, J. A. 2009. "The Impact of the Global Financial Crisis on Latin America." Cepal Review 97:9-32.

Odotei, I. K., and A. K. Awedoba, eds. 2006. Chieftancy in Ghana: Culture, Governance and Development. Legon, Ghana: Sub-Saharan Publishers.

Odum, E. P. 1969. "The Strategy of Ecosystem Development." Science 164 (3877): 262-70.

Okali, C., and S. Berry. 1983. "Alley Farming in West Africa in Comparative Perspective." Discussion Paper 11. African-American Issues Center, Boston University.

Okereke, C., and K. Dooley. 2009. "Principles of Justice in Proposals and Policy Approaches to Avoided Deforestation: Towards a Post-Kyoto Climate Agreement." Global Environmental Change-Human and Policy Dimensions 20 (1): 82-95.

Oliveira, P. J. C., G. P. Asner, D. E. Knapp, A. Almeyda, R. Galván-Gildemeister, S. Keene, R. F. Raybin, and R. C. Smith. 2007. "Land-Use Allocation Protects the Peruvian Amazon." Science 317 (31 Aug): 1233-36.

Oliveira, P. S., and R. J. Marquis. 2002. The Cerrados of Brazil: Ecology and Natural History of a Neotropical Savanna. New York: Columbia University Press.

Olschewski, R., and P. C. Benitez. 2005. "Secondary Forests as Temporary Carbon Sinks? The Economic Impact of Accounting Methods on Reforestation Projects in the Tropics." Ecological Economics 55 (3): 380-94.

Olsen, K. H. 2007. "The Clean Development Mechanism's Contribution to Sustainable Development: A Review of the Literature." Climatic Change 84 (1): 59-73.

Openlands. n.d. Openlands Lakeshore Preserve. Available from http://www.openlands .org/policy.asp?pgid=342.

Orozco, M. 2002. "Globalization and Migration: The Impact of Family Remittances in Latin America." Latin American Politics and Society 44 (2): 41-66.

Organismo de Formalización de la Propiedad Informal (COFOPRI). 2007 (March 30). "Ministro de Vivienda entregó 976 títulos de propiedad en Ucayali [Housing Minister delivered 976 titles in Ucayali]." Retrieved from http://www.cofopri.gob.pe/ pren_ver_detnot.asp?cod=78&img=1.

Ortiz, A. 1969. The Tewa World : Space, Time, Being, and Becoming in a Pueblo Society. Chicago: University of Chicago Press.

Orwell, G. (1934) 1987. Burmese Days: A Novel. New York: Harper & Bros.

Ostrom, E., and H. Nagendra. 2006. "Insights on Linking Forests, Trees, and People from the Air, on the Ground, and in the Laboratory." Proceedings of the National Academy of Sciences 103 (51): 19224.

Pacheco, P. 2006. "Agricultural Expansion and Deforestation in Lowland Bolivia: The Import Substitution versus the Structural Adjustment Model." Land Use Policy 23 (3): 205-25.

———. 2009. "Smallholder Livelihoods, Wealth and Deforestation in the Eastern Amazon." Human Ecology 37 (1): 27-41.

Pacheco, P., D. Barry, P. Cronkleton, and A. Larson. 2008. "From Agrarian to Forest Tenure Reforms in Latin America: Assessing Their Impacts for Local People and Forests." Paper presented at the 12th Biennial Conference of the International Association for the Study of the Commons (IASC). Cheltenham, England, July 14–18, 2008.

Padmanabhan, M. 1998. "Conflict in a National Park." The Hindu, August 2.

Padoch, C. 1999. Várzea: Diversity, Development, and Conservation of Amazonia's Whitewater Floodplains. Advances in Economic Botany vol. 13. Bronx, NY: New York Botanical Garden Press.

Padoch, C., E. Brondizio, S. Costa, M. Pinedo-Vasquez, R. R. Sears, and A. Siqueira. 2008. "Urban Forest and Rural Cities: Multi-Sited Households, Consumption Patterns, and Forest Resources in Amazonia." Ecology and Society 13 (2): 2.

Padoch, C., and W. de Jong. 1990. "Santa Rosa: The Impact of the Forest Products Trade on an Amazonian Place and Population." In New Directions in the Study of Plants and People, edited by G. T. P. a. M. J. B, 8:151–58. Bronx: Advances in Economic Botany.

Padoch, C., and M. Pinedo-Vasquez. 1996. "Smallholder Forest Management: Looking Beyond Non-Timber Forest Products." In Current Issues in Non-Timber Forest Products Research, edited by M. Ruiz Pérez and J. E. M. Arnold, 103. Bogor: Center for International Forestry Research (CIFOR).

———. 2006. "Concurrent Activities and Invisible Technologies: An Example of Timber Management in Amazonia." In Human Impacts on Amazonia: The Role of Traditional Ecological Knowledge in Conservation and Development, edited by D. A. Posey and M. J. Balick, 172–80. New York: Columbia University Press.

Paganoto, F. 2007. "Reflexões Sobre Novas Tendências Migratórias no Brasil Contemporâneo: A Ascensão do Fluxo Urbano–Rural [Reflections on New Migratory Tendencies in Contemporary Brazil: The Rise of Rural–Urban Flow]." Paper presented at Proceedings of the 12th Encontro da Associação Nacional de Pós-graduação e pesquisa em planejamento urbano e regional. Belém, 21–25 May.

Pagiola, S., E. Ramìrez, J. Gobbi, C. de Haan, M. Ibrahim, E. Murgueitio, and J. P. Ruìz. 2007. "Paying for the Environmental Services of Silvopastoral Practices in Nicaragua." Ecological Economics 64 (2): 374–85.

Pal, P. 1978. The Classical Tradition in Rajput Painting. New York: Pierpont Morgan Library.

Palo, M. 1999. Forest Transitions and Carbon Fluxes: Global Scenarios and Policies. Helsinki, Finland: UNU World Institute for Development Economics Research.

Palomino, C. 2006. "Peruvian Center for Strategic Forestry Information (CIEF), Ministerio de Agricultura, INRENA, Peru."

Panagariya, A. 2008. India: The Emerging Giant. New York: Oxford University Press.

Panwar, H. S. 1978. "Decline and Restoration Success of the Central Indian Barasinga (*Cervus duvauceli branderi*)." Paper presented at Proceedings of a Working Meeting of the Deer Specialist Group of the Species Survival Commission, Morges.

Papworth, S., J. Rist, L. Coad, and E. Milner-Gulland. 2009. "Evidence for Shifting Baseline Syndrome in Conservation." Conservation Letters 2 (2): 93–100.

Park, M. (1799) 2000. Travels in the Interior Districts of Africa. Durham, NC: Duke University Press.

Parrish, S. S. 2006. American Curiosity: Cultures of Natural History in the Colonial British Atlantic World. Chapel Hill: University of North Carolina Press.

Parry, L., C. A. Peres, B. Day, and S. Amaral. 2010. "Rural–Urban Migration Brings Conservation Threats and Opportunities to Amazonian Watersheds." Conservation Letters 3 (4): 251–59.

Pärssinen, M., A. Ranzi, S. Saunaluoma, and A. Siiriäinen. 2003. "Geometrically Patterned Ancient Earthworks in the Rio Blanco Region of Acre, Brazil: New Evidence of Ancient Chiefdom Formations in Amazonian Interfluvial Terra Firme Environments." In Western Amazonia-Amazônia Ocidental: Multidisciplinary Studies on Ancient Expansionistic Movements, Fortifications, and Sedentary Life, edited by M. Pärssinen and A. Korpisaari, 135–72. Helsinki: Renvall Institute for Area and Cultural Studies, University of Helsinki.

Participatory Geographic Information Systems (PGIS). 2008. Participatory Avenues, the Gateway to Community Mapping, PGIS & PPGIS. Available from http://ppgis .iapad.org/about.htm.

Patel, R. 2008. Stuffed and Starved: The Hidden Battle for the World Food System. New York: Melville House.

———. 2010. The Value of Nothing. New York: Picador.

Paz-Rivera, C., and F. E. Putz. 2009. "Anthropogenic Soils and Tree Distributions in a Lowland Forest in Bolivia." Biotropica 41 (6): 665–75.

Peet, J. R. 1969. "The Spatial Expansion of Commercial Agriculture in the Nineteenth Century: A Von Thunen Interpretation." Economic Geography 45:283–301.

Peet, R., and M. Watts, eds. 2004. Liberation Ecologies: Environment, Development, Social Movements. New York: Routledge.

Peeters, L. Y. K., L. Soto-Pinto, H. Perales, G. Montoya, and M. Ishiki. 2003. "Coffee Production, Timber, and Firewood in Traditional and Inga-Shaded Plantations in Southern Mexico." Agriculture Ecosystems and Environment 95 (2–3): 481–93.

Peluso, N. 1996. "Fruit Trees and Family Trees in an Anthropogenic Forest: Ethics of Access, Property Zones, and Environmental Change in Indonesia." Comparative Studies in Society and History 38 (03): 510–48.

———. 2009. "Rubber Erasures, Rubber Producing Rights: Making Racialized Territories in West Kalimantan, Indonesia." Development and Change 40 (1): 47–80.

Peluso, N. L. 1992. Rich Forests, Poor People: Resource Control and Resistance in Java. Berkeley: University of California Press.

———. 2003a. "Territorializing Local Struggles for Resource Control: A Look at Environmental Discourses and Politics in Indonesia." In Nature in the Global South: Environmental Projects in South and Southeast Asia, edited by P. Greenough and A. L. Tsing, 231–52. Durham, NC: Duke University Press.

———. 2003b. "Weapons of the Wild: Strategic Uses of Wildness and Violence in West Kalimantan." In In Search of the Rainforest, edited by C. Slater, 204–45. Berkeley: University of California Press.

Peluso, N. L., and E. Harwell. 2001. "Territory, Custom, and the Cultural Politics of Ethnic War in West Kalimantan, Indonesia." In Violent Environments, edited by N. L. Peluso and M. Watts, 83–116. Ithaca: Cornell University Press.

Peluso, N. L., and P. Vandergeest. 2001. "Genealogies of the Political Forest and Customary Rights in Indonesia, Malaysia, and Thailand." Journal of Asian Studies 60 (3): 761–812.

Peluso, N. L., P. Vandergeest, and L. Potter. 1995. "Social Aspects of Forestry in Southeast Asia: A Review of Postwar Trends in the Scholarly Literature." Journal of Southeast Asian Studies 26 (1): 196–218.

Peluso, N. L., and M. Watts, eds. 2001. Violent Environments. Ithaca: Cornell University Press.

Perfecto, I., I. Armbrecht, S. M. Philpott, L. Soto-Pinto, and T. V. Dietsch. 2007. "Shaded Coffee and the Stability of Rainforest Margins in Northern Latin America." Stability of Tropical Rainforest Margins: Linking Ecological, Economic and Social Constraints of Land Use and Conservation: 227–63.

Perfecto, I., and J. Vandermeer. 2010. "The Agroecological Matrix as Alternative to the Land-Sparing/Agriculture Intensification Model." Proceedings of the National Academy of Sciences of the United States of America 107 (13): 5786–91.

Perkins, D. H. 1905. The Metropolitan Park System. Report of the Special Commission to the City of Chicago.

Persson, R. 1974. World Forest Resources. Research Notes no. 17. Stockholm: Royal College of Forestry.

Perz, S. G. 2007. "Grand Theory and Context-Specificity in the Study of Forest Dynamics: Forest Transition Theory and Other Directions." Professional Geographer 59 (1): 105–14.

Perz, S. G., and D. L. Skole. 2003. "Secondary Forest Expansion in the Brazilian Amazon and the Refinement of Forest Transition Theory." Society and Natural Resources 16 (4): 277–94.

Peters, C. M. 2000. "Precolumbian Silviculture and Indigenous Management of Neotropical Forests." In Imperfect Balance: Landscape Transformations in the Precolumbian Americas, edited by D. L. Lentz, 203–23. New York: Columbia University Press.

Peters, C. M., M. J. Balick, F. Kahn, and A. B. Anderson. 1989. "Oligarchic Forests of Economic Plants in Amazonia—Utilization and Conservation of an Important Tropical Resource." Conservation Biology 3 (4): 341–49.

Petersen, J. B., E. G. Neves, and M. J. Heckenberger. 2001. "Gift from the Past: Terra Preta and Prehistoric Amerindian Occupation in Amazonia." In The Unknown Amazon: Culture in Nature in Ancient Brazil, edited by C. Mcewan, C. Barretos, and E. G. Neves, 86–105. London: British Museum Press.

Petryna, A. 2002. Life Exposed: Biological Citizens after Chernobyl. Princeton, NJ: Princeton University Press.

Pfaff, A., and R. Walker. (2010). "Regional Interdependence and Forest 'Transitions': Substitute Deforestation Limits the Relevance of Local Reversals." Land Use Policy 27:119–29

Phanvilay, K. 2010. "Impacts of Land-Use and Land-Cover Transitions on People's Livelihoods in the Uplands of Northern Laos: Case Studies from Bokeo and Louang Namtha Provinces." PhD diss., University of Hawaii, Manoa.

Pickett, S., M. Cadenasso, and S. Bartha. 2001. "Implications from the Buell-Small Succession Study for Vegetation Restoration." Applied Vegetation Science 4 (1): 41–52.

Pickett, S. T. A., J. Kolasa, J. J. Armesto, and S. L. Collins. 1989. "The Ecological Concept of Disturbance and Its Expression at Various Hierarchical Levels." Oikos 54 (2): 129–36.

Pinedo-Vasquez, M., J. Barletti Pasqualle, D. del Castillo Torres, and K. Coffey. 2002. "A Tradition of Change: The Dynamic Relationship between Biodiversity and Society in Sector Muyuy, Peru." Environmental Science and Policy 5:43–53.

Pinedo-Vasquez, M., and C. Padoch. 2009. "Urban, Rural and in-Between: Multi-Sited Households Mobility and Resource Management in the Amazon Flood Plain." In Mobility and Migration in Indigenous Amazonia: Contemporary Ethnoecological Perspectives, edited by M. Alexiades, 86. Oxford: Berghahn Books.

Pinedo-Vasquez, M., and F. Rabelo. 2002. "Sustainable Management of an Amazonian Forest for Timber Production: A Myth or Reality?" In Cultivating Biodiversity: Understanding, Analyzing, and Using Agricultural Diversity, edited by H. Brookfield, C. Padoch, H. Parsons, and M. Stocking, 186–93. London: United Nations University.

Pinedo-Vasquez, M., R. R. Sears, I. Cardama, and M. Pinedo Panduro. 2005. "A Hybrid Concept for Understanding the Functionality of Seed Systems in Smallholder Societies of the Peruvian Amazonia." In Seed Systems and Crop Genetic Diversity on-Farm, edited by D. Jarvis, R. Sevilla-Panizo, J.-L. Chavez-Servia, and T. Hodgkin, 37–48. Proceedings of a Workshop, September 16–20, 2003, Pucallpa, Peru. Rome, Italy: International Plant Genetic Resources Institute.

Pinedo-Vasquez, M., D. Zarin, K. Coffey, C. Padoch, and F. Rabelo. 2001. "Post-Boom Logging in Amazonia." Human Ecology 29 (2): 219–39.

Pinto Parada, R. 1987. Pueblo de Leyenda. Trinidad: Tiempo del Bolivia.

Piperno, D. R., and D. M. Pearsall. 1998. The Origins of Agriculture in the Lowland Neotropics. San Diego: Academic Press.

Piperno, D. R., A. J. Ranere, I. Holst, and P. Hansell. 2000. "Starch Grains Reveal Early Root Crop Horticulture in the Panamanian Tropical Forest." Nature 407 (6806): 894–97.

Poffenberger, M. 1990. Keepers of the Forest: Land Management Alternatives in Southeast Asia. West Hartford, CT: Kumarian Press Inc.

Politis, G. G. 1996. "Moving to Produce: Nukak Mobility and Settlement Patterns in Amazonia." World Archaeology 27 (3): 492–511.

Pollak, H., M. Mattos, and C. Uhl 1997. "A Profile of Palm Heart Extraction in the Amazon Estuary." Human Ecology 23 (3): 357–85.

Pomeranz, K. 2000. The Great Divergence: Europe, China, and the Making of the Modern World Economy. Princeton, NJ: Princeton University Press.

Ponsart-Dureau, M. C. 1986. Le Pays Kissi de Guinée Forestière: Contribution a la Connaissance Du Milieu; Problematique de Dévéloppement [Country Kissi of Guinea Forestry: Contribution to Knowledge from the Middle]. Montpellier: Ecole Superieure d'Agronomie Tropicale.

Poole, P. 2006. "Is There Life after Tenure Mapping? Participatory Learning and Action 54." Retrieved from http://www.iied.org/NR/agbioliv/pla_notes/pla_backissues/54.html.

Porritt, V. L. 2004. The Rise and Fall of Communism in Sarawak, 1940-1990, vol. no. 59. Clayton: Monash Asia Institute.

Portères, R. 1965. "Les Noms Des Riz en Guinée: Vi—Les Noms Des Variétés de Riz Chez les Toma [The Names of Rice in Guinea: The Vi-Names Rice Varieties in Toma]." Journal d'Agriculture Tropicale et de Botanique appliquée 12:687-728.

Portes, A., C. Escobar, and A. W. Radford. 2007. "Immigrant Transnational Organizations and Development: A Comparative Study." International Migration Review 41 (1): 242-81.

Posey, D. A. 2004. Indigenous Knowledge and Ethics: A Darrell Posey Reader, K. Plenderleith ed. Studies in Environmental Anthropology vol. 10. New York: Routledge.

Posey, D. A., and W. L. Balée. 1989. Resource Management in Amazonia: Indigenous and Folk Strategies. Advances in Economic Botany vol. 7. Bronx, NY: New York Botanical Garden.

Posey, D. A., and M. J. Balick. 2006. Human Impacts on Amazonia: The Role of Traditional Ecological Knowledge in Conservation and Development. New York: Columbia University Press.

Potapov, P., A. Yaroshenko, S. Turubanova, M. Dubinin, L. Laestadius, C. Thies, D. Aksenov, et al. 2008. "Mapping the World's Intact Forest Landscapes by Remote Sensing." Ecology and Society 13 (2). Retrieved from http://lib.icimod.org/record/13548/files/4240.pdf.

Potter, L. M. 2003. "Forests versus Agriculture: Colonial Forest Services, Environmental Ideas and the Regulation of Land-Use Change in Southeast Asia." In The Political Ecology of Tropical Forests in Southeast Asia: Historical Perspectives, edited by T.-P. Lye, W. de Jong, and K.-I. Abe, 29-71. Melbourne: Trans Pacific Press.

Povinelli, E. A. 2002. The Cunning of Recognition: Indigenous Alterities and the Making of Australian Multiculturalism. Durham, NC: Duke University Press.

Poulgrain, G. 1998. The Genesis of Konfrontasi: Malaysia, Brunei and Indonesia, 1945-1965. Kuala Lumpur, Malaysia: Crawford House Publishing.

Prasad, A. 2003. Against Ecological Romanticism: Verrier Elwin and the Making of an Anti-Modern Tribal Identity. New Delhi: Three Essays Collective.

Prater, S. H. 1940. "The Number of Tigers Shot in the Reserved Forests in India and Burma During the Year 1938-39." Journal of the Bombay Natural History Society 41:881-89.

Pred, A. 2000. Even in Sweden: Racisms, Racialized Spaces, and the Popular Geographical Imagination. Berkeley: University of California Press.

Primack, R. B., and T. E. Lovejoy. 1995. Ecology, Conservation, and Management of Southeast Asian Rainforests. New Haven, CT: Yale University Press.

Programme d'Aménagement de Bassins Versants Haute Guinée (Project Kan II). n.d. Etude sociologique [*Sociological Study*]. Report. Conakry, Républisque de Guinée: Author.

Programme D'aménagement Des Bassins Versants Haute Guinée (Kan II). 1992. Plan D'opération [Plan of Operation]. Conakry: République de Guinée; Programme d'Aménagement des Hauts Bassins du Fleuve Niger.

Programme D'aménagement Des Hauts Bassins Du Fleuve Niger. n.d. Etude Sociologique [*Sociological Study*]. Conakry: République de Guinée; Programme d'Aménagement des Hauts Bassins du Fleuve Niger.

Putz, F. E., and K. H. Redford. 2010. "The Importance of Defining 'Forest': Tropical Forest Degradation, Deforestation, Long-Term Phase Shifts, and Further Transitions." Biotropica 42 (1): 10–20.

Putzel, L. 2010. "The Tree That Held up the Forest: Shihuahuaco and the Chinese Timber Trade." PhD diss., Department of Biology, City University of New York.

Putzel, L., C. Padoch, and M. Pinedo-Vasquez. 2008. "The Chinese Timber Trade and the Logging of Peruvian Amazonia." Conservation Biology 22 (6): 1659–61.

Pyne, S. J. 1998. "Forged in Fire: History, Land, and Anthropogenic Fire." In Advances in Historical Ecology, edited by W. L. Balée, 62–103. New York: Columbia University Press.

Rachman, A. 1970. Sejarah Singkat Kodam Xii Tanjungpura, Kalimantan Barat [A Brief History of Military Command Xii Tanjungpura, West Kalimantan]. Pontianak.

Radhakrishna, M. March 2008. "Unsettled People, Unsettled Issues: State, Civil Society and Nomadic Communities." Presented at the Workshop on Political Ecology, Department of Sociology. Delhi School of Economics.

Raffles, H. 2002. In Amazonia: A Natural History. Princeton, NJ: Princeton University Press.

Raffles, H., and A. Winkler-Prins. 2003. "Further Reflections on Amazonian Environmental History: Transformations of Rivers and Streams." Latin American Research Review 38 (3): 165–87.

Raintree, J. 2001. "Human Ecology and Rural Livelihood in Lao PDR." In National Human Development Report, Lao PDR 2001, 71–95. Vientiane: United Nations Development Programme.

Rajagopalan, G., R. Sukumar, R. Ramesh, B. K. Pant, and G. Rajagopalan. 1997. "Late Quarternary Vegetational and Climatic Changes from Tropical peats in Southern India—An Extended Record up to 40,000 Years BP." Current Science 73:60–63.

Ramankutty, N., E. Heller, and J. Rhemtulla. 2010. "Prevailing Myths About Agricultural Abandonment and Forest Regrowth in the United States." Annals of the Association of American Geographers 100 (3): 502–12.

Ramanujan, A. K., and Tolkappiyar. 1985. Poems of Love and War: From the Eight Anthologies and the Ten Long Poems of Classical Tamil. New York: Columbia University Press.

Rambo, A. T. 1996. "The Composite Swiddening Agroecosystem of the Tay Ethnic Minority of the Northwestern Mountains of Vietnam." In Montane Mainland

Southeast Asia in Transition, edited by B. Rerkasem, 43–64. Chiang Mai: Chiang Mai University Consortium.

Ramnath, M. 2008. "Surviving the Forest Rights Act: Between Scylla and Charybdis." Economic and Political Weekly 43 (9): 37.

Randhawa, M. S. 1965. Flowering Trees. New Delhi: National Book Trust.

Rangan, H. 2000. Of Myth and Movements: Rewriting Chipko into Himalayan History. Brooklyn, NY: Verso.

Rangarajan, M. 1996. "The Politics of Ecology: The Debate on Wildlife and People in India, 1970–95." Economic and Political Weekly 31 (35): 2391–409.

———. 1999. Fencing the Forest: Conservation and Ecological Change in India's Central Provinces, 1860–1914. Delhi: Oxford University Press.

———. 2001a. "From Princely Symbol to Conservation Icon: A Political History of the Lion in India." In The Unfinished Agenda, Nation Building in South Asia, edited by M. Hasan and N. Nakazato, 399–442. Delhi: Manohar.

———. 2001b. India's Wildlife History: An Introduction. Delhi: Permanent Black in association with Ranthambhore Foundation: Distributed by Orient Longman.

———. 2002. "Polity, Ecology and Landscape: Fresh Writing on South Asia's Past." Studies in History 17:135–48.

———. 2006. "Ideology, the Environment and Policy: Indira Gandhi." India International Centre Quarterly 33:50–64.

———. 2012. "Introduction." In India's Environmental History: From Ancient Times to the Colonial Period, edited by M. Rangarajan and K. Sivaramakrishnan, 1–34. Delhi, India: Oxford University Press.

Rangarajan, M., and G. Shahabuddin. 2006. "Displacement and Relocation from Protected Areas: Towards a Biological and Historical Synthesis." Conservation and Society 4 (3): 359.

Ranjitsinh, M. K. 1997. Beyond the Tiger: Portraits of Asian Wildlife. Delhi: Brijbasi.

Ranzi, A., and R. Aguiar. 2004. Geoglifos da Amazônia: Perspectiva Aérea [Geoglyphs of the Amazon: Aerial Perspective]. Florianópolis: Faculdades Energia.

Rathbone, R. 2000. Nkrumah and the Chiefs: The Politics of Chieftaincy in Ghana, 1951–1960. Oxford: James Currey.

Rattray, R. S. 1927. Religion and Art in Ashanti. Oxford: Clarendon Press.

Rauch, J. H. 1869. Public Parks: Their Effects Upon the Moral, Physical and Sanitary Condition of the Inhabitants of Large Cities; with Special Reference to the City of Chicago. Chicago: S. C. Griggs.

Raychaudhuri, T., and I. Habib. 1982. "Population." In The Cambridge Economic History of India, vol. I, C. 1200–1750, edited by T. Raychaudhuri and I. Habib, 163–70. Cambridge: Cambridge University Press.

Reardon, T., and G. Escobar. 2001. "Rural Nonfarm Employment and Incomes in Latin America: Overview and Policy Implications." World Development 29 (3): 395–409.

Record, S. J. 1930. "Forestry in British Honduras." Tropical Woods 24:6–15.

Redford, K. H. 1991. "The Ecologically Noble Savage." Talking about People. Readings in Contemporary Cultural Anthropology 15 (1): 11–13.

Redford, K. H., and S. E. Sanderson. 2000. "Extracting Humans from Nature." Conservation Biology 14 (5): 1362–64.

Redford, K. H., and A. M. Stearman. 1993. "Forest-Dwelling Native Amazonians and the Conservation of Biodiversity—Interests in Common or in Collision." Conservation Biology 7 (2): 248–55.

Redman, C. L. 1999. Human Impact on Ancient Environments. Tucson: University of Arizona Press.

Reij, C., G. Tappan, and A. Belemvire. 2005. "Changing Land Management Practices and Vegetation on the Central Plateau of Burkina Faso (1968–2002)." Journal of Arid Environments 63 (3): 642–59.

Reij, C., G. Tappan, and M. Smale. 2009. Agroenvironmental Transformation in the Sahel: Another Kind of "Green Revolution." Washington, DC: International Food Policy Research Institute (IFPRI).

Reij, C. P., and E. M. A. Smaling. 2008. "Analyzing Successes in Agriculture and Land Management in Sub-Saharan Africa: Is Macro-Level Gloom Obscuring Positive Micro-Level Change?" Land Use Policy 25 (3): 410–20.

Reimer, P. J., M. G. L. Baillie, E. Bard, A. Bayliss, J. W. Beck, C. J. H. Bertrand, P. G. Blackwell, et al. 2004. "Intcal04 Terrestrial Radiocarbon Age Calibration, 0–26 Cal Kyr Bp." Radiocarbon 46:1029–58.

Reno, W. 1998. Warlord Politics and African States. Boulder, CO: Lynne Rienner Publishers.

République de Guinée, J. D. Bourque, and R. Wilson. 1990. Guinea Forestry Biodiversity Study—Ziama and Diecke Reserves. Gland, Switzerland: IUCN for République de Guinée.

République de Guinée, and B. Jean. 1989. La Gestion Des Ressources Naturelles [Natural Resources Management]. Conakry: République de Guinée; Ministère d'Agriculture et des Ressources Animales.

Ribbentrop, B. 1900. Forestry in British India. Calcutta: Office of the Superintendent of Government Printing.

Ribot, J. C. 2001. "Reframing Deforestation. Global Analysis and Local Realities—Studies in West Africa." Development and Change 32 (1): 181–82.

Richards, J. F. 2003. The Unending Frontier: An Environmental History of the Early Modern World. Berkeley: University of California Press.

Richards, J. F., J. R. Hagen, and E. S. Haynes. 2008. "Changing Land Use in Bihar, Punjab and Haryana, 1850–1970." Modern Asian Studies 19 (03): 699–732.

Richards, J. F., and R. P. Tucker. 1988. World Deforestation in the Twentieth Century. Duke Press Policy Studies. Durham, NC: Duke University Press.

Rigg, J. 2003. "Evolving Rural-Urban Relations and Livelihoods in Southeast Asia." In Southeast Asia Transformed: A Geography of Change, edited by C. L. Sien, 231–56. Singapore: Institute of Southeast Asian Studies.

———. 2005. Living with Transition in Laos: Market Integration in Southeast Asia. London: Routledge.

———. 2006. "Land, Farming, Livelihoods, and Poverty: Rethinking the Links in the Rural South." World Development 34 (1): 180–202.

Rigg, J., and S. Nattapoolwat. 2001. "Embracing the Global in Thailand: Activism and Pragmatism in an Era of Deagrarianization." World Development 29 (6): 945–60.

Rival, L. M. 2002. Trekking through History: The Huaorani of Amazonian Ecuador. New York: Columbia University Press.

Robbins, P. 2000. "The Rotten Institution: Corruption in Natural Resource Management." Political Geography 19 (4): 423–43.

Robertson, J. 1799. General View of the Agriculture in the County of Perth. Perth: Printed by order of the Board of Agriculture for J. Morison.

Robin, C. 1999. "Towards an Archaeology of Everyday Life: Maya Farmers of Chan Noòhol, Belize." PhD diss., Department of Anthropology, University of Pennsylvania.

Robin, L. 1998. Defending the Little Desert: The Rise of Ecological Consciousness in Australia. Carlton: Melbourne University Press.

Robbins, P. 2001. "Tracking Invasive Landcovers in India, or Why Our Landscapes Have Never Been Modern." Annals of the Association of Human Geographers 91 (4): 637–59.

Rocheleau, D. 1987. "Women, Trees and Tenure: Implications for Agroforestry Research and Development." In Land, Trees and Tenure, edited by J. Raintree, 79–121. Madison: University of Wisconsin Land Tenure Center/ICRAF.

Rocheleau, D. E., B. P. Thomas-Slayter, and E. Wangari. 1996. Feminist Political Ecology: Global Issues and Local Experiences. New York: Routledge.

Rodger, A. 1921. A Handbook of the Forest Products of Burma. Rangoon: Government Printing.

Rodgers, W. A., and H. S. Panwar. 1988. Planning a Protected Area Network in India. Volume I—the Report. Dehadrun, India: Wildlife Institute of India.

Roe, E. M. 1991. "Development Narratives, or Making the Best of Blueprint Development." World Development 19 (4): 287–300.

Roe, E. M., and L. Fortmann. 1982. Season and Strategy: The Changing Organization of the Rural Water Sector in Botswana. Ithaca: Rural Development Committee, Center for International Studies, Cornell University.

Roosevelt, A. C. 1991. Moundbuilders of the Amazon: Junqueira Geophysical Archaeology on Marajo Island, Brazil. San Diego: Academic Press.

———.1996. "Paleo-Indian Cave Dwellers in the Amazon: The Peopling of the Americas." Science 272:373–84.

———. 1999. "The Development of Prehistoric Complex Societies: Amazonia, a Tropical Forest." Archeological Papers of the American Anthropological Association 9 (1): 13–33.

Roosevelt, A. C., M. Lima da Costa, C. Lopes Machado, M. Michab, N. Mercier, H. Valladas, J. Feathers, W. Barnett, M. Imazio da Silveira, and A. Henderson. 1996. "Paleoindian Cave Dwellers in the Amazon: The Peopling of the Americas." Science 272 (5260): 373.

Rosa, H., S. Kandel, and L. Dimas. 2005. Compensation for Ecosystem Services and Rural Communities: Lessons from the Americas. El Salavdor: PRISMA (Programa Salvadoreneo de Investigacion sobre Desarrollo y Medio Ambiente).

Rosencranz, A., and S. Lele. 2008. "Supreme Court and India's Forests." Economic and Political Weekly 43 (5):11–14.

Ross, M. L. 2001. Timber Booms and Institutional Breakdown in Southeast Asia. Cambridge Cambridge University Press.

——. 2004. "How Do Natural Resources Influence Civil War? Evidence from Thirteen Cases." International Organization 58 (01): 35–67.

——. 2006. "A Closer Look at Oil, Diamonds, and Civil War." Political Science 9 (1): 265.

Rosset, P. 2006. Food Is Different: Why We Must Get the WTO out of Agriculture. Zed Books.

Rostain, S. 1999. "Secuencia Arqueológica en Montículos del Valle del Upano en la Amazonía Ecuatoriana [Sequence of Archaeological Mounds in Upano Valley in the Ecuadorian Amazon]." Bulletin de l'Institut Français de Ètudes Andines 28 (1): 53–89.

Rothman, H. 1992. On Rims and Ridges: The Los Alamos Area since 1880. Lincoln: University of Nebraska Press.

Rouvinen, S., and J. Kouki. 2008. "The Natural Northern European Boreal Forests: Unifying the Concepts, Terminologies, and Their Application." Silva Fennica 42 (1): 135–46.

Rudel, T., and J. Roper. 1997. "The Paths to Rain Forest Destruction: Crossnational Patterns of Tropical Deforestation, 1975–90." World Development 25 (1):53–65.

Rudel, T. K. 2002a. "Paths of Destruction and Regeneration: Globalization and Forests in the Tropics." Rural Sociology 67 (4):622–36.

——. 2002b. "Transnational Peasants: Migrations, Networks, and Ethnicity in Andean Ecuador." Rural Sociology 67 (3):486–88.

Rudel, T. K., D. Bates, and R. Machinguiashi. 2002. "A Tropical Forest Transition? Agricultural Change, Out-Migration, and Secondary Forest in the Ecuadorian Amazon." Annals of the American Association of Geographers 92 (1): 87–102.

Rudel, T. K., O. T. Coomes, E. Moran, F. Achard, A. Angelsen, J. C. Xu, and E. Lambin. 2005. "Forest Transitions: Towards a Global Understanding of Land Use Change. Global Environmental Change 15 (1): 23–31.

Rudel, T. K., R. Defries, G. P. Asner, and W. F. Laurance. 2009. Changing Drivers of Deforestation and New Opportunities for Conservation. Conservation Biology 23 (6): 1396–1405.

Rudel, T. K., and B. Horowitz. 1993. Tropical Deforestation : Small Farmers and Land Clearing in the Ecuadorian Amazon. New York: Columbia University Press.

Rudel, T. K., M. Perez-Lugo, and H. Zichal. 2000. When Fields Revert to Forest: Development and Spontaneous Reforestation in Post-War Puerto Rico. Professional Geographer 52 (3): 386–397.

Ruf, F. 1995. "From 'Forest Rent' to 'Tree Capital': Basic 'Laws' of Cocoa Supply. In Cocoa Cycles: The Economics of Cocoa Supply, edited by F. Ruf and P. S. Siswoputranto, 1–54. Cambridge: Woodhead.

Russell, E. 2001. War and Nature: Fighting Humans and Insects with Chemicals from World War I to Silent Spring. Cambridge: Cambridge University Press.

Rustad, S. C. A., J. K. Rod, W. Larsen, and N. P. Gleditsch. 2008. "Foliage and Fighting: Forest Resources and the Onset, Duration, and Location of Civil War." Political Geography 27 (7): 761–82.

Ruyssen, B. 1957. "Le Karité au Soudan [Shea of the Sudan]." Agronomie tropicale
12:143–72.

Rzedowski, J. 1981. Vegetación de México [Vegetation of Mexico]. Mexico: Editorial
Limusa.

Saavedra, O. 2006. "El Sistema Agrícola Prehispánico de Camellones en la Amazonía
Boliviana [Pre-Hispanic Agricultural System in the Bolivian Amazon Camellones]."
In Agricultura Ancestral Camellones y Albarradas: Contexto Social, Usos y Retos
del Passado y del Presente: Coloquio Agricultura Prehisp·Nica Sistemas Basados
en El Drenaje y en la Elevaciûn de los Suelos Cultivados, edited by F. Valdez, 295.
Quito: Editorial Abya-Yala.

Saberwal, V. K. 1999. Pastoral Politics: Shepherds, Bureaucrats, and Conservation in
the Western Himalaya. Oxford: Oxford University Press.

Sahgal, B., and S. Thiyagarajan. 2008. "The New Delhi Television Tiger Campaign."
Sanctuary Asia, 16–23.

Saikia, A. 2005. Jungles, Reserves, Wildlife: A History of Forests in Assam. Guwahati:
Wildlife Areas Development and Welfare Trust.

Salafsky, N., and E. Wollenberg. 2000. "Linking Livelihoods and Conservation: A
Conceptual Framework and Scale for Assessing the Integration of Human Needs
and Biodiversity." World Development 28 (8): 1421–38.

Salve, H. 2008. Speech at WWF Felicitation Ceremony. Delhi.

Sanderson, E. W., M. Jaiteh, M. A. Levy, K. H. Redford, A. V. Wannebo, and G. Wool-
mer. 2002. "The Human Footprint and the Last of the Wild." Bioscience 52 (10):
891–904.

Sanderson, M. R., and J. D. Kentor. 2009. "Globalization, Development and International Migration: A Cross-National Analysis of Less-Developed Countries,
1970–2000." Social Forces 88 (1): 301–36.

Sanford Jr., R. L., J. Saldarriaga, K. E. Clark, C. Uhl, and R. Herrera. 1985. "Amazon
Rain-Forest Fires." Science 227 (4682): 53.

Sang, J. 1998. "Tainted Tumbleweeds Concern Hanford." Tri-City Herald, December 27.

Sanghal, B., ed. 2008. Lest We Forget: Kailash Sankhala's India. Mumbai: Sanctuary
Asia.

Sankhala, K. 1977. Tiger!: The Story of the Indian Tiger. London: Simon & Schuster.

Santos-Granero, F., and F. Barclay. 2000. Tamed Frontiers: Economy, Society, and
Civil Rights in Upper Amazonia. Boulder, CO: Westview Press.

Saravanan, V. 2008. "Economic Exploitation of Forest Resources in South India during
the Pre-Forest Act Colonial Era, 1793–1882." International Forestry Review 10 (1):
65–73.

Sastri, K. A. N. 1975. A History of South India, 4th ed. Madras, India: Oxford University Press.

Satterthwaite, D., and C. Tacoli. 2002. "Seeking an Understanding of Poverty That
Recognizes Rural-Urban Differences and Rural-Urban Linkages." In Urban Liveli-
hoods: A People-Centred Approach to Reducing Poverty, edited by C. Rakodi and T.
Lloyd Jones, 52–70. London: Earthscan Publications.

Saul, M., J. M. Ouadba, and O. Bognounou. 2003. "The Wild Vegetation Cover of Western Burkina Faso: Colonial Policy and Post-Colonial Development." In *African Savannas: Global Narratives and Local Knowledge of Envrionmental Change*, edited by T. J. Bassett and D. Crummey, 121–60. Oxford: James Currey.

Savur, M. 2003. *And the Bamboo Flowers in the Indian Forests: What Did the Pulp and Paper Industry Do?* Delhi: Manohar.

Sayer, J. 1992. "Development Assistance Strategies to Conserve Africa's Rainforests." In *Conservation of West and Central African Forests*, edited by K. Cleaver, M. Munashighe, M. Dyson, N. Egli, A. Peuker, and F. Wencélius, 3–9. Washington, DC: World Bank.

Scarascia-Mugnozza, G., H. Oswald, P. Piussi, and K. Radoglou. 2000. "Forests of the Mediterranean Region: Gaps in Knowledge and Research Needs." *Forest Ecology and Management* 132 (1): 97–109.

Scarborough, V., F. Valdez Jr., S. Luzzadder-Beach, T. Beach, and J. G. Jones. 1999. "Temple Mountains, Sacred Lakes, and Fertile Fields: Ancient Maya Landscapes in Northwestern Belize." *Antiquity* 72:650–60.

Schabas, M. 2005. *The Natural Origins of Economics*. Chicago: University of Chicago Press.

Schama, S. 1995. *Landscape and Memory*. New York: A.A. Knopf, distributed by Random House.

Schell, J. 1982. *The Fate of the Earth*, 1st ed. New York: Knopf.

Schiff Natural Lands Trust Inc. 2008. *About Schiff*. Available from http://www.schiff naturepreserve.org/.

Schimmel, A., B. K. Waghmar, and C. Atwood. 2004. *The Empire of the Great Mughals: History, Art and Culture*. London: Reaktion Books.

Schlager, E., and E. Ostrom. 1992. "Property-Rights Regimes and Natural Resources: A Conceptual Analysis." *Land Economics* 68 (3): 249–62.

Schlawin, J. R., and R. A. Zahawi. 2008. "'Nucleating' Succession in Recovering Neotropical Wet Forests: The Legacy of Remnant Trees." *Journal of Vegetation Science* 19 (4): 485–92.

Schmidt, P. M., and M. J. Peterson. 2009. "Biodiversity Conservation and Indigenous Land Management in the Era of Self-Determination." *Conservation Biology* 23 (6): 1458–66.

Schmink, M. 2004. "Communities, Forests, Markets, and Conservation." In *Working Forests in the Neotropics: Conservation through Sustainable Management?*, edited by D. J. Zarin, J. R. R. Alavalapatti, F. E. Putz, and M. Schmink, 119–29. New York: Columbia University Press.

Schneeberger, W. F. 1979. *Contributions to the Ethnology of Central Northeast Borneo: Parts of Kalimantan, Sarawak and Sabah*. Studia Ethnologia Bernensia vol. 2. Berne: University of Berne, Institute of Ethnology.

Schroeder, R. A. 1999. *Shady Practices: Agroforestry and Gender Politics in the Gambia*. Berkeley: University of California Press.

Schroeder, R., and K. Suryanata. (1996) 2004. "Gender and Class Power in Agroforestry Systems: Case Studies from Indonesia and West Africa." In *Liberation*

Ecologies: Environment, Development, Social Movements, edited by R. Peet and M. Watts, 299–315. New York: Routledge.

Schwartz, S. I. 1998. Atomic Audit: The Costs and Consequences of US Nuclear Weapons since 1940. Washington, DC: Brookings Institution Press.

Scoones, I., ed. 1994. Living with Uncertainty: New Directions for Pastoral Development in Africa. London: IT Publications.

Scott, C. W. 1945. "Burma Teak Today." Wood 10 (4): 81–84.

Scott, J. C. 1998. Seeing Like a State: How Certain Schemes to Improve the Human Condition Have Failed. New Haven, CT: Yale University Press.

———. 2009. The Art of Not Being Governed: An Anarchist History of Upland Southeast Asia. New Haven, CT: Yale University Press.

Scott Parrish, S. 2006. American Curiosity: Cultures of Natural History in the Colonial British Atlantic World. Williamsburg: University of North Carolina Press.

Sears, R. R. 2003. "New Forestry on the Floodplain: The Ecology and Management of *Calycophyllum spruceanum* (Rubiaceae)." PhD diss., Ecology, Evolution and Environmental Biology, Columbia University, New York.

Sears, R. R., C. Padoch, and M. Pinedo-Vasquez. 2007. "Amazon Forestry Transformed: Integrating Knowledge for Smallholder Timber Management." Human Ecology 35:697–707.

Secretariat of the Convention on Biological Diversity. 2001. Global Biodiversity Outlook. Montreal: Secretariat of the Convention on Biological Diversity.

Sedjo, R. A. 1992. "Temperate Forest Ecosystems in the Global Carbon Cycle." Ambio 21 (4): 274–77.

Sedjo, R. A., and M. Clawson. 2000. "Global Forests." In The Resourceful Earth, edited by J. L. Simon and H. Kahn, 128–71. Oxford: Basil Blackwell.

Sedjo, R. A., and A. M. Solomon. 1989. "Climate and Forests." In Greenhouse Warming: Abatement and Adaptation, edited by N. J. Rosenberg, W. E. Easterling, P. R. Crosson, and J. Darmstadter, 105–19. Washington, DC: Resources for the Future.

Sellato, B. 1994. Nomads of the Borneo Rainforest: The Economics, Politics,and Ideology of Settling Down. Honolulu: University of Hawaii Press.

Sethi, N. 2008. "Sariska Tiger Project Hits Traffic Roadblock." The Times of India, May 11.

Seymour, G. L. 1859–1860. "The Journal of the Journey of George L. Seymour to the Interior of Liberia: 1858." New York Colonization Journal IX, no. 12, vol. X, nos. 6 and 8: 108–09.

Sezen, U. U., R. L. Chazdon, and K. E. Holsinger. 2005. "Genetic Consequences of Tropical Second-Growth Forest Regeneration." Science 307 (5711): 891.

———. 2007. "Multigenerational Genetic Analysis of Tropical Secondary Regeneration in a Canopy Palm." Ecology 88 (12): 3065–75.

Shahabuddin, G., R. Kumar, and M. Shrivastava. 2007. "Creation of Inviolate Space: Lives, Livelihoods and Conflict in Sariska Tiger Reserve." Economic and Political Weekly 42 (20): 1855.

Shahi, S. P. 1977. Backs to the Wall: Saga of Wildlife in Bihar, India. New Delhi: Affiliated East-West Press.

Shandra, J. M., T. Rudel, and B. London. 2008. Non-Governmental Organizations, Counter Coalitions, and Conserving Biodiversity: A Cross-National Analysis. Stony Brook: State University of New York.

Shetler, J. B. 2007. Imagining Serengeti: A History of Landscape Memory in Tanzania from Earliest Times to the Present. Athens: Ohio University Press.

Sikor, T. 2006. "Analyzing Community-Based Forestry: Local, Political and Agrarian Perspectives." Forest Policy and Economics 8:339–49.

Simon, D. Urban. 2008. "Urban Environments: Issues on the Peri-Urban Fringe." Annual Review of Environmental Resources 33:167–85.

Simmonds, P. L. 1885. "The Teak Forests of India and the East, and Our British Imports of Teak." Journal of the Society of Arts 33:345–59.

Sims, J. L. 1859–1860. "The Journal of a Journey in the Interior of Liberia by James L. Sims, of Monrovia. Scenes in the Interior of Liberia: Being a Tour through the Countries of the Dey, Goula, Pessah Barlain, Kpellay, Suloang, and the King Boatswain's Tribes, in 1858." New York Colonization Journal IX, no. 12, vol. X, nos. 6 and 8.

Sinclair, J. 1791–99. The Statistical Account of Scotland. 21 vols. Aberdeen: William Creech.

Sinclair, J., and Board of Agriculture. 1795. General View of the Agriculture of the Northern Counties and Islands of Scotland: Including the Counties of Cromarty, Ross, Sutherland and Caithness and the Islands of Orkney and Shetland: With Observations on the Means of Their Improvement. London: Printed by C. Macrae.

Singh, A. 1993. The Legend of the Maneater. Delhi: Ravi Dayal.

Singh, K. 1969. Hints on Tiger Shooting. Jaico Shikar Series. Bombay: Jaico Pub. House.

———. 2008. "No Room to Roam." Sanctuary Asia, 4–6.

Sioh, M. 1998. "Authorizing the Malaysian Rainforest: Configuring Space, Contesting Claims and Conquering Imaginaries." Cultural Geographies 5 (2): 144.

———. 2004. "An Ecology of Postcoloniality: Disciplining Nature and Society in Malaya, 1948–1957." Journal of Historical Geography 30 (4): 729–46.

Siqueira, A. D. 2006. "Mullheres, relações de gênero e tomadas de decisão em unidades caboclas do estuário amazônico [Women, Gender Relations and Decision-Making Units in the Amazon Estuary Caboclas]." In Sociedades Caboclas Amazônicos, edited by C. Adams, R. S. S. Murrieta, and W. A. Neves, 135–236. Sao Paolo: Anna Blume.

Sivaramakrishnan, K. 1997. "A Limited Forest Conservancy in Southwest Bengal, 1864–1912." Journal of Asian Studies 56 (1): 75–112.

———. 1999. Modern Forests: Statemaking and Environmental Change in Colonial Eastern India. Stanford: Stanford University Press.

———. 2009. Forests and the Environmental History of Modern India." Journal of Peasant Studies 36 (2) (April): 299–324.

Sivaramakrishnan, K., and Arun Agrawal. 2000. Agrarian Environments: Resources, Representations, and Rule in India. Raleigh, NC: Duke University Press.

Slade, H. 1896. "Too Much Fire Protection in Burma." Indian Forester 22 (5): 172–76.

Slater, C. 1996. "Amazonia as Edenic Narrative." In Uncommon Ground: Rethinking the Human Place in Nature, edited by W. Cronon, 114–31. New York: W.W. Norton and Co.

———, ed. 2003. In Search of the Rain Forest. Durham, NC: Duke University Press.

Slocum, M. G., and C. C. Horvitz. 2000. "Seed Arrival under Different Genera of Trees in a Neotropical Pasture." Plant Ecology 149 (1): 51–62.

Smit, W. 1998. "The Rural Linkages of Urban Households in Durban, South Africa." Environment and Urbanization 10 (1): 77.

Smith, A. 1976. An Inquiry into the Nature and Causes of the Wealth of Nations. Edited by R. H. Campbell, A. S. Skinner, and W. B. Todd, vol. 2. Oxford: Oxford University Press.

Smith, A. M. 1982. Jacobite Estates of the Forty-Five. Edinburgh: John Donald.

Smith, B. K. 1991. "Classifying Animals and Humans in Ancient India." Man 26 (3): 527–48.

Smith, J. 1798. General View of the Agriculture of the County of Argyll; with Observations on the Means of Its Improvement. Edinburgh: Printed by Mundell & Son; sold by G. Nicol, London; and by Messrs. Robinson; J. Sewell; Cadell & Davies; William Creech, Edinburgh; and John Archer, Dublin.

Smith, M. J. 1999. Burma: Insurgency and the Politics of Ethnicity, 2nd ed. London: Zed Books.

Smith, N. 1984. Uneven Development: Nature, Capital, and the Production of Space. New York: Blackwell.

Smith, N., A. Serraõ, P. Alvim, and I. Falesi. 1995. Amazonia: Resiliency and Dynamism of a Land and Its People. Tokyo: United Nations Press.

Smith, N. J. H. 1980. "Anthrosols and Human Carrying Capacity in Amazonia." Annals of the Association of American Geographers 70 (4): 553–66.

———. 1999. The Amazon River Forest: A Natural History of Plants, Animals, and People. New York: Oxford University Press.

Smout, T. C. 2000. Nature Contested: Environmental History in Scotland and Northern England since 1600. Edinburgh: Edinburgh University Press.

———. 2003. People and Woods in Scotland: A History. Edinburgh: Edinburgh University Press.

Smout, T. C., A. R. Macdonald, and F. J. Watson. 2005. A History of the Native Woodlands of Scotland, 1500–1920. Edinburgh: Edinburgh University Press.

Soemadi. 1974. Peranan Kalimantan Barat Dalam Menghadapi Subversi Komunis Asia Tenggara: Suatu Tinjauan Internasional Terhadap Gerakan Komunis Dari Sudut Pertahanan Wilayah, Khususnya Kalimantan Barat [Facing Role in West Kalimantan Communist Subversion of Southeast Asia: An Overview of International Communist Movement Against Defense of Angles Area, particularly West Kalimantan]. Pontianak: Yayasan Tanjungpura.

Soepardi Poerwokoesoemo, R. 1974. Hutan Dan Kehutanan Dalam Tiga Jaman [Forests and Forestry in Three Times]. Jakarta: Perum Perhutani.

Somchai, P. 2006. Civil Society and Democratization: Social Movements in Northeast Thailand, vol. 99. Copenhagen: Netherlands Institute of Advanced Studies (NIAS).

Sommer, A. 1976. "Attempt at an Assessment of the World's Tropical Forests." Unasylva 28 (112): 5–25.

Sommerfelt, S. 1999. "Shares and Sharing: Dynamics of Exchange, Identity, Rank and Power in a Gambian Town." MA thesis, Department and Museum of Anthropology, University of Oslo.

Soto-Pinto, L., M. Anzueto, J. Mendoza, G. J. Ferrer, and B. de Jong. 2010. "Carbon Sequestration through Agroforestry in Indigenous Communities of Chiapas, Mexico." Agroforestry Systems 78 (1): 39–51.

Soulé, M. E., and G. Lease. 1995. Reinventing Nature?: Responses to Postmodern Deconstruction. Washington, DC: Island Press.

Spence, M. D. 1999. Dispossessing the Wilderness: Indian Removal and the Making of the National Parks. New York: Oxford University Press.

Spencer, L. 2003. "Deer Dilemma: Too Much of a Good Thing?" Chicago Wilderness Magazine 7:7–9.

Sprugel, D. G. 1991. "Disturbance, Equilibrium, and Environmental Variability: What Is 'Natural' Vegetation in a Changing Environment?" Biological Conservation 58 (1): 1–18.

Stab, S., and J. Arce. 2000. "Pre-Hispanic Raised-Field Cultivation as an Alternative to Slash-and Burn Agriculture in the Bolivian Amazon: Agroecological Evaluation of Field Experiments." In Biodiversidad, Conservación y Manejo en la Región de la Reserva de la Biosfera Estación Biológica del Beni, Bolivia, edited by O. Herrera-Macbryde, F. Dallmeier, B. Macbryde, J. A. Comiskey, and C. Miranda, 317. Washington, DC: Smithsonian Institution, SI/MAB Biodiversity Program.

Stahl, P. W. 1991. "Arid Landscapes and Environmental Transformations in Ancient Southwestern Ecuador." World Archaeology 22 (3): 346–59.

———. 1996. "Holocene Biodiversity: An Archaeological Perspective from the Americas." Annual Review of Anthropology 25 (1): 105–26.

———. 2000. "Archaeofaunal Accumulation, Fragmented Forests, and Anthropogenic Landscape Mosaics in the Tropical Lowlands of Prehispanic Ecuador." Latin American Antiquity 11 (3): 241–57.

———. 2006. "Microvertebrate Synecology and Anthropogenic Footprints in the Forested Neotropics." In Time and Complexity in Historical Ecology: Studies in the Neotropical Lowlands, edited by W. L. Balée and C. L. Erickson, 127. New York: Columbia University Press.

Standley, P. C., and B. E. Dahlgren. 1930. Flora of Yucatan. Publication/Field Museum of Natural History vol. 279. Chicago: Field Museum of Natural History.

Stang, John. 1998. "Tainted Tumbleweeds Concern Hanford." Tri-City Herald. December 27.

Stark, B. L., and A. Ossa. 2007. "Ancient Settlement, Urban Gardening, and Environment in the Gulf Lowlands of Mexico." Latin American Antiquity 18 (4): 385–406.

Starr, F., ed. 1912. Narrative of the Expedition Despatched to Musahdu by the Liberian Government under Benjamin K. Anderson Esq. In 1874. Monrovia: College of West Africa Press.

Stearman, A. M. 1985. Camba and Kolla: Migration and Development in Santa Cruz, Bolivia. Orlando: University of Central Florida Press.

Stebbing, E. P. 1947. "The Teak Forests of Burma." Nature 160:818–20.

Steininger, M. K. 2000. "Secondary Forest Structure and Biomass Following Short and Extended Land-Use in Central and Southern Amazonia." Journal of Tropical Ecology 16 (05): 689–708.

Stevens, W. K. 1995. Miracle under the Oaks: The Revival of Nature in America. New York: Pocket Books.

Steward, A. S. 2008. "Changing Lives, Changing Fields: Diversity in Agriculture and Economic Strategies in Two Caboclo Communities in the Amazon Estuary." PhD diss., Department of Plant Sciences, City University of New York.

Steward, J. H. 1948. Handbook of South American Indians vol. 3: Tropical Forest Tribes. 7 vols. Bulletin/Smithsonian Institution, Bureau of American Ethnology vol. 143. Washington, DC: Smithsonian Institution Bureau of American Ethnology.

Stibig, H. J., F. Achard, and S. Fritz. 2004. "A New Forest Cover Map of Continental Southeast Asia Derived from Spot-Vegetation Satellite Imagery." Applied Vegetation Science 7 (2): 153–62.

Stibig, H. J., and J. P. Malingreau. 2003. "Forest Cover of Insular Southeast Asia Mapped from Recent Satellite Images of Coarse Spatial Resolution." Ambio 32 (7): 469–75.

Stickler, C. M., D. C. Nepstad, M. T. Coe, D. G. Mcgrath, H. O. Rodrigues, W. S. Walker, B. S. Soares, and E. A. Davidson. 2009. "The Potential Ecological Costs and Cobenefits of REDD: A Critical Review and Case Study from the Amazon Region." Global Change Biology 15 (12): 2803–24.

Stieglitz, F. V. 1990. Exploitation Forestière Rurale et Réhabilitation Des Forêts: Premièrs Résultats D'un Projet de Recherche Interdisciplinaire en Haute-Guinée [*Logging and Rural Rehabilitation of Forests: First Results of an Interdisciplinary Research Project in Upper Guinea*]. Berlin: République de Guinée.

Stoian, D. 2005. "Making the Best of Two Worlds: Rural and Peri-Urban Livelihood Options Sustained by Nontimber Forest Products from the Bolivian Amazon." World Development 33 (9): 1473–90.

———. 2006. La Economía Extractivista de la Amazonia Boliviana [The Extractive Economy of the Bolivian Amazon]. Bogor: CIFOR.

Stoler, A. L. 2008. "Imperial Debris: Reflections on Ruins and Ruination." Cultural Anthropology 23 (2): 191–219.

Stott, P. A., and S. Sullivan. 2000. Political Ecology: Science, Myth and Power. London: Oxford University Press.

Stratford, J. A., and W. D. Robinson. 2005. "Gulliver Travels to the Fragmented Tropics: Geographic Variation in Mechanisms of Avian Extinction." Frontiers in Ecology and the Environment 3 (2): 85–92.

Stubbs, R. 1989. Hearts and Minds in Guerrilla Warfare: The Malayan Emergency, 1948–1960. New York: Oxford University Press.

Stuiver, M., and G. W. Pearson. 1993. "High Precision Bidecadal Calibration of the Radiocarbon Time Scale, AD 1950–500 BC and 2500–6000 BC." Radiocarbon 35 (1): 1–23.

Stump, R. W. 2005. "Religion and the Geographies of War." In The Geography of War and Peace: From Death Camps to Diplomats, edited by C. R. Flint, 149–73. New York: Oxford University Press.

Sturgeon, J. C. 2005. Border Landscapes : The Politics of Akha Land Use in China and Thailand. Seattle: University of Washington Press.

———. 2010. "Governing Minorities and Development in Xishuangbanna, China: Akha and Dai Rubber Farmers as Entrepreneurs." Geoforum 41 (2): 318–28.

———. 2011. "Rubber Transformations: Post-Socialist Livelihoods and Identities for Akha and Dai Farmers in Xishuangbanna, China." In *Moving Mountains:Ethnicity and Livelihoods in Highland China, Vietnam, and Laos*, edited by Jean Michaud and Tim Forsyth, chapter 9. Vancouver: UBC Press.

Sturgeon, J. C., and N. K. Menzies. 2008. "Ideological Landscapes: Rubber in Xishuangbana." Asian Geographer 25:21–37.

Sukumar, R. 2003. The Living Elephants: Evolutionary Ecology, Behavior, and Conservation. New York: Oxford University Press.

Sukumar, R., R. Ramesh, R. K. Pant, and G. Rajagopan. 1993. "A Delta 13 C Record of Late Quaternary Climate Change from tropical peats in southern India." Nature 364 (19): 703–06.

Sundberg, J. 2009. "Cat Fights on the Rio, Diabolic Caminos in the Desert: The Nature of Boundary Enforcement in the United States-Mexico Borderlands." Paper presented at Luce Project on Green Governance, University of California, Berkeley, Workshop on Environmental Politics.

Sunderlin, W. D., A. Angelsen, B. Belcher, P. Burgers, R. Nasi, L. Santoso, and S. Wunder. 2005. "Livelihoods, Forests, and Conservation in Developing Countries: An Overview." World Development 33 (9): 1383–402.

Sunderlin, W. D., J. Hatcher, and M. Liddle. 2008. From Exclusion to Ownership? Challenges and Opportunities in Advancing Forest Tenure Reform. Washington, DC: Rights and Resource Initiative.

Sutton, M. Q., and E. N. Anderson. 2004. Introduction to Cultural Ecology. Walnut Creek, CA: AltaMira Press.

Svenning, J. C. 2002. "Non-Native Ornamental Palms Invade a Secondary Tropical Forest in Panama." Palms 46:81–86.

Tacoli, C. 2002. "Changing Rural-Urban Interactions in the Sub-Saharan Africa and Their Impact on Livelihoods: A Summary." Working Paper Series on Rural-Urban Interactions and Livelihood Strategies. London: International Institute for Environment and Development (IIED).

———. 2009. "Crisis or Adaptation? Migration and Climate Change in a Context of High Mobility." Environment and Urbanization 21:513–25.

Tambiah, S. 1985. Ritual, Performance, and Thought. Cambridge, MA: Harvard University Press.

Teak and Mahogany. 2005. "Outdoor Furniture, Outdoor Lifestyle" (Advertisement). Expat Living, January/February, 79.

Teak Decking Limited. 2007. Available from http://www.teak-decking.co.uk.

Teichgraeber, R. F. 1987. "'Less Abused Than I Had Reason to Expect': The Reception of the Wealth of Nations in Britain, 1776–90." The Historical Journal 30 (02): 337–66.

Terborgh, J. 1999. Requiem for Nature. Washington, DC: Island Press.

Terpend, M. N. 1982. La Filière Karité; Produit de Cueillette, Produit de Luxe [*The Die Shea, Product Collection, Luxury Products*]. Paris: Les Dossiers Faim-Développement.

Thapar, R. 2001. "Perceiving the Forest: Early India." Studies in History 17 (1): 1.

———. 2012. "Perceiving the Forest: Early India." In India's Environmental History: From Ancient Times to the Colonial Period , edited by M. Rangarajan and K. Sivaramakrishnan, 105–26. Delhi: Oxford University Press.

Thapar, V. 2005. "Note of Dissent." In Joining the Dots: The Report of the Tiger Task Force, 163–80. New Delhi: Ministry of Environment and Forests, Government of India.

———. 2006. The Last Tiger: Struggling for Survival. New York: Oxford University Press.

Thapar, V., K. U. Karanth, J. D. Nichols, J. Seidenstricker, E. Dinerstein, J. L. D. Smith, C. Mcdougal, A. J. T. Johnsingh, and R. S. Chundawat. 2003. "Science Deficiency in Conservation Practice: The Monitoring of Tiger Populations in India." Animal Conservation 6 (02): 141–46.

Thapar, V., and F. S. Rathore. 1999. The Secret Life of Tigers. New York: Oxford University Press.

The Vplants Project. n.d. Vplants: A Virtual Herbarium of the Chicago Region. Available from http://www.vplants.org.

Thomas, D. 2005. Developing Watershed Management Organizations in Pilot Sub-Basins of the Ping River Basin. Bangkok: Office of Natural Resources and Environmental Policy and Planning, Ministry of Natural Resources and Environment.

Thomas, D. E., B. Ekasingh, M. Ekasingh, L. Lebel, H. M. Ha, L. Ediger, S. Thongmanivong, X. Jianchu, C. Sangchyoswat, and Y. Nyberg. 2008. Comparative Assessment of Resource and Market Access of the Poor in Upland Zones of the Greater Mekong Region. Chiang Mai: World Agroforestry Center.

Thompson, D. F., ed. 1996. The Oxford Compact English Dictionary. Oxford: Oxford University Press.

Thompson, E. P. 1975. Whigs and Hunters: The Origin of the Black Act. London: Allen Lane.

Thompson, J. E. S. 1954. The Rise and Fall of Maya Civilization. Norman: University of Oklahoma Press.

Thomson, P. A. B. 2004. Belize: A Concise History. Oxford: Macmillan.

Thongchai, W. 1994. Siam Mapped: A History of the Geo-Body of a Nation. Honolulu: University of Hawaii Press.

Thongmanivong, S., and Y. Fujita. 2006. "Recent Land Use and Livelihood Transitions in Northern Laos." Mountain Research and Development 26 (3): 237–44.

Thongmanivong, S., Y. Fujita, and J. Fox. 2005. "Resource Use Dynamics and Land-Cover Change in Ang Nhai Village and Phou Phanang National Reserve Forest, Lao PDR." Environmental Management 36 (3): 382–93.

Tilly, C. 1984. Big Structures, Large Processes, Huge Comparisons. New York: Russell Sage Foundation.

Toledo, M., and J. Salick. 2006. "Secondary Succession and Indigenous Management in Semideciduous Forest Fallows of the Amazon Basin." Biotropica 38 (2): 161–70.

Tomber, R. 2008. Indo-Roman Trade. New York: Duckworth.

TPCG, and Kalpavkriksh. 2005. Securing India's Future: Final Technical Report of the National Biodiversity Strategy and Action Plan Process. Pune/Delhi: NBSAP Technical and Policy Group.

Trac, C. J., S. Harrell, T. M. Hinckley, and A. C. Henck. 2007. "Reforestation Programs in Southwest China: Reported Success, Observed Failure, and the Reasons Why." Journal of Mountain Science 4 (4): 275–92.

Trautmann, T. R. 1982. "Elephants and the Mauryas." In India, History and Thought: Essays in Honour of Al Basham, ed. S. N. Mukherjea, 254–81. Calcutta: Firma L Mukhopadhyay.

Trentmann, F. 2004. "Beyond Consumerism: New Historical Perspectives on Consumption." Journal of Contemporary History 39 (3): 373–401.

Trouillot, M. R. 1991. "Anthropology and the Savage Slot: The Poetics and Politics of Otherness." In Recapturing Anthropology: Working in the Present, edited by R. Fox, 17–44. Santa Fe: School of American Research.

Tsing, A. L. 1999. "Becoming a Tribal Elder, and Other Green Development Fantasies." In Transforming the Indonesian Uplands, edited by T. Li, 159–202. London: Harwood Academic Publishers.

———. 2005. Friction: An Ethnography of Global Connection. Princeton, NJ: Princeton University Press.

Tucker, C. J., and J. R. G. Townshend. 2001. "Strategies for Monitoring Tropical Deforestation Using Satellite Data." International Journal of Remote Sensing 21 (6): 1461–71.

Tucker, R. P., and J. F. Richards. 1983. Global Deforestation and the Nineteenth-Century World Economy. Duke Press Policy Studies. Durham, NC: Duke University Press.

Turner, M. D. 2004. "Political Ecology and the Moral Dimensions of 'Resource Conflicts': The Case of Farmer-Herder Conflicts in the Sahel." Political Geography 23 (7): 863–89.

US Bureau of the Census. 1913. Thirteenth Census of the United States Taken in the Year 1910, Volume II, Population. Washington, DC: Government Printing Office.

US Department of Energy. 1995b. Estimating the Cold War Mortgage: The 1995 Baseline Environmental Management Report. Washington, DC: US Government Printing Office.

———. 1995a. Closing the Circle on the Splitting of the Atom. Washington, DC: US Government Printing Office.

———. 1999. Richardson Announces 1,000 Acres at Los Alamos National Laboratory to Protect Wildlife. News Release, October 30.

———. 2001. Long-Term Stewardship Study. Washington, DC: US Government Printing Office.

US Department of Energy, Idaho National Engineering and Environmental Laboratory. 1999. "Energy Department, Bureau of Land Management Create Sagebrush Steppe Reserve." News Release, July 19.

US Department of Energy, Pacific Northwest National Laboratory. 1999. "Ecology Reserve: A Haven for Plants and Animals." Backgrounders, August 1–2.

US Department of Energy, Savannah River Operation Office. 1999. "Department of Energy Teams with State on Dedication and Management of 10,000-Acre Crackerneck Wildlife Management Area and Ecological Reserve." News Release, June 24.

US Forest Service. Midewin National Tallgrass Prairie. Available from http://www .fs.fed.us/mntp/.

Ubink, J. M. 2008. In the Land of the Chiefs: Customary Law, Land Conflicts, and the Role of the State in Peri-Urban Ghana. Leiden: Leiden University Press.

Uhlig, H. 1984. Spontaneous and Planned Land Settlement in Southeast Asia. Hamburg: Institute of Asian Affairs; Giesssener Geographische Schriften.

Ulrich, R. S. 1984. "View through a Window May Influence Recovery from Surgery." Science 224 (4647): 420–21.

United Nations. 1999. "World Urbanization Prospects: 1999 Revision." Retrieved from http://www.un.org/esa/population/pubsarchive/urbanization/urbanization.pdf.

United Nations Environment Programme (UNEP), the Amazon Cooperation Treaty Organization (ACTO) in collaboration with the Research Center of the Universidad del Pacífico (CIUP). 2009. Environment Outlook in Amazonia: GEO Amazonia. Panama City: UNEP and ACTO.

Valdez, F. 2006. Agricultura Ancestral, Camellones y Albarradas : Contexto Social, Usos y Retos del Passado y del Presente [*Ancestral Agriculture, Camellones and Albarradas: Social Context, Applications and Challenges Passado and Present*]. 1ra ed. Colección Actas & Memorias del Ifea vol. 3. Quito: Editorial Abya-Yala.

Valmiki. 1984. The Ramayana of Valmiki: An Epic of Ancient India. Edited by R. P. Goldman, S. J. Sutherland, and R. Lefeber. Princeton, NJ: Princeton University Press.

Van de Sandt, J., and A. MacKinven. 2007. "Mapping Indigenous Territories. National Committee of the Netherlands, Part A: Overview and Synthesis of Project Cluster." Small Grants Programmes, Knowledge Management series. No. 1. IUCN.

Van der Geest, S. 2000. "Funerals for the Living: Conversations with Elderly People in Kwahu, Ghana." African Studies Review 43 (3): 103–29.

Van Dijk, C. 1981. Rebellion under the Banner of Islam: The Darul Islam in Indonesia. The Hague: Martinus Nijhoff; Rijksuniversiteit te Leiden.

Van Gemerden, B. S., H. Olff, M. P. E. Parren, and F. Bongers. 2003. "The Pristine Rain Forest? Remnants of Historical Human Impacts on Current Tree Species Composition and Diversity." Journal of Biogeography 30 (9): 1381–90.

Van Haaften, E. H., and F. J. R. Van de Vijver. 2003. "Human Resilience and Environmental Degradation: The Eco-Cultural Link in the Sahel." International Journal of Sustainable Development and World Ecology 10 (2): 85–99.

Vandergeest, P. 1996. "Mapping Nature: Territorialization of Forest Rights in Thailand." Society and Natural Resources 9 (2): 159–75.

———. 2003. "Racialization and Citizenship in Thai Forest Politics." Society and Natural Resources 16 (1): 19–37.

Vandergeest, P., and N. Peluso. 1995. "Territorialization and State Power in Thailand." Theory and Society 24 (3): 385–426.

Vandergeest, P., and N. L. Peluso. 2006a. "Empires of Forestry: Professional Forestry and State Power in Southeast Asia, Part 1." Environment and History 12 (1): 31–64.

———. 2006b. "Empires of Forestry: Professional Forestry and State Power in Southeast Asia, Part 2." Environment and History 12 (1): 359–93.

Vandermeer, J., I. G. de la Cerda, I. Perfecto, D. Boucher, J. Ruiz, and A. Kaufmann. 2004. "Multiple Basins of Attraction in a Tropical Forest: Evidence for Nonequilibrium Community Structure." Ecology 85 (2): 575–79.

Vandermeer, J., M. A. Mallona, D. Boucher, K. Yih, and I. Perfecto. 1995. "Three Years of Ingrowth Following Catastrophic Hurricane Damage on the Caribbean Coast of Nicaragua: Evidence in Support of the Direct Regeneration Hypothesis." Journal of Tropical Ecology 11 (03): 465–71.

Vandermeer, J., and I. Perfecto. 2007. "The Agricultural Matrix and a Future Paradigm for Conservation." Conservation Biology 21 (1): 274–77.

VanWey, L. K., E. Ostrom, and V. Meretsky 2005. "Theories Underlying the Study of Human-Environment Interactions." In Seeing the Forest and the Trees: Human-Environment Interactions in Forest Ecosystems, edited by E. F. Moran and E. Ostrom, 23–56. Cambridge, MA: MIT Press.

Vardhan, H. 1978. "Tiger Lacks Lobby—a Case Study of the Sariska and Ranthambhore Tiger Reserves for Shifting of Villages." Paper presented at Forest International Symposium on the Tiger, Delhi.

Vasanthy, G. 1988. "Pollen Analysis of Late Quaternary Sediments: Evolution of Upland Savanna in Sandynallah (Nilgiris, South India)." Review of Palaeobotany and Palynology 55 (1–3): 175–92.

Vaughan, D. A., K. Kadowaki, A. Kaga, and N. Tomooka. 2008. "On the Phylogeny and Biogeography of the Genus *Oryza*." Breeding Science 55:113–22.

Vicentini, Y. 2004. Cidade e História na Amazônia [*City and History of the Amazon*]. Curitiba: Editoria UFPR.

Vincke, C., I. Diedhiou, and M. Grouzis. 2010. "Long Term Dynamics and Structure of Woody Vegetation in the Ferlo (Senegal)." Journal of Arid Environments 74 (2): 268–76.

Vogel, C. 2007. Illinois State Land Conservation Funding. Illinois Environmental Council Education Fund; the Nature Conservancy; Trust for Public Land.

Von Benda-Beckmann, F. 1999. "A Functional Analysis of Property Rights with Special Reference to Indonesia." In Property Rights and Economic Development; Land and Natural Resources in South-East Asia and Oceania, edited by T. Van Meijl and F. Von Benda-Beckmann, 15–56. London: Kegan Paul.

Vongsay, P. 2004. "Hat Yao Making Most of para Rubber." Vientiane Times, 11.

Vormisto, J., J. C. Svenning, P. Hall, and H. Balslev. 2004. "Diversity and Dominance in Palm (Arecaceae) Communities in Terra Firme Forests in the Western Amazon Basin." Journal of Ecology 92 (4): 577–88.

Waddell, D. A. G., and Royal Institute of International Affairs. 1961. British Honduras: A Historical and Contemporary Survey. London: Oxford University Press for the Royal Institute of International Affairs.

Waghorne, J. P. 1994. The Raja's Magic Clothes: Re-Visioning Kingship and Divinity in England's India. Philadelphia: Pennsylvania State University Press.

Wagley, C. 1953. Amazon Town: A Study of Man in the Tropics. New York: Macmillan.

Wagner, P. L. 1964. "Natural Vegetation of Middle America." In Natural Environment and Early Culture, edited by R. Wauchope, 216–64. Austin: University of Texas Press.

Wainwright, J., and M. Robertson. 2003. Territorialization, Science and the Colonial State: The Case of Highway 55 in Minnesota. Cultural Geographies 10 (2): 196.

Walker, A. 1999. "Regimes of Regulation: 1800–1988." In The Legend of the Golden Boat: Regulation, Trade and Traders in the Borderlands of Laos, Thailand, China and Burma, edited by A. Walker, 25–63. Honolulu: University of Hawai'i Press.

———. 2004. "Seeing Farmers for the Trees: Community Forestry and the Arborealisation of Agriculture in Northern Thailand." Asia Pacific Viewpoint 45 (3): 311–24.

Walker, H. C. 1908. "Fire Protection in Burma." Indian Forester 34:339–49.

Walker, J. 1808. "An Economical History of the Hebrides and Highlands of Scotland." Edinburgh: Printed for Longman, Hurst, Rees for Guthrie & Anderson.

Walker, J., and M. M. Mckay. 1980. The Rev. Dr. John Walker's Report on the Hebrides of 1764 and 1771. Edinburgh: John Donald.

Walker, J. H. 2004. Agricultural Change in the Bolivian Amazon. University of Pittsburgh Memoirs in Latin American Archaeology vol. 13. Pittsburgh: University of Pittsburgh.

Walker, J. S. 2000. Permissible Dose: A History of Radiation Protection in the Twentieth Century. Berkeley: University of California Press.

Walker Papers La III. n.d. Edinburgh University Library, 352/3/4.

Walker, R., J. Browder, E. Arima, C. Simmons, R. Pereira, M. Caldas, R. Shirota, and S. de Zen. 2009. "Ranching and the New Global Range: Amazonia in the 21st Century." Geoforum 40 (5): 732–45.

Washington, S. H. 2005. Packing Them In : An Archaeology of Environmental Racism in Chicago, 1865–1954. Lanham: Lexington Books.

Watts, M. 2004. "Resource Curse? Governmentality, Oil and Power in the Niger Delta, Nigeria." Geopolitics 9 (1): 50–80.

Webb, J. L. A. 2002. Tropical Pioneers: Human Agency and Ecological Change in the Highlands of Sri Lanka, 1800–1900. Athens: Ohio University Press.

Weisgall, J. M. 1994. Operation Crossroads: The Atomic Tests at Bikini Atoll. Annapolis: Naval Institute Press.

Westoby, J. C. 1987. The Purpose of Forests: Follies of Development. New York: Blackwell.

Westphal, L. M. 2003. "Urban Greening and Social Benefits: A Study of Empowerment Outcomes." Journal of Arboriculture 29 (3): 137–47.

Westphal, L. M., M. Longoni, C. L. Leblanc, and A. Wali. 2008. "Anglers' Appraisals of the Risks of Eating Sport-Caught Fish from Industrial Areas: Lessons from Chicago's Calumet Region." Human Ecology Review 15 (1): 46.

White, A., and A. Martin. 2002. Who Owns the World's Forests?: Forest Tenure and Public Forests in Transition. Washington, DC: Forest Trends; Center for International Environmental Law.

White, J., and M. H. Madany. 1978. "Classification of Natural Communities in Illinois." In Illinois Natural Areas Inventory Technical Report, Volume 1: Survey Methods and Results, edited by J. White, 310–505. Urbana: Illinois Natural Areas Inventory.

White, L. J. T., and J. F. Oates. 1999. "New Data on the History of the Plateau Forest of Okomu, Southern Nigeria: An Insight into How Human Disturbance Has Shaped the African Rain Forest." Global Ecology and Biogeography 8 (5): 355–61.

Whitmore, T. C., and C. P. Burnham. 1975. Tropical Rain Forests of the Far East. Oxford: Clarendon Press.

Whitmore, T. C., and D. F. R. P. Burslem. 1988. "Major Disturbances in Tropical Rainforests." In Dynamics of Tropical Communities, edited by D. M. Newbery, H. H. T. Prins, and N. D. Brown, 549–66. Oxford: Blackwell Science Ltd.

Whittaker, M. 1999. "Preserving Open Space on the Rural-Urban Fringe: The Role of Land Trusts." In Contested Countryside: The Rural-Urban Fringe in North America, edited by O. Furuseth and M. Lapping, 263–87. Aldershot: Ashgate Publishers.

Whittaker, R. H. 1957. "Recent Evolution of Ecological Concepts in Relation to the Eastern Forests of North America." American Journal of Botany 44 (2): 197–206.

Whitten, R. G. 1979. "Comments on the Theory of Holocene Refugia in the Culture History of Amazonia." American Antiquity 44:238–51.

Wiersum, K. F. 1984. "Introduction: Towards a Global Forestation Strategy." Paper presented at Proceedings of an International Symposium on Strategies and Designs for Afforestation, Reforestation and Tree Planting, Wageningen.

Wiggers, R. 2000. "Classic Restorations: Fermilab." Chicago Wilderness Magazine 3:4–9.

Wilcox, B. P., M. K. Owens, W. A. Dugas, D. N. Ueckert, and C. R. Hart. 2006. "Shrubs, Streamflow, and the Paradox of Scale." Hydrological Processes 20 (15): 3245–59.

Williams, J. H. 1950. Elephant Bill. London: Hart-Davis.

Williams, M. 1989. "Deforestation: Past and Present." Progress in Human Geography 13 (2): 176.

———. 2003. Deforesting the Earth: From Prehistory to Global Crisis. Chicago: University of Chicago Press.

Wily, A. 2008. "Custom and Commonage in Africa: Rethinking the Orthodoxies." Land Use Policy 25 (1): 43–52.

WinklerPrins, A. 2002. "House-Lot Gardens in Santarém, Pará, Brazil: Linking Rural with Urban." Urban Ecosystems 6 (1): 43–65.

———. 2006. "Urban House-Lot Gardens and Agrodiversity in Santarém, Pará, Brazil: Spaces of Conservation That Link Urban with Rural." In Globalization and New Geographies of Conservation, edited by K. S. Zimmerer, 121–40. Chicago: University of Chicago Press.

WinklerPrins, A., and P. de Souza. 2005. "Surviving the City: Urban Home Gardens and the Economy of Affection in the Brazilian Amazon." Journal of Latin American Geography 4 (1): 107–26.

Wisconsin Department of Natural Resources. n.d. Common Buckthorn (Rhamnus Cathartica). Available from http://dnr.wi.gov/invasives/fact/buckthorn_com.htm.

Wiseman, F. M. 1978. "Agricultural and Historical Ecology of the Maya Lowlands." In Pre-Hispanic Maya Agriculture, edited by P. D. Harrison and B. L. Turner II, 63–115. Albuquerque: University of New Mexico Press.

Wolfson, R. 1993. Nuclear Choices: A Citizen's Guide to Nuclear Technology. Rev. ed. Cambridge, MA: MIT Press.

Woodman, G. R. 1996. Customary Land Law in the Ghanaian Courts. Accra: Ghana Universities Press.

Woods, W. I., and J. M. McCann. 1999. "The Anthropogenic Origin and Persistence of Amazonian Dark Earths." Paper presented at Yearbook, Conference of Latin American Geography.

World Commission on Environment and Development. 1991. Our Common Future. Oxford: Oxford University Press.

World Resources Institute. 2007. Annual Report 2006–2007. Washington, DC: World Resources Institute.

Worster, D. 1996. "The Two Cultures Revisited: Environmental History and the Environmental Sciences." Environment and History 2 (1): 3–14.

Wri. 2007. Earthtrends. Washington, DC: World Resources Institute.

Wright, S. J., and H. C. Muller-Landau. 2006. "The Future of Tropical Forest Species." Biotropica 38 (3): 287–301.

Wright, S. J., K. E. Stoner, N. Beckman, R. T. Corlett, R. Dirzo, H. C. Muller-Landau, G. Nuñez-Iturri, C. A. Peres, and B. C. Wang. 2007. "The Plight of Large Animals in Tropical Forests and the Consequences for Plant Regeneration." Biotropica 39 (3): 289–91.

Wrigley, E. A. 1988. Continuity, Chance and Change: The Character of the Industrial Revolution in England. Cambridge: Cambridge University Press.

———. 1988. "The Limits of Growth: Malthus and the Classical Economists." Population and Development Review 14:30–48.

Wu, J. 1990. 中国地方志茶叶历史资料。北京。农业出版社 [The Chinese Local Tea Historical Data]. Beijing: Agriculture Press.

Wu, Z. L., H. M. Liu, and L. Y. Liu. 2001. "Rubber Cultivation and Sustainable Development in Xishuangbanna, China." International Journal of Sustainable Development and World Ecology 8 (4): 337–45.

Wulf, A. 2008. The Brother Gardeners: Botany, Empire and the Birth of an Obsession. London: William Heinemann.

Wunder, S. 2001. "Poverty Alleviation and Tropical Forests—What Scope for Synergies?" World Development 29 (11): 1817–33.

Wüst, I., and C. Barreto. 1999. "The Ring Villages of Central Brazil: A Challenge for Amazonian Archaeology. Latin American Antiquity 10 (1): 3–23.

Wyatt-Smith, J. 1947. "Save the Belukar." Malayan Forester 11:24–26.

———. 1949. "Regrowth in Cleared Areas." Malayan Forester 13:83–86.

Wynter-Blyth, M. A. 1956. "The Lion Census of 1955." Journal of the Bombay Natural History Society 53 (4): 527–36.

Xu, J. 2006. "The Political, Social, and Ecological Transformation of a Landscape: The case of rubber in Xishuangbanna, China. Mountain Research and Development 26 (3): 254–62.

Xu, J., E. T. Mai, D. Tashi, Y. Fu, Z. Lu, and D. Melick. 2005. "Integrating Sacred Knowledge for Conservation: Cultures and Landscapes in Southwest China." Ecology and Society 10 (2): 7.

Xu, J. C., J. Fox, X. Lu, N. Podger, S. Leisz, and X. H. Ai. 1999. "Effects of Swidden Cultivation, State Policies, and Customary Institutions on Land Cover in a Hani Village, Yunnan, China." Mountain Research and Development 19 (2): 123–32.

Yang, B. 2004. "Horses, Silver, and Cowries: Yunnan in Global Perspective." Journal of World History 15 (3): 281–322.

Yashar, D. J. 2005. Contesting Citizenship in Latin America: The Rise of Indigenous Movements and the Postliberal Challenge. Cambridge Studies in Contentious Politics. Cambridge: Cambridge University Press.

Yunnan Province Bureau of Reclamation. 2003. Commemoration of the 50th Anniversary of the Establishment of the First State Farm in Xishuangbanna. DVD.

Zarin, D. J., M. J. Ducey, J. M. Tucker, and W. A. Salas. 2001. "Potential Biomass Accumulation in Amazonian Regrowth Forests." Ecosystems 4 (7): 658–68.

Zarin, D. J., V. F. G. Pereira, H. Raffles, F. G. Rabelo, M. Pinedo Vasquez, and R. G. Congalton. 2001. "Landscape Change in Tidal Floodplains near the Mouth of the Amazon River." Forest Ecology and Management 154:383–93.

Zerouki, B. 1993. Etude Relative au Feu Auprès Des Populations Des Bassins Versants Types Du Haut Niger [*Relative Study of the Fire Populations in the Watershed Types of Haut Niger*]. Conakry: République de Guinée; Programme d'Amenagement des Bassins Versants Types du Haut Niger.

Zhang, Y. 2006. 中国普洱茶古六大茶山的过去和现在 ["Chinese Tea Ancient Dasan Six Past and Present"]. In 普洱茶经典文选 [*Classic Tea Anthology*], edited by W. Meijin. Kumming: Yunnan Meishu Chubanshe.

Zimmerer, K. S. 2000. "The Reworking of Conservation Geographies: Nonequilibrium Landscapes and Nature-Society Hybrids." Annals of the Association of American Geographers 90 (2): 356–69.

Zimmerer, K. S., and K. R. Young. 1998. Nature's Geography: New Lessons for Conservation in Developing Countries. Madison: University of Wisconsin Press.

Zimmerman, J. K., T. M. Aide, M. Rosario, M. Serrano, and L. Herrera. 1995. "Effects of Land Management and a Recent Hurricane on Forest Structure and Composition in the Luquillo Experimental Forest, Puerto Rico." Forest Ecology and Management 77 (1–3): 65–76.

Zimmermann, F. 1987. The Jungle and the Aroma of Meats: An Ecological Theme in Hindu Medicine. Berkeley: University of California Press.

Zoomers, A. 2010. "Globalisation and the Foreignisation of Space: Seven Processes Driving the Current Global Land Grab." Journal of Peasant Studies 37 (2): 429–47.

CONTRIBUTORS

Fredrik Albritton Jonsson
Department of History
University of Chicago
Chicago, IL 60637
USA

Braulio Vilchez Alvarado
Escuela de Ingeniería Forestal
Instituto Tecnológico de Costa Rica
Cartago 159-7050
Costa Rica

Deborah Barry
Ministry of Environment and Natural
   Resources, MARN
San Salvador
El Salvador

Huw Barton
School of Archaeology and Ancient His-
   tory
University of Leicester
Leicester LE1 7RH
United Kingdom

Sara Berry
Department of History
Johns Hopkins University
Baltimore, MD 21218
USA

Eduardo S. Brondizio
Department of Anthropology and An-
   thropological Center for Training and
   Research on Global Environmental
   Change (ACT)
Indiana University
Bloomington, IN 47405-7100
USA

Raymond L. Bryant
Department of Geography
King's College London
London WC2R 2LS
United Kingdom

Judith Carney
Department of Geography and Institute
   of the Environment and Sustainability
University of California, Los Angeles
Los Angeles, CA 90095-1524
USA

Robin L. Chazdon
Department of Ecology and Evolutionary
   Biology
University of Connecticut
Storrs, CT 06269-3043
USA

Peter Crane
School of Forestry and Environmental
   Studies
Yale University
New Haven, CT 06511
USA

Marlène Elias
Bioversity International
Selangor Darul Ehsan 43400
Malaysia

Clark L. Erickson
Department of Anthropology
University of Pennsylvania
Philadelphia, PA 19104-6398
USA

James Fairhead
Department of Anthropology
University of Sussex
Brighton, BN1 9RH
United Kingdom

Carlos Fausto
Department of Anthropology
Museu Nacional
Universidade Federal do Rio de Janeiro
Rio de Janeiro, RJ 20940-040
Brazil

Jefferson Fox
East-West Center
Honolulu, Hawai'i 96848
USA

Bruna Franchetto
Department of Anthropology
Museu Nacional
Universidade Federal do Rio de Janeiro
Rio de Janeiro, RJ 20940-040
Brazil

Yayoi Fujita Lagerqvist
School of Geosciences
University of Sydney
New South Wales 2006
Australia

Alan Grainger
School of Geography
University of Leeds
Leeds LS2 9JT
United Kingdom

Susanna B. Hecht
Institute of the Environment and School
of Public Affairs
University of California, Los Angeles
Los Angeles, CA 90095-1656
USA

Michael J. Heckenberger
Department of Anthropology
University of Florida
Gainesville, FL 32611-7305
USA

Liam Heneghan
Department of Environmental Science
and Studies
DePaul University
Chicago, IL 60604
USA

Monica Janowski
Department of Anthropology and Sociol-
ogy
School of Oriental and African Studies
(SOAS)
United Kingdom

Samantha Jones
School of Geography, Archaeology and
Paleoecology
Queens University Belfast
Belfast, Northern Ireland BT7 1NN
United Kingdom

Afukaka Kuikuro
Associação Indígena Kuikuro do Alto
Xingu,
Parque Indígena do Xingu, Mato Grosso
Brazil

Urissap'a Tabata Kuikuro
Associação Indígena Kuikuro do Alto
Xingu
Parque Indígena do Xingu, Mato Grosso
Brazil

Brian Lane
Department of Anthropology
University of Hawai'i at Manoa
Honolulu, Hawai'i 96822-2223
USA

Melissa Leach
Institute of Development Studies
University of Sussex
Brighton, BN1 9RH
United Kingdom

David L. Lentz
Department of Biological Sciences
University of Cincinnati
Cincinnati, OH 45221-0006
USA

Susan G. Letcher
School of Natural and Social Sciences
Purchase College
Purchase, NY 10577
USA

Mark T. Lycett
Department of Anthropology and Pro-
    gram on the Global Environment
University of Chicago
Chicago, IL 60637
USA

Joseph Masco
Department of Anthropology
University of Chicago
Chicago, IL 60637
USA

Ruth Meinzen-Dick
International Food Policy Research
    Institute
Washington, DC 20006-1002
USA

Nicholas K. Menzies
Asia Institute
University of California, Los Angeles
Los Angeles, CA 90095-1487
USA

Kathleen D. Morrison
Department of Anthropology and Pro-
    gram on the Global Environment
University of Chicago
Chicago, IL 60637
USA

Francie Muraski-Stotz
Chicago Zoological Society
Brookfield, IL 60513
USA

Roderick P. Neumann
Department of Global and Cultural Studies
Florida International University
Miami, Florida 33199
USA

Christine Padoch
Center for International Forestry Re-
    search (CIFOR)
Situ Gede Bogor Barat 16115
Indonesia

Nancy Lee Peluso
Department of Environmental Science,
    Policy, and Management
University of California, Berkeley
Berkeley, CA 94720-3114
USA

Ivette Perfecto
School of Natural Resources and Environ-
    ment
University of Michigan
Ann Arbor, MI 48109-1041
USA

Miguel Pinedo-Vasquez
Center for Environmental Research and
    Conservation
Columbia University
New York, NY 10027-5557
USA
and
Center for International Forestry Re-
    search (CIFOR)
Situ Gede Bogor Barat 16115
Indonesia

Melinda Pruett-Jones
Chicago Wilderness Alliance
Chicago, IL 60603
USA

Louis Putzel
Center for International Forestry Research (CIFOR)
Situ Gede Bogor 16115
Indonesia

Mahesh Rangarajan
Nehru Memorial Museum & Library
New Delhi – 110011
India

Chris Reij
VU University
1081 HV Amsterdam
The Netherlands

Laurel Ross
Division of Science and Education
The Field Museum
Chicago, IL 60605
USA

Medardo Miranda Ruiz
Administracion y Negocios Internacionales
Universidad Alas Peruanas
Pucallpa
Perú

J. Christian Russell
Land-Use and Environmental Change Institute
University of Florida
Gainesville, FL 32611
USA

Morgan Schmidt
Department of Geography
University of Florida
Gainesville, FL 32611
USA

Robin R. Sears
The School for Field Studies
Beverly, MA 01915
USA

U. Uzay Sezen
Center for Applied Genetic Technologies, Georgia Research Alliance
University of Georgia
Atlanta, GA 30303
USA

Andrea D. Siqueira
Anthropological Center for Training and Research on Global Environmental Change (ACT)
Indiana University
Bloomington, IN 47405-7100
USA

Angela Steward
Research Group in Agroecology
Mamirauá Institute for Sustainable Development
69470-000 Tefé (AM)
Brazil

Peter Vandergeest
Department of Geography
York University
Toronto, Ontario M3J 1P3
Canada

John Vandermeer
Department of Ecology and Evolutionary Biology
University of Michigan
Ann Arbor, MI 48109-1048
USA

Nathan Vogt
Anthropological Center for Training and Research on Global Environmental Change (ACT)
Indiana University
Bloomington, IN 47405-7100
USA
and
National Institute for Space Research
Av. Dos Astronautas, 1.758
Jd. Granja – CEP: 12227-010
São José dos Campos – SP
Brazil

Alaka Wali
Division of Science and Education
The Field Museum
Chicago, IL 60605
USA

Amanda Wendt
Department of Ecology and Evolutionary
    Biology
University of Connecticut
Storrs, CT 06269-3043
USA

Lynne Westphal
USDA Forest Service
Northern Research Station
Evanston, Illinois 60201-3172
USA

INDEX